PERFORMANCE EVALUATION
OF NUMERICAL SOFTWARE

IFIP TC-2 Working Conference on
Performance Evaluation of Numerical Software
Baden, Austria, 11-15 December 1978

organized by
IFIP Technical Committee 2.5,
Programming (Numerical Software)
International Federation for Information Processing

Program Committee
L. Fosdick (Chairman), Th. Dekker, B. Einarsson, B. Ford,
T. Hull, C. Lawson, J. Reid

NORTH-HOLLAND PUBLISHING COMPANY
AMSTERDAM•NEW YORK•OXFORD

PERFORMANCE EVALUATION OF NUMERICAL SOFTWARE

Proceedings of the IFIP TC 2.5 Working Conference on
Performance Evaluation of Numerical Software

edited by

Lloyd D. FOSDICK

Department of Computer Science
University of Colorado at Boulder
USA

1979

NORTH-HOLLAND PUBLISHING COMPANY
AMSTERDAM•NEW YORK•OXFORD

ISBN: 0 444 85330 8

Published by:
NORTH-HOLLAND PUBLISHING COMPANY—AMSTERDAM • NEW YORK • OXFORD

Sole distributors for the U.S.A. and Canada
ELSEVIER NORTH-HOLLAND, INC.
52 Vanderbilt Avenue
New York, N.Y. 10017

Library of Congress Cataloging in Publication Data

IFIP TC 2.5 Working Conference on Performance Evaluation
    of Numerical Software, Baden, Austria, 1978.
    Performance evaluation of numerical software.

    "Organized by IFIP Technical Committee 2.5, Pro-
gramming (Numerical Software), International Federation
for Information Processing."
    Includes index.
    1. Numerical analysis--Computer programs--Evaluation
--Congresses. I. Fosdick, Lloyd Dudley. II. Inter-
national Federation for Information Processing. Technical
Committee 2.5, Programming (Numerical Software)
III. Title.
QA297.I18 1978   519.4'028'5425   79-13317

ISBN 0-444-85330-8

PRINTED IN THE NETHERLANDS

PREFACE

Our ability to design and build complex systems such as nuclear reactors and gas turbine engines, to analyze large masses of data as are gathered in oil exploration, and to forecast the behaviour of a complex system such as the earth's atmosphere, is limited by our ability to perform the necessary computations. The matter of determining the right software for such computations, of being assured the software is reliable and effective, and of establishing principles for measuring and comparing software is the subject of these proceedings.

The papers are divided into two areas, one of very broad scope which is concerned with general aspects of performance evaluation, and one with a much narrower scope which is concerned with the performance evaluation of numerical software in three subjects - linear algebra, ordinary differential equations, and optimization. A novel and important topic addressed by several papers in this conference is the warranty of numerical software, and another special feature is a panel discussion on the use of mathematical software outside the mathematical community.

The working conference recorded here was held at the Hotel Gutenbrunn in Baden, Austria, from 11 December to 15 December 1978. It was the first working conference to be organized by Working Group 2.5 (Numerical Software) for Technical Committee 2 (Programming) of the International Federation of Information Processing Societies. Seventy-three invitees participated.

The conference chair was L. Fosdick. The program committee was: L. Fosdick (chair), Th. Dekker, B. Einarsson, B. Ford, T. Hull, C. Lawson, and J. Reid. The local arrangements committee was: H. Stetter (chair), and C. Reinsch. During the conference the local arrangements committee was assisted by F. Macsek, J. Schneid, C. Ueberhuber, and E. Weinmueller, all of the Technical University, Vienna. Financial support for the conference was provided by: Austrian Ministry for Science and Research, Control Data Corporation, Honeywell-Bull, IBM Austria, and the U.S. Army Research Office. Finally, secretarial support was provided by H. Ortiz of the University of Colorado, M. Orchard and E. Capanni of the Numerical Algorithms Group Ltd., and B. Katzberger of the Technical University, Vienna.

Lloyd D. Fosdick
Oxford

PARTICIPANTS

E.L. Battiste, 212 Edinburgh Drive, Cary, NC 27511, U.S.A.

B.R. Benjamin, South Australian Institute of Tech., Ingle Farm, South Australia

M.A. Bossavit, 1 Ave. du General de Gaulle, 92141 Clamart, France

J.M. Boyle, Appl. Math., Argonne Nat. Lab., Argonne, Illinios 60439, U.S.A.

R.P. Brent, Australian Nat. University, Canberra, ACT 2600, Australia

W.S. Brown, Bell Laboratories, Murray Hill, New Jersey 07974, U.S.A.

R. Bulirsch, TU München, D-8, München 2, West Germany

J.R. Bunch, Dept. of Math., University of Calif. SD, La Jolla, Calif. 92093, U.S.A.

J. Bus, Mathem. Centrum, Tw. Boerhaavestraat 49, Amsterdam, Netherlands

W.R. Cowell, Appl. Math., Argonne Nat. Lab., Argonne, Illinois 60439, U.S.A.

A.R. Curtis, Bldg. 8.9, A.E.R.E., Harwell, Didcot, Oxford OX11 0RA, G.B.

J.L. de Jong, Dept. of Math., Eindhoven Univ. of Tech., Eindhoven, Netherlands

E. de Doncker, University of Leuven, Haverlee 3030, Belgium

Th. Dekker, van Utrechtlaan 25, Castricum, Netherlands

L.M. Delves, Dept. of CSS, University of Liverpool, Liverpool, G.B.

I.S. Duff, Bldg. 8.9, A.E.R.E., Harwell, Didcot, Oxford OX11 0RA, G.B.

B. Einarsson, Nat. Defense Res. Institute, FOA 252, S-14700, Tumba, Sweden

W.M. Enright, Dept. of Comp. Science, Univ. of Toronto, Toronto M5S 1A7, Canada

A.M. Erisman, Boeing Computer Services, 11006 S.E. 24th Pl., Bellevue 98004, U.S.A.

B. Ford, NAG, 7 Banbury Road, Oxford OX2 6NN, G.B.

L.D. Fosdick, NAG, 7 Banbury Road, Oxford OX2 6NN, G.B.

P. Fox, Old Short Hills Road, Short Hills, New Jersey 07078, U.S.A.

C.W. Gear, Dept. of Comp. Science, Univ. of Illinois, Urbana, IL 61801, U.S.A.

W.M. Gentleman, Comp. Science Dept., Univ. of Waterloo, Waterloo N2L 3G1, Canada

J. Gergely, Comp. & Autom. Institute, Kende ucta 13-17, Budapest XI, Hungary 1502

A.J. Geurts, Dept. of Math., Univ. of Tech., Eindhoven, Netherlands

P.E. Gill, Div. of Numer. Anal., NPL, Teddington, Middlesex, England

I. Gladwell, Dept. of Math., University of Manchester, Manchester M13 9PL, G.B.

G.K. Gupta, Dept. of Comp. Science, Monash Univ., Clayton, Vic.3168, Australia

P.W. Hemker, Mathem. Centrum, Tw. Boerhaavestraat 49, Amsterdam, Netherlands

S. Hitotumatu, Res. Institute of Math. Science, Kyoto Univ., Kyoto, Japan 606

W. Hoffman, Dept. of Math., University of Amsterdam, Amsterdam 1018, Netherlands

T.E. Hull, Dept. of Comp. Science, University of Toronto, Toronto M5S 1A7, Canada

T.L. Jordan, C-3, MS265, Los Alamos Scientific Lab., Los Alamos, NM 87545, U.S.A.

W. Kahan, University of Calif., Berkeley 94720, U.S.A.

R.W. Klopfenstein, David Sarnoff Res. Center, RCA, Princeton, NJ 08540, U.S.A.

C.L. Lawson, MS 125-128, Jet Propulsion Lab., Pasadena, California 91103, U.S.A.

F.A. Lootsma, Dept. of Math., Univ. of Tech., Delft AJ 2600, Netherlands

J. Lyness, Appl. Math. Div., Argonne Nat. Lab., Argonne, IL 60439, U.S.A.

B. Meyer, EDF-DER Service IMA, 1 Ave. du Gen. de Gaulle, 92141 Clamart, France

I.N. Molchanov, Institute Kibernetiki an UKR SSR, Kiev SU-252207, U.S.S.R.

J.J. Moré, Dept. of Appl. Math. & Theor. Physics, Silver Street, Cambridge, G.B.

W. Murray, DNACS, Nat. Physical Lab., Teddington, Middlesex, G.B.

J. Nelder, Statistics Dept., Rothamsted Exp. Station, Harpenden, Herts, G.B.

B. Niblett, Dept. of Comp. Science, Univ. College of Swansea, Swansea SA2 8DE, G.B.

L.J. Osterweil, Dept. of Comp. Science, Univ. of Colorado, Boulder, CO 80309, U.S.A.

M.H.C. Paardekooper, Katholieke Hogeschool, Tilburg, Netherlands

S.J. Polak, ISA, VN573, Philips M.V., Eindhoven, Netherlands

J.K. Reid, Bldg. 8.9, A.E.R.E. Harwell, Didcot, Oxford OX11 0RA, G.B.

C. Reinsch, Leibniz-Rechenzentrum, 8 München 2, Barerstrasse 21, West Germany

J.R. Rice, Math. Sciences, Purdue Univ., West Lafayette, Indiana 47907, U.S.A.

I. Robinson, Div. Appl. Math., Catholic Univ. of Leuven, Heverlee 3030, Belgium

A. Ruhe, Dept. of Information Processing, Univ. Umea, S-90187, Umea, Sweden

R.D. Russell, c/o Tom Brown, 4378 Arure Road, Blaine, WA, U.S.A.

D. Sayers, NAG, 7 Banbury Road, Oxford OX2 6NN, G.B.

R.B. Schnabel, Dept. of Comp. Science, Univ. of Colorado, Boulder, CO 80309, U.S.A.

K. Schittowski, Inst. f. Angew. Math., Universität, Wurzburg 87, West Germany

J.L. Schonfelder, Comp. Center, Univ. of Birmingham, Birmingham B15 2TT, G.B.

K.R. Schwarz, Angew. Math., Univ. Zurich, CH-8032, Zurich, Switzerland

L.F. Shampine, Numer. Math. Div. Sandia Labs., Albuquerque, NM 87185, U.S.A.

A.H. Sherman, Dept. of Comp. Science, Univ. of Texas, Austin, Texas 78712, U.S.A.

B.T. Smith, Appl. Math. Div., Argonne Nat. Lab., Argonne, Illinois 60439, U.S.A.

E. Spedicato, Via Firenze, Paderno, Dagnano, Milano 20037, Italy

H.J. Stetter, Inst. f. Numer. Math., TU Wien, Wien 1040, Austria

G.W. Stewart, Dept. of Comp. Science, Univ. of Maryland, College Park, ML 20742

C.F. Tapper, Magdalen College, Oxford Univ., Oxford OX1 4AU, G.B.

C. Ueberhuber, Inst. f. Numer. Math., TU Wien, Wien 1070, Austria

P.J. van der Houwen, Mathem. Centrum, Tw. Boerhaaverstraat 49, Amsterdam, NL

H.A. van der Vorst, Academic Comp. Center, Budapestlaan 6, Utrecht-de-Uithof, NL

J.G. Verwer, Mathem. Centrum, Tw Boerhaaverstraat 49, Amsterdam 1091, Netherlands

P.A. Wedin, Inst. f. Information Processing, University, Umea 90187, Sweden

J.H. Wilkinson, Dept. Num. Anal., Nat. Physical Lab., Teddington, Middlesex, G.B.

N.N. Yanenko, Institute of Theor. & Appl. Math., Novosibirsk, 630090, U.S.S.R.

CONTENTS

SESSION 1 : GENERAL ASPECTS OF PERFORMANCE
EVALUATION

Performance Evaluation of Numerical Software, Fosdick (ed.)
© IFIP, North-Holland Publishing Company, 1979

CORRECTNESS OF NUMERICAL SOFTWARE

T.E. Hull
Department of Computer Science
University of Toronto
Toronto, Ontario, Canada

Current research in program verification relies heavily on the
use of assertions, as well as on abstract data types, and, of
course, structured programming. Present trends seem to be
emphasizing attempts to automate the verification process.
However, interest appears to be concerned exclusively with the
correctness of nonnumerical programs.

We consider how these ideas and techniques, particularly the use
of assertions, might be adapted to proving the correctness of
numerical programs (i.e., of programs that involve roundoff or
truncation errors in a significant way).

Structured programming is of course helpful in making any kind of
program easier to understand, and hence easier to prove correct.
We conclude that the use of assertions and abstract data types
might also help in organizing proofs in some limited situations,
but that otherwise the techniques of program verification are not
likely to be helpful with numerical programs.

The most useful technique of all is, of course, careful mathematical
analysis. Beyond that, there are at least three major improvements
that could be made in our programming language facilities and that
would greatly facilitate proving the correctness of numerical
programs. These are (1) clean and well-defined floating-point
arithmetic, (2) precision control, and (3) certain symbol manipulation facilities.

A brief discussion of various interpretations of what we might mean
by program correctness is also included.

INTRODUCTION

We are of course all in favour of programs being correct. We may not have a clear
idea of what we mean by the "correctness" of a program, or we may find that different individuals have different interpretations of the idea. Nevertheless, it is
a "good thing" to be in favour of correctness!

During the last decade or so there has been a considerable amount of research done
in an area which is usually known as "program verification". This research is
concerned with concepts and techniques which are at least related to whatever we
might mean by program correctness, and it would seem reasonable to ask if any of
the ideas and results from this area can be useful to those of us who are interested in numerical software. Unfortunately, the work is concerned exclusively with
nonnumerical calculations, and moreover is often quite formal. Nevertheless, it
seems to be concerned with the general area of correctness and we should see if
any of it can be helpful to us.

A few years ago I attempted a similar study. This led to a paper (by Hull,
Enright and Sedgwick [1972]) which attempted to show how some of the techniques of

program verification, particularly the idea of using assertions, could be adapted to programs for doing numerical calculations. The essential ideas will be summarized in later sections.

In preparation for this talk, I have studied some of the more recent work on program verification in the hope of answering the following two questions:

(1) Are there any new techniques?

(2) If so, are they of any help to us?

At the same time, I have also tried to provide preliminary answers to two further questions, namely:

(3) What might we mean by correctness?

(4) What new programming facilities would be helpful in the development of correct programs?

(For surveys of recent work on program verification, see London [1977a, 1977b] and Luckham [1977].)

Before providing answers to these questions, a general, but still rather preliminary discussion of correctness will be introduced in the next section. Then three sections will be devoted to examples whose purpose is to illustrate the use of program verification techniques in numerical calculations. (Emphasis is placed on the use of assertions, abstract data types, and structured programming.) Then in the next three sections we return to the questions stated above. These are in turn followed by a section in which we summarize our conclusions.

A PRELIMINARY VIEW OF CORRECTNESS

It has been my experience that the subject of program correctness almost always evokes an initial response that consists of two parts.

First of all, it is taken for granted (not surprisingly) that testing alone is not enough. It is of course recognized that at least some testing is needed, and various reasons for testing can easily be identified. For example, one reason is simply to try to catch relatively minor errors, even as "minor" as a missing comma. Another reason for testing is to obtain measures of efficiency, perhaps in comparison with other programs for solving a similar class of problems. A third is to help delineate the domain over which the program can be used safely. A fourth is that some test runs may be an integral part of the "proof of correctness" in certain situations.

The second part of the initial response is very often a statement to the effect that "it is not possible to _really_ prove the correctness of a program." I am sympathetic to the feeling behind such a statement, but I am convinced that such statements are misleading and actually miss an important point about program correctness.

I believe that it is much more helpful to consider proofs about programs to be completely analogous to proofs in mathematics. Then not _really_ being able to prove the correctness of a program is analogous to not _really_ being able to prove a theorem in mathematics. Strictly speaking, the latter is true. After all, there are famous examples in mathematics when "proofs" turned out to be invalid. One example was Euler's proof of Fermat's last theorem. Another involves the Four Colour problem; it was believed for a thirteen year period from 1879 to 1892 that this conjecture had been proven. There are also doubts about the foundations of mathematics in the minds of many mathematicians. But this uncertainty does not deter us from making successful use of mathematical results, and considering them to be _really_ proven.

Admittedly, there are differences. As a trivial example, a missing comma in a
mathematical theorem may cause no trouble when we make use of the theorem, whereas
a missing comma in a computer program can easily cause havoc. This only means
that there are some differences in detail. But the main significance of the analogy remains the same, which is that we can prove programs correct in the sense
that we can prove mathematical results are correct.

The question of exactly what should be proven is one that will be discussed in more
detail in a later section. For the time being it will be convenient to concentrate
our attention on a special case. This is the case where a function has been
defined and we consider a program to be correct if it evaluates the function. For
example, the value of the function might be the greatest common divisor of two
positive integers. Or the domain of the function might consist of the set of all
n-tuples of real numbers which are machine representable, and the value of the
"sort" function for a particular n-tuple would then be a rearrangement of that
n-tuple into nondecreasing order.

As we shall see later there are other ways in which we can view the correctness
of a program. However, this way is quite a useful one for the purposes of the
examples in the next three sections. One of our aims in the next three sections
is to show how this way of viewing functions can be applied in numerical contexts,
where roundoff and truncation errors must be taken into account when we define
the functions to be evaluated by our programs.

A SIMPLE EXAMPLE

Much of the stimulus for work on program verification was provided by a paper of
Floyd's [1967]. The simple example he used to illustrate the basic ideas is
described by the flowchart in Fig.1. The example is a program for finding the sum
of n numbers, $a_1, a_2, \ldots, a_n$. (Although we use flowcharts in this and the following
examples, it is intended that our discussion be directly applicable to programs
in a particular programming language.)

For our purposes, the key idea is the introduction of assertions, which are statements about the circumstances that prevail at the designated points in the computation. In particular, the assertion A_i in this example will contain the following
requirement:

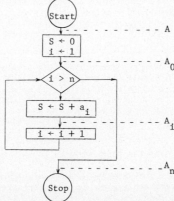

Figure 1.   Outline of a program for finding the sum of
n numbers $a_1, a_2, \ldots, a_n$. The A's are assertions.

$$S = \sum_{j=1}^{i} a_j,$$

which merely asserts that S is the i-th partial sum.

Assertions are used so that the proof of the correctness of the entire program can be broken down into a number of lemmas. The key lemma in this example is the inductive one of showing that A_i remains true after each time around the loop, i.e., that A_i is invariant. It is also necessary to prove that A_0 and $1 \le n$, followed by $S \leftarrow S + a_0$ implies A_0. (Of course the assertion A_0 includes the requirements $S = 0$ and $i = 1$.) Two other lemmas are also needed, as well as a proof that the procedure terminates.

Since the publication of Floyd's paper, a great deal about program verification has developed. In particular, the use of assertions has been firmly established. In fact, a number of programming languages make explicit provision for the inclusion of assertions which can be checked at run time if the user so wishes.

Progress has also been made towards more general formalizations of verification procedures, and also in attempts to automate, at least partially, these verification procedures. All of these developments have fitted in nicely with the general acceptance of "structured programming" that has taken place during the same period of time. These developments have also been complemented by the more recent interest in "abstract data types". The example considered in this section is too simple to provide a good illustration of structured programming or abstract data types, but we will return to these points in later examples. For the time being, the point being made is that the use of assertions as illustrated in this example, along with structured programming and abstract data types, are all contributing to the development of research in the general area of program verification.

On the negative side, as has already been stated, none of this development seems to have been concerned in any way with programs for doing numerical computations. However, with the simple example of this section, it is not difficult to provide modifications which take the numerical aspects into account.

For example, anyone interested in the technique of "backward error analysis" would find it natural to replace the partial sum in A_i with

$$S = \sum_{j=1}^{i} a_j^{(i)}, \quad \text{where } a_j^{(i)} = a_j(1+\varepsilon)^{i-j+1},$$

which asserts that S is the i-th partial sum, as before, except that now it is the partial sum of $a_j^{(i)}$ rather than a_j, but the former differ from the latter by no more than i-j+1 rounding errors. From this point of view, the correctness of the program now is not that it finds the sum of the original n numbers, but rather that it finds the sum of n numbers that differ by no more than some known number of rounding errors from the original n numbers. (Of course, a full statement of what the program does would be qualified by statements about what happens in case of overflow or underflow, but the main part of the result is the one described here.)

Despite being able to modify the point of view in this way, and apparently to adapt the use of assertions to numerical calculations, there are still a number of difficulties to consider. For example, correctness in the sense of a backward error analysis is not necessarily what one might require of this program. As an alternative, it could be that one is only interested in an error bound for the final result. Such a bound can be derived from the backward analysis. It happens to take a particularly simple form when the a_i are all of the same sign. (In the general case, the program could be modified to produce, a posteriori, a bound on the error in the computed result.) As another alternative, it could be that one is interested in a statement about how errors would propagate through the calculation, provided one could assume that the individual errors were drawn at random from a particular probability distribution. This last alternative may not be very important in practice, but it does help to point out the fact that there can be a considerable choice about what one might want to prove regarding a program for doing numerical calculations. It may have been obvious what was meant by a correct program in the nonnumerical case. But in the numerical case, we have a choice as to what we mean by the correctness of a particular program. The same program can be proven correct in one sense, or in some other sense; perhaps it would be better to speak of proving different properties of a program, rather than proving it correct in different senses.

We have been emphasizing the way in which the presence of roundoff error can affect our views of program correctness. So far it would appear that "program verification" techniques might be adapted to take account of this factor. However, we have considered only a very simple example and it should not be taken for granted that more complicated programs can be handled in a similar way. And it must also be kept in mind that we have not yet considered any programs that involve truncation errors. These topics will be illustrated in the next two sections.

Before concluding this section, we would like to mention something to do with the historical origin of assertions. We have pointed out that Floyd's 1967 paper provided much of the impetus for using assertions and the ensuing growth of interest in program verification. In fact, such assertions are often referred to as Floyd assertions. However, it should be acknowledged that their use goes back at least to the 1940's when stored-program computers were first being constructed. Assertions were used, for example, by Goldstine and von Neumann [1947], and by Turing [1949]. They were called "snapshots" in a later paper by Naur [1966].

APPLICATIONS IN LINEAR ALGEBRA

It happens that the backward error analysis can be applied to many programs for solving problems in linear algebra. In fact, this was an important aspect of the tremendous advances in numerical linear algebra that began to develop so rapidly in the 1950's. The work is associated primarily with the name of Wilkinson, but also with many others as well.

This work was not developed directly in terms of programs, with assertions attached to specific points in those programs. However, it is only a matter of shifting our point of view slightly and expressing the mathematical results in terms of assertions, in order to present those results in a form that is completely analogous to the simple example of the preceding section.

The details for a method of solving linear equations are given in the paper by Hull, Enright and Sedgwick [1972] that was mentioned earlier. We will therefore present here only a brief outline. In Fig.2 a high level description of a program for solving a system of linear equations is given. The first step is to perform a triangular decomposition, and this is followed by two more steps, each involving the solution of a triangular system. (For simplicity, no reference has been made to pivoting, which in any event does not affect the error bounds. For the same reason, we have omitted details about overflow and underflow; counters have also not been mentioned; needed restrictions on n and u have also been omitted.) The main point about Fig.2 is that the key parts of the appropriate assertions have been included. The details of the results are those given by Forsythe and Moler [1967].

The first step in the high level description of Fig.2 is simply a statement that A is to be decomposed into L and U, and that certain bounds are to be satisfied. If the specified action is carried out, it can be shown that the assertion that follows is true. However, just how this specified action is carried out must be

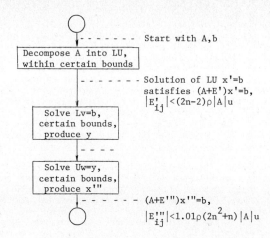

Figure 2.   High level outline of a program for the
approximate solution of Ax=b, with appropriate
assertions. Here ρ is the usual growth factor,
$|A|=\max|A_{ij}|$, and u is the relative roundoff
error bound.

explained in a refinement of the first step of Fig.2. Such a refinement is outlined in Fig.3, along with appropriate assertions.

Refinements are also needed for the other steps in Fig.2, to complete the second
level of description of the program for solving linear equations. Then of course
still further refinements of the major steps at this level are required, and so on,
until the level of refinement finally reaches the level of the programming
language itself.

For further details, see the paper by Hull, Enright and Sedgwick [1972]. The point
we have intended to emphasize is that a backward error analysis can be incorporated
into the kind of framework used for proving program correctness by those who rely
on the use of assertions for program verification. However, it must also be
admitted that the adaptation is not nearly as easy to make with more complicated
programs such as linear equation solvers as it was with the simple program of the
preceding section. To complicate matters still further, it should be noted that
allowing for overflow and underflow is non-trivial, and may not fit into the overall scheme very well. For example, following Forsythe and Moler again, it can be
shown that the occurrence of overflow means there is a singular matrix B such that

$$|A_{ij}-B_{ij}| < 1.01(2n^2+n)\rho|A|u + n|b|/M$$

where M is the largest floating point number the machine can store.

On the other hand, it may be that the effort to merge the results of a backward
error analysis with the assertion-verification point of view is worthwhile. This
would be true, for example, if the latter turned out to be a good way to help
organize the proofs of correctness for programs in such areas as linear algebra.
We will return to this point in a later section.

To conclude this section it should be pointed out that the example considered here
also provides a good illustration of the need for abstract data types. In this

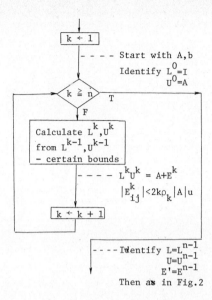

Figure 3. A refinement of the first step in the
program outlined in Fig.2, along with
appropriate assertions.

case the data types are matrices, and the need for abstraction arises because the
separate refinements refer to individual matrices, and it is essential that the
operations performed on these matrices be the same for each refinement. It could
not be tolerated, for example, if one refinement expected a matrix to be stored by
rows, while another expected it to be stored by columns.

APPLICATIONS TO ODEs AND OTHER AREAS

In the paper by Hull, Enright and Sedgwick, it is also shown how assertions can be
used with a program for solving ordinary differential equations. In this situation, truncation error is dominant and, for the sake of simplicity, roundoff error
can be assumed negligible. An outline of the example is given in Fig.4.

At the high level of Fig.4 it is merely stated that a value of the step-size h
satisfying certain properties must be found. One of those properties is that h be
bounded away from zero, and this of course can be used to prove termination. The
other property about the error estimate being small enough will not be discussed
in detail here. However, there are two points to make about it. One is that
"small enough" means small enough so that a bound on the true error can be
guaranteed, and the assertion can then be established. (Note that the assertion
is in a form that would be expected for a backward error analysis, namely, that the
program has produced the true solution of a slightly different problem.)

The other point shows up more and more clearly as further refinements of the high
level description in Fig.4 are made and is, not surprisingly, that the requirement
of "small enough" can only be met for very limited classes of problems. Such
results were originally established by Hull [1968, 1970] for the case where
$f(x,y) = Ay$, A being a constant matrix. Enright [1972] established corresponding
results for stiff systems, and Sedgwick [1973] was able to handle a limited class

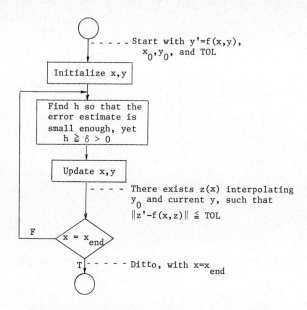

Figure 4.   Outline of a program for solving ordinary differential
equations. The correctness is in terms of finding the
exact solution of a slightly different problem,
analogous to the backward error analyses of linear
algebra. However, such results can be established
only for very limited classes of problems.

of non-linear equations. Later Jackson [1974] followed through an analysis of the
linear case for a realistic library program; roundoff error was included, and so
were all the details of the actual program that had not been tailored in any way
for the purposes of proving correctness.

It is clear that the use of assertions in these cases is limited in two important
ways. One is that it is limited to proofs which are valid only for very restricted
classes of problems. The other is that a considerable effort is needed to incorporate the necessary assertions. Whether the effort is worthwhile is not completely clear. As with programs in linear algebra, it may be that the final form is a
good way to organize a proof of correctness, although in the case of ODEs the
proofs are only for very limited classes of problems. (It is worth mentioning that
Jackson's attempts to prove lemmas that were as strong as possible under the
circumstances led to minor modifications being made in the program.)

With regard to areas other than matrix calculations and ODEs, one can imagine the
sorts of results that can be obtained, for example, in the case of zero finders
and quadrature.

The only other attempt to prove correctness of a real program that I have been
closely associated with was carried out by Steele [1974]. The program was a pre-
processor for DEFT (a Disciplined Extension of Fortran) and was therefore not a
program for doing numerical calculations. However, the work demonstrated very
clearly how important it is to work with well-structured programs, and how closely
the assertions can be tied to the program during the course of developing the
program. These lessons are certainly relevant to numerical programs as well,

although the second point about developing the assertions along with the program is probably not as applicable in numerical cases as it is in nonnumerical ones.

ARE THE TECHNIQUES OF "PROGRAM VERIFICATION" HELPFUL?

We return now to the first two questions raised in the introduction about the techniques of "program verification" and whether or not they are helpful in proving correctness of programs for carrying out numerical calculations.

We identify these techniques as being, first of all, the use of assertions, which is what we have emphasized in the preceding sections. Second, there is the discipline of structured programming, which of course is by no means restricted to those who are doing research in program verification. Third, there is the use of abstract data types. And fourth, there has been a great deal of effort devoted to attempts to formalize and to automate the process of program verification (theorem proving, proof checkers, automatic management of the collection of lemmas, and so on).

Let us begin with assertions. On the positive side, it appears that much of what has been accomplished in the analysis of roundoff error can be adapted to presentations of programs and appropriate assertions. And it may be that this is a good way to help organize, in a systematic way, the proof of the correctness of programs for doing algebraic calculations, at least when the correctness is expressed in terms that are related to a backward error analysis.

However, there are points to be made on the negative side as well. One is that the backward analysis is not always applicable, for example, with matrix inversion, and the appropriate assertions would probably appear to be less natural in these circumstances. Another point is that it may be that we are interested in proving correctness in terms other than what is provided by a backward analysis; for example, we may want to prove something about a bound on the final error, and it may not be clear what are appropriate assertions in such cases.

The situation becomes even less favourable with calculations that involve truncation error, such as arise in connection with ordinary differential equations or quadrature. Here the domain over which there is any hope of proving correctness appears to be relatively very limited. Even so, with the differential equations we considered only questions concerned with local error control, and did not consider the more difficult question of global error control.

Altogether it seems that assertions might be helpful in organizing proofs in some cases. But they do not seem to provide a particularly useful or powerful tool in general cases. It seems unlikely that the use of assertions can be "scaled up" in a profitable way from simple programs, such as the one for finding a sum, to library programs for standard calculations involving matrices, ODEs, and so on. It also seems particularly apparent that assertions would not be used, as they are supposed to be used by program verifiers, to help motivate the development of programs. It would seem to be particularly unlikely, for example, that one would start with the invariant assertion of a loop, and then try to derive the program from that starting point.

As for structured programming, there can be no disagreement that we would all like to have programs that are nicely structured in some sense or other, and thereby, among other things, to be able to prove properties of the programs more easily. However, it is perhaps worthwhile making one point about this topic, that has to do with a difficulty often found in writing numerical programs. The difficulty is that overflow or underflow, and other exceptional conditions, arise more often in numerical programs and make it necessary to have convenient facilities for escaping from intermediate stages in the course of a calculation. We cannot afford to be too rigid about what we allow in well-structured programs.

With regard to abstract data types, it is likely that the main contribution is to help identify and conceptualize what is already taken for granted by anyone who writes programs for doing numerical calculations. Perhaps, in the long run, better programming language facilities will become available as a result of the work on program verification so that the various data structures can be handled more conveniently. In the meantime, it is obvious that a particular operation performed on a matrix in the refinement of one part of a program must be identical to the same operation when it turns up in a refinement of another part of the same program. This means that it might be useful to isolate the specification of the matrix (for example, a particular sparse matrix) and the operations to be performed on it, from other parts of the program. This is what is done already, in any event, but it might be helpful to formalize the concept a bit more and, as already stated, to have more appropriate language facilities for the necessary implementations. However, in numerical calculations, it would also be of primary importance to have complete control of any aspect that might affect the efficiency of the program.

The final aspect of program verification activities that we identified has to do with attempts to formalize and automate parts of the process. It is easy to get the impression that relatively little progress is being made in this area. If this is so, it seems unlikely that what is being done here is going to be of help with numerical programs. This is particularly so if the use of assertions is of only limited use anyway. Moreover, the kinds of assertions needed ("there exists an ε such that..." or "there exists a $z(x)$ such that...") seem to be more complex than what the automated systems are designed to handle. Scaling up to programs of a realistic size and complexity is also a formidable barrier.

Altogether, it appears that there is not a great deal to be gained by trying to apply the techniques of program verification to programs for numerical calculations. However, we will summarize our conclusions more specifically in the final section.

WHAT CAN "CORRECTNESS" MEAN?

It is time to consider the third question raised in the introduction regarding what we might mean by the correctness of a program.

Up until now we have taken the point of view that we had a particular function in mind, and that a program was to be considered correct if it evaluated that function. This seems to be the point of view in most of the work on program verification.

One of the points that has come up in previous sections has been that with numerical calculations it is quite possible to have different functions in mind for the same program. For example, we may have in mind the kind of mapping we associate with a backward error analysis, or, for the same program, we may have in mind a statement about a bound on the error in the final result. We could also have in mind statements involving the efficiency with which certain results are obtained. (Of course, proofs about efficiency can be made complicated in major ways by factors other than the program itself, such as the way in which subscripts are handled by the compiler.)

But while the idea of evaluating a function correctly is, strictly speaking, always a valid way of interpreting the correctness of a program, it may not always be the most useful way. For example, it may be helpful to identify other properties when discussing the correctness of an operating system, or of a simulation.

To pursue this idea just a little further, we might ask what is meant by the correctness of a program for playing chess. There are cases, and this may be one of them, when we mean a program is correct when it is identical to another, the other being of course described in another "programming" language (such as English). Alternatively, we may mean a program to be correct when it is merely equivalent to another program.

These alternatives must be kept in mind in dealing with numerical programs as
well. To conclude with just one example. We often mean "identical" to another
program, or at least "equivalent", in connection with numerical programs. This
arises when we have carefully analyzed a particular process and we then wish to
write a program to carry out that process "correctly". The function aspect has
already been dealt with in the analysis, and the correctness of the program
applies only to the proper coding of the procedure.

WHAT NEW FACILITIES ARE NEEDED?

We turn now to the fourth question raised in the introduction. If we find the
techniques of program verification to be at best of somewhat limited usefulness in
helping prove the correctness of numerical programs, we ask the question as to
whether there are other facilities that would be of help.

It seems to me that there are at least three major improvements that are feasible,
and that would provide a great deal of help. Each is in the nature of making our
programming language so much easier to use and the resulting programs so much
easier to understand, that their correctness (whatever we may mean by that) is
very much easier to establish. Each improvement is a subject in itself, so we
will restrict ourselves in this section to merely mentioning each of the three
improvements.

The first of these is clean arithmetic. The development of sensible and well-
defined standards for floating-point arithmetic is long overdue. We should not
have to put up any longer with the annoyance of small differences between
different kinds of machines, and very nearly inexplicable special properties for
individual machines. We should obtain properly rounded floating-point results on
the machines we use, with simple provision for handling exceptions such as overflow and underflow. This would help immeasurably in the design and analysis of
numerical programs. The programs would be portable, and more easily proven correct. Further discussion, and some specific proposals, are given in a paper by
Hull [1979]. See also Reinsch [1979].

A second improvement would be to have control over the precision of our calculations. There are many ways in which this would help simplify programming, as well
as broadening the kinds of programs we can develop, and also make it much easier
to analyze the numerical processes being carried out. This is somewhat more
difficult to accomplish in the hardware than is the clean arithmetic alone, but
the two are related. This topic, along with numerous examples, are also discussed
in more detail in the paper just mentioned (Hull [1979]), and in the references
given there.

The third improvement that, in my opinion, is most needed is to have suitable
symbol manipulation facilities incorporated in our programming language. There
often are occasions when we need to do such manipulations in the course of a numerical calculation. For example, we may need to find the Jacobian of a vector-
valued function, or we may need to manipulate power series. Of course, systems
for doing such things exist. But they are not incorporated into the programming
languages we use in a convenient way, and we are therefore not able to write the
necessary programs in a simple and easily understood form, i.e., in a form that is
easily shown to be correct!

CONCLUSION

In summary, it seems to me that some of the techniques and ideas used by persons
interested in "program verification" could be of some limited usefulness to those
of us interested in proving the correctness of numerical programs. Structured
programming, at least in a general sense (and as long as sufficient provision is

made for the necessary "escapes" that arise in numerical calculations), is of course very helpful in making programs easily understood, and hence more easily proven correct.

We have also seen how assertions can be adapted to numerical programs. Along with abstract data types, they might help us, at least in some situations, to organize a proof of correctness in a systematic way.

However, the advantages seem to me to be rather limited. And I am especially doubtful about gaining any benefit from what I perceive as the present trend towards formalizing and automating the verification process. For one thing, the emphasis is still almost completely on nonnumerical programs. But what may be even more important, it is difficult to see how the automatic techniques can be scaled up to handle programs of a realistic size and complexity.

The really significant requirement in proving program correctness is the need for careful mathematical analysis, such as has been so dramatically illustrated, for example, in numerical linear algebra since the 1950's. (Programs, and quite substantial ones at that, were being proven correct long before the use of assertions, etc., became popular.)

It also seems to me that this process of proving numerical programs correct would be very substantially assisted by the development of better programming language facilities, such as those mentioned in the preceding section regarding clean floating-point arithmetic, precision control, and symbol manipulation.

Perhaps it would also be worthwhile paying more attention to the question of exactly what we do mean by the correctness of particular programs. Some discussion has been used to show that this question can be interpreted in a variety of ways.

REFERENCES

[1] Wayne Enright (1972) Studies in the numerical solution of stiff ordinary differential equations. Ph.D. thesis, University of Toronto.

[2] R.W. Floyd (1967) Assigning meanings to programs. Proceedings Symposium in Applied Mathematics, AMS, 19, 19-32. (Edited by J.T. Schwartz.)

[3] G.E. Forsythe and C.B. Moler (1967) Computer solution of linear algebraic systems. Prentice-Hall, Englewood Cliffs.

[4] H.H. Goldstine and J. von Neumann (1947) Planning and coding of problems for an electronic computing instrument. Report prepared for U.S. Army Ordinance Dept., Part 2, vol.1. Reprinted in John von Neumann, Collected Works, 5, 80-151. The MacMillan Company, New York. (Edited by A.H. Taub.)

[5] T.E. Hull (1968) The numerical integration of ordinary differential equations. Proceedings IFIP Congress 1968, 1 (1969), 40-53. North-Holland, Amsterdam.

[6] T.E. Hull (1970) The effectiveness of numerical methods for ordinary differential equations. Studies in Numerical Analysis 2, 114-121. SIAM, Philadelphia. (Edited by J.N. Ortega and W.C. Rheinboldt.)

[7] T.E. Hull (1979) Desirable floating-point arithmetic and elementary functions for numerical computation. To appear in SIGNUM Newsletter.

[8] T.E. Hull, W.H. Enright and A.E. Sedgwick (1972) The correctness of numerical algorithms. Proceedings SIGPLAN Symposium on Proofs of Assertions about Programs, Las Cruces, New Mexico, 66-73.

[9] Kenneth R. Jackson (1974) Proving properties about subroutines that solve

ODEs numerically. M.Sc. thesis, University of Toronto.

[10] R.L. London (1977a) Perspectives on program verification. Current Trends in
Programming Methodology, II, Program Validation, 151-172. Prentice-Hall,
Englewood Cliffs. (Edited by R.T. Yeh.)

[11] R.L. London (1977b) Program verification. To appear in Research Directions
in Software Technology. (Edited by Peter Wegner.)

[12] D.C. Luckham (1977) Program verification and verification oriented programming. Information Processing 77, Proceedings IFIP Congress 77, 783-793.
North-Holland, Amsterdam. (Edited by B. Gilchrist.)

[13] P. Naur (1966) Proof of algorithms by general snapshots. BIT 6, 4, 310-316.

[14] Christian H. Reinsch (1979) Principles and preferences for computer arithmetic.
To appear in SIGNUM Newsletter.

[15] Arthur Sedgwick (1973) An effective variable order variable step Adams method.
Ph.D. thesis, University of Toronto.

[16] C.A. Steele (1974) Problems arising in proving the correctness of a computer
program. M.Sc. thesis, University of Toronto.

[17] A. Turing (1949) Checking a large routine. Conference on High Speed Automatic
Calculating Machines, 67-68. Cambridge University Mathematics Laboratory,
Cambridge.

Performance Evaluation of Numerical Software, Fosdick (ed.)
© IFIP, North-Holland Publishing Company, 1979

Some Fundamentals of Performance Evaluation for Numerical Software

W. Stanley Brown

Bell Laboratories
Murray Hill, New Jersey 07974

1. Introduction

Although people often use the word "performance" in discussing the speed of a program (measured in specified operations or in time on a specified computer), it might be wise to ask first how well the program performs, and indeed whether it performs at all. Is there a procedure for installing it on your computer? Will it produce correct results for all valid inputs?

Many numerical programs depend explicitly on the *environment parameters* that define the precision and range of numbers on the host computer. Have the values of these parameters for your computer been determined? Are they conveniently available?

If claims are made concerning the accuracy of a numerical program, then the result cannot be considered correct unless it supports those claims. Whether the claims are *a priori* or *a posteriori,* and whether they involve bounds or statistical expectations, they are likely to be expressed in terms of the environment parameters mentioned above, whose values are therefore crucial to any meaningful assessment of numerical performance.

Overflow and underflow are also involved in the issue of performance. Can overflow or underflow (or anomalies near the limits of the range) cause a loss of accuracy not anticipated in the error analysis? Can roundoff error cause overflow or underflow when the exact computation is known to remain in range?

To develop a portable program for which the answers to these questions are satisfactory, one must be sure that all assumptions about the host computer are explicit, and that all machine dependencies are expressed so that they can easily be found, understood, and changed. Even in non-portable programs, these measures are desirable for readability, reliability, and maintainability.

With these requirements in mind, the author has developed a model of floating-point computation[1] intended as a basis for the development and analysis of efficient and highly portable programs for a wide variety of mathematical algorithms. After reviewing this model in Section 2, we show in Section 3 how it can be used in the development and analysis of a simple algorithm to compute the mean of the components of a vector. The extension of these ideas to other, more elaborate algorithms is briefly discussed in Section 4.

In practice, the use of the model depends on the availability of correct values for its environment parameters. A test program to assist in the evaluation of these parameters is described in Section 5. Since this program tests the numerical performance of the host computer, it is a fundamental tool for the performance evaluation of numerical software. The paper concludes in Section 6 with some observations about other tools.

2. Review of the Model

The purpose of the model is to capture the fundamental concepts of floating-point computation in a small set of parameters, which characterize the numerically most important attributes of a computing environment, and in a small set of axioms, which describe floating-point computation in terms of these parameters. In writing portable software, one focuses on the model and avoids thinking about the idiosyncrasies of any particular machine. Since the parameters of the model may be used freely, a program tailors itself to each environment in which it is installed, and a high degree of efficiency is possible. In writing about portable software, one also focuses on the model and its parameters, and thus the documentation is portable too.

The model includes a few basic parameters and a few derived parameters for each floating-point number system that is supported by the host computer. The basic parameters are

1. the *base*, $b \geqslant 2$
2. the *precision*, $p \geqslant 2$.
3. the *minimum exponent*, $e_{min} < 0$
4. the *maximum exponent*, $e_{max} > 0$

These define a system of *model numbers* consisting of zero and all numbers of the form

$$x = fb^e \tag{1}$$

where

$$f = \pm (f_1 b^{-1} + \cdots + f_p b^{-p})$$
$$f_1 = 1,...,b-1$$
$$f_2, \ldots, f_p = 0,...,b-1 \tag{2}$$

and

$$e_{min} \leqslant e \leqslant e_{max} \tag{3}$$

The parameters must be chosen so that these model numbers are exactly representable in the machine, and so that operations on them behave according to a few simple and desirable axioms. Operations on other *machine numbers,* if any, are bounded by the corresponding operations on the neighboring model umbers.

On some computers, the model numbers may coincide exactly with the machine numbers. However, anomalies in the behavior of a computer may require that the parameters be penalized to reduce the purported range or precision. Such penalties also reduce the size of the set of model numbers, and thus create a set of machine numbers that are not model numbers and need not behave quite so well in computation. Any anomalies that cannot be accommodated by modest penalties are usually recognized by the manufacturer as design errors, and repaired in due course.

Since the model numbers with a given exponent, e, are equally spaced on an absolute scale, the relative spacing decreases as the magnitude of the fraction-part, f, increases. For error analysis, the maximum relative spacing

$$\epsilon = b^{1-p} \tag{4}$$

is of critical importance. Also of interest throughout this paper are the smallest positive model number

$$\sigma = b^{e_{min}-1} \tag{5}$$

and the largest model number

$$\lambda = b^{e_{max}}(1-b^{-p}). \tag{6}$$

To provide at least a minimal range for any given precision (enough for the algorithm discussed in Section 3 and for many others, but far less than one might like), we propose requiring that

$$\sigma < \epsilon^2 \tag{7}$$

and

$$\lambda > \epsilon^{-2}. \tag{8}$$

These inequalities may be restated as

$$e_{min} \leqslant 2-2p \tag{9}$$

and

$$e_{max} \geqslant 2p-1. \tag{10}$$

While the double-precision numbers on Honeywell 6000 series computers meet these inequalities with very little to spare, we find it difficult to imagine a floating-point format in which the exponent gets a smaller fraction of the total space. Assuming the same word size (72 bits) and the same split (8 bits for the signed exponent and 64 bits for the signed fraction-part), one could use an implicit normalization bit to increase the precision to 64. Plausible exponent ranges (after setting one exponent aside for the number zero) would then be $[-127, +127]$ or $[-126, +128]$. The proposed inequalities are the tightest that would permit both of these plausible systems. Of course, (9) or (10) could fail in a high-precision number system with a range that is both small and asymmetric, but machine designers seem to agree that this would not be an attractive combination of attributes. If necessary, a failure of (9) or (10) could be overcome by penalizing the claimed precision.

Since (7) and (8) appear to be necessary for a useful number system, while the equivalent (9) and (10) appear to be universally valid but barely safe for existing and plausible computers, we tentatively propose that these inequalities be incorporated into the model. A simple corollary is that

$$\sigma < \epsilon^4\lambda. \tag{11}$$

Turning to arithmetic, the model admits the possibility that a machine number x may be *extra-precise* (between adjacent nonzero model numbers) or *out-of-range* ($0 < |x| < \sigma$ or $|x| > \lambda$). If out-of-range numbers are avoided, it is shown in [1] that the model supports conventional forward error analysis with only minor qualifications. However, for the purposes of this paper we need to strengthen the axioms to permit computation with the result of an underflow, and to restate some of the rules of error analysis in terms of absolute rather than relative errors.

If a floating-point operation underflows and the exact mathematical result, x, is positive, the axioms require that the computed value, $\tilde{x}$, be in the interval $[0,\sigma]$. Thus the absolute error satisfies

$$|x - \tilde{x}| < \sigma . \tag{12}$$

While this error may be serious or even catastrophic, we would like to strengthen the axioms by removing the restriction against using such a result in subsequent computation. Since the present paper illustrates the need for the stronger axioms, and since our earlier fears of unusable unnormalized tiny numbers appear to have been unwarranted, we tentatively propose that the restrictions against operands in the underflow range be removed, except that no such operand may be used as a divisor.

To conclude this section, we present the rules for error analysis that will be needed in Section 3. These rules can be derived from the rules presented in Table 1 of [1], but the reader will obtain more insight by deriving them directly from the axioms of the model. Let

$$\tilde{x} = x + \xi$$
$$\tilde{y} = y + \eta$$
$$\tilde{z} = z + \zeta \tag{13}$$

where x, y, z are the exact mathematical values of some desired quantities; $\tilde{x}$, $\tilde{y}$, $\tilde{z}$ are the corresponding approximate computed values; and ξ, η, ζ are the absolute errors. Also, let n be an exactly known model number, and let v be the absolute error in the computed value of its reciprocal, n^{-1}. Table 1 gives exact error formulas for the relevant arithmetic operations.

Operation	Error Formula
$x = y \pm z$	$\xi = \eta \pm \zeta + \delta\tilde{x}$
$x = y \cdot z$	$\xi = \eta z + \zeta\tilde{y} + \delta\tilde{x}$
$x = y/n$	$\xi = \eta/n + \delta\tilde{x}$
$x = y \cdot (n^{-1})$	$\xi = \eta/n + \delta\tilde{x} + v\tilde{y}$

Table 1.
Error formulas for the relevant arithmetic operations.

In the formulas for addition, subtraction, and multiplication, there are two terms to account for the error in the operands, and a third term $\delta\tilde{x}$, with $|\delta| < \epsilon$, to account for the error introduced by rounding or chopping the result. Instead of considering division in full generality, we restrict our attention to the case where the divisor is an exactly known model number. The table gives two rules for division. The first assumes that division is implemented by a single machine instruction, while the second assumes that it requires a reciprocation followed by a multiplication.

There is one operation that is vital to the algorithm in Section 3 but omitted from the table; namely scaling by a power of the base. Since this is exact for model numbers, and since the model effectively treats other machine numbers as model intervals, it may be considered exact for the purposes of error analysis.

3. An Algorithm to Compute the Mean of a Vector

To explore the issues involved in using the model, we present an accurate, efficient, and portable algorithm to compute the mean of the components of a vector $\mathbf{x} = (x_1, \ldots, x_n)$. By appropriate scaling, the algorithm avoids all overflow. However, it tolerates underflow if the resulting loss of accuracy is either negligible (relative to some x_i) or unavoidable.

The algorithm is accurate in the sense that the simple (and realistic) error bound is not affected by underflow unless the computed mean itself underflows.

The algorithm is efficient in the sense that it makes only one pass over the vector, and uses only a single accumulator. The overhead due to scaling consists of at most $n+1$ comparisons and at most $n+2$ multiplications.

Finally, the algorithm is portable in the sense that it depends on the host computer only through the environment parameters and arithmetic axioms of the model. Thus a program implementing the algorithm can be used without change on any computer that supports the model, providing only that the parameters receive correct values by some mechanism such as a compiler, a preprocessor, or a set of library functions.

We are now prepared to present the algorithm, followed by some comments and a careful analysis:

$$
\begin{aligned}
&\textbf{procedure } mean\,(x,n) \\
&check: \\
&\quad \text{assert } (0 < n < \epsilon^{-1}) \\
&small: \\
&\quad S := 0 \\
&\quad \text{for } i := 1,\ldots,n \\
&\qquad \text{if } (|x_i| < b\sigma/\epsilon^2) \; S := S + \epsilon^{-1}x_i \\
&\qquad \text{else go to large} \\
&\quad \text{return } ((\epsilon S)/n) \\
&large: \\
&\quad S := \epsilon^2 S \\
&\quad \text{for } i := i,\ldots,n \\
&\qquad S := S + \epsilon x_i \\
&return: \\
&\quad S := S/(neb) \\
&\quad \text{if } (S \geqslant \lambda/b) \text{ return } (\lambda) \\
&\quad \text{else if } (S \leqslant -\lambda/b) \text{ return } (-\lambda) \\
&\quad \text{else return } (Sb) \\
&\quad \text{end}
\end{aligned}
$$

To permit proofs, we need a bound on n, and to avoid excessive loss of accuracy, we need $n\epsilon \ll 1$, though we require only that $n\epsilon < 1$. The algorithm begins (see *check)* by checking this condition.

For efficiency, the algorithm makes only one pass over the vector, and uses only a single accumulator. To avoid destructive underflow, it scales away from zero (see *small)* as long as all the components are *small* (i.e., less than $b\sigma/\epsilon^2$ in magnitude). To avoid overflow, it scales toward zero (see *large)* starting with the first *large* (i.e., at least $b\sigma/\epsilon^2$ in magnitude) component. Note that the boundary, $b\sigma/\epsilon^2$, and the scale factors, ϵ and ϵ^{-1}, are all powers of the base, b. Although the exact mathematical mean cannot overflow, there is a danger that the computed mean might overflow because of accumulated roundoff. To avoid this, the computation of the mean from the scaled sum is performed with great care (see *return)* whenever at least one large component has been encountered.

In analyzing the error in the computed mean, we consider only the error added by the algorithm, assuming that all the x_i are model numbers. A separate analysis of the effects of errors in the x_i is also required, and if any x_i is not a model number, the error interval assigned to it must include both of the neighboring model numbers. We can safely require that any such x_i has two neighbors, because it is essential for portability to ensure that $|x_i| \leqslant \lambda$ for $i = 1,...,n$.

For simplicity, we first consider a simplified version of the algorithm in which no scaling is done, and we assume that no underflow or overflow occurs. Let

$$S_i = x_1 + \cdots + x_i$$
$$S = x_1 + \cdots + x_n = S_n$$
$$M = S/n, \tag{14}$$

and

$$S^+ = \max(|\tilde{S}_i|)$$
$$S^* = \begin{cases} \max(S^+, \sigma), & \text{if if all components are small,} \\ \max(S^+, b\sigma/\epsilon^2), & \text{otherwise} \end{cases}$$
$$M^* = S^*/n \ , \tag{15}$$

where $\tilde{S}_i$ is the computed value of S_i. From Table 1, the incremental error introduced in computing $S_i = S_{i-1} + x_i$ *(for $i=2,...,n$)* is bounded by $\epsilon|\tilde{S}_i|$, and therefore

$$|S - \tilde{S}| < \epsilon \sum_{i=2}^{n} |\tilde{S}_i| \leqslant (n-1)\epsilon S^* \ . \tag{16}$$

It follows from the division rules in Table 1 that

$$|M - \tilde{M}| < \begin{cases} n\epsilon M^*, & \text{if division is basic,} \\ (n+1)\epsilon M^*, & \text{if division is composite.} \end{cases} \tag{17}$$

While the relative error in $\tilde{M}$ may be very large, these bounds imply that rounding errors in the initial data could yield a comparable (or perhaps much larger) error in the result.

Turning to the general case, it is clear that scaling and unscaling do not affect (17) unless they cause underflow. Suppose x_i is the first large component, and an underflow occurs when we rescale the scaled sum $\epsilon^{-1}\tilde{S}_{i-1}$. Then the exact mathematical value of

$\epsilon \tilde{S}_{i-1}$ is less than σ in magnitude, while its computed value (the result of the underflow) is less than or equal to σ in magnitude. Furthermore, if the computed value is nonzero, it has the same sign as the exact value. On the other hand, since x_i is large, we have $\epsilon |x_i| \geqslant b\sigma/\epsilon$. It follows that the exact and computed values of the sum, $\epsilon \tilde{S}_{i-1} + \epsilon x_i$, are within one unit in the last place of ϵx_i, and are also within one unit in the last place of each other. Hence the combined error of the rescaling and the addition is bounded by ϵS^*, and the underflow has no effect on (17). A similar argument shows that we need not be concerned about the possibility of underflow due to the scaling of subsequent small components.

Let us now consider the possibility of underflow during addition. Let α_i be the scale factor (ϵ or ϵ^{-1}) applied to x_i. The error introduced in computing the sum $\alpha_i S_i = \alpha_i S_{i-1} + \alpha_i x_i$ is bounded by σ if underflow occurs, and by $\epsilon \alpha_i S^*$ otherwise. We shall prove that $\sigma \leqslant \epsilon \alpha_i S^*$ for all i, and hence any underflow in this context has no effect on (17). First,

$$S^* \geqslant \begin{cases} \sigma, & \text{if all components are small,} \\ \sigma/\epsilon^2, & \text{otherwise.} \end{cases} \tag{18}$$

Next,

$$\alpha_i \geqslant \alpha_n = \begin{cases} \epsilon^{-1}, & \text{if all components are small,} \\ \epsilon, & \text{otherwise.} \end{cases} \tag{19}$$

Multiplying these inequalities, we see that $\alpha_i S^* \geqslant \sigma/\epsilon$, as was to be shown.

It remains to bound the error, μ, introduced into the mean by computing it from the scaled sum. For simplicity, we first consider the case where division is a basic operation.

If all the components of x are small, and if an underflow occurs in computing $(\epsilon S)/n$, then $|\mu| < \sigma$. If underflow does not occur, then $|\mu| < \epsilon M^*$ as before. In either case, $|\mu| < \max(\epsilon M^*, \sigma)$.

On the other hand, if at least one component of x is large, and if no underflow occurs in computing $S/(n\epsilon b)$, then $|\mu| < \epsilon M^*$ as before. If underflow does occur, then $|\mu| < \sigma b < \sigma b/n\epsilon = \epsilon(\sigma b/\epsilon^2)/n \leqslant \epsilon S^*/n = \epsilon M^*$. In either case, $|\mu| < \epsilon M^*$.

Whether or not x has large components, we have $|\mu| < \max(\epsilon M^*, \sigma)$ whenever division is a basic operation. If division is a composite operation, the reciprocation of n (or $n\epsilon b$, which differs from n only in the exponent field) may involve roundoff, but it cannot underflow. The error in M due to this roundoff is bounded by ϵM^*, and the rest of the analysis is unaffected. Hence $|\mu| < \epsilon M^* + \max(\epsilon M^*, \sigma)$. Combining these bounds with the contribution from (16), we conclude that

$$|M - \tilde{M}| < \begin{cases} (n-1)\epsilon M^* + \max(\epsilon M^*, \sigma), & \text{if division is basic,} \\ n\epsilon M^* + \max(\epsilon M^*, \sigma), & \text{if division is composite.} \end{cases} \tag{20}$$

In the usual case when M does not underflow, this bound can be replaced by (17).

In considering the possibility of overflow, we need two simple lemmas, which follow easily from [1]. These describe rare exceptions to the general rule that tight mathematical inequalities may be reversed by computational roundoff.

Lemma 1

Let n be an integer with $0 < n \leqslant b^p$, let B be a power of the base with $\sigma \leqslant B$ and $nB \leqslant \lambda$, and let $u_1, \ldots, u_n$ be machine numbers with $|u_i| \leqslant B$. Then the magnitude of the computed value of $u_1 + \cdots + u_n$ does not exceed nB.

Lemma 2

Let x be a machine number, y a model number, and B a power of the base. Suppose B and y/B are both in range, and hence also model numbers. Also suppose $|x| \leqslant \lambda$. If the computed value of the relation $x \geqslant y/B$ is *false*, then in fact $x \leqslant y/B$ and the computed value of Bx does not exceed y.

In the *small* section of the algorithm, S is a sum of terms, each of which is bounded by $b\sigma/\epsilon^3$. Hence by Lemma 1, $|S| \leqslant nb\sigma\epsilon^{-3}$. Since $n < \epsilon^{-1}$ and $\sigma \leqslant \epsilon^2/b$, it follows that $|S| < \epsilon^{-2} < \lambda$, so no overflow can occur.

In the *large* section of the algorithm, S is a sum of terms, each of which is bounded by $\epsilon\lambda < \epsilon b^{e_{max}}$. Hence by Lemma 1, $|S| \leqslant n\epsilon b^{e_{max}}$. Since $n \leqslant \epsilon^{-1} - 1$, it follows that $|S| \leqslant (1-\epsilon)b^{e_{max}} < \lambda$, so no overflow can occur.

It remains to prove that the precautions taken in the *return* section of the algorithm are sufficient to avoid overflow. Upon entering this section, $|S| \leqslant n\epsilon b^{e_{max}}$ as noted above, and the magnitude of the mean $S/(n\epsilon)$, if computed directly, could overflow by one or two rounding errors. Hence the algorithm computes $S/(n\epsilon b)$, which is perfectly safe, and then multiplies carefully by b. Actually the multiplication operation is not invoked unless the computed value of the relation $S \geqslant \lambda/b$ is *false*, and in this case Lemma 2 implies that the computed value of Sb does not exceed λ. By a similar argument, one can also show that the computed value of Sb is not less than $-\lambda$, and hence no overflow can occur.

4. Other Numerical Algorithms

The preceding algorithm demonstrates a very high standard of provable performance, and the reader may wonder whether a complete numerical library can or should be built to that standard. To explore this question, I have studied various algorithms for computing the Euclidean norm of a vector and the scalar product of two vectors.

For the Euclidean norm, J. L. Blue[2] has presented a portable and efficient algorithm that goes beyond our standard by avoiding underflow as well as overflow. Although Blue's algorithm makes only one pass over the vector, it uses three accumulators, and many numerical analysts consider it unreasonably complicated.

A simpler algorithm that tolerates nondestructive underflow has been presented by C. L. Lawson.[3] Although Lawson did not attempt a rigorous analysis of his algorithm, one can easily recast it along the lines of the preceding section of this paper, and prove that it always achieves the desired accuracy and never overflows.

For the scalar product, there are several approaches that yield a suitable level of accuracy and can be guaranteed not to overflow. However, I have not been able to develop an algorithm that is both efficient and straightforward. Essentially one must either accept a minimum of three scale factors for the components of each vector (and hence at least five for the terms of the scalar product), or use multiple accumulators or some other scheme that effectively provides an extended range.

So far, we have implicitly accepted the constraint of having only a single floating-point number system. If an auxiliary *long real* system is available, providing both extended precision and extended range, then it is very easy to implement all of these algorithms so that overflow and destructive underflow are avoided and the results are accurate to within a single rounding error. For example, the following algorithm computes the mean of the components of a short real vector, x, to within σ if the result underflows, or a single unit in the last place otherwise. As before, there is a danger that the computed mean might overflow because of accumulated roundoff. If so, the overflow will occur when the final result, S/n, is demoted to short. In this case, we assume that the *short* function can be made to return $\pm\lambda$ (with the correct sign) instead of putting the program into an error state.

```
procedure mean (x, n)
integer n
short real array(n) x
long real S
S := 0
for  i := 1,...,n
   S := S + x_i
return (short(S/n))
end
```

Formally, this approach requires separate environment parameters for the short and long number systems, and a new axiom for conversions from one system to the other.

To avoid overflow in our new algorithm for the mean, we require only that $n\lambda_1 \leqslant \lambda_2$, where the subscripts refer to the short and long number systems, respectively. In the corresponding algorithms for the Euclidean norm and the scalar product, we would require that $n\lambda_1^2 \leqslant \lambda_2$. Since $n < \epsilon_1^{-1} < \lambda_1$, it suffices for these three algorithms to have $\lambda_1^3 \leqslant \lambda_2$. For most practical purposes it should suffice to have 2 extra bits in the exponent field $(\lambda_2 \approx \lambda_1^4)$, but for still greater safety we recommend 3 extra bits $(\lambda_2 \approx \lambda_1^8)$ or one extra decimal digit $(\lambda_2 \approx \lambda_1^{10})$.

To avoid destructive underflow in our algorithm for the mean, we need only that $n\sigma_2 < \epsilon_1\sigma_1$, which can be guaranteed by requiring that $\sigma_2 \leqslant \sigma_1^2$. In the corresponding algorithm for the Euclidean norm, only the squaring of components threatens underflow, and this can be avoided by requiring that $\sigma_2 \leqslant \sigma_1^2$. Finally, to avoid destructive underflow in computing the scalar product, we would need $n\sigma_2 < \epsilon_1\sigma_1^2$, which can be guaranteed by requiring that $\sigma_2 \leqslant \sigma_1^3$. This requirement is sufficient for all three of these algorithms. However, for greater safety we again recommend 3 extra bits in the exponent field $(\sigma_2 \approx \sigma_1^8)$ or one extra decimal digit $(\sigma_2 \approx \sigma_1^{10})$.

Finally, to guarantee that our new algorithm for the mean is accurate to within a relative error of ϵ_1, we require only that $n\epsilon_2 < \epsilon_1$. For the scalar product the requirement is the same, while for the Euclidean norm it is weaker by about a factor of 2, since the square root in the definition halves the relative error of the final result. In all three cases, 20 extra bits or 6 extra decimal digits in the fraction field will suffice for all $n \leqslant 10^6$.

Once we have a long real number system at our command, we will want to have library procedures for computing with long real numbers, and for those procedures we will need another auxiliary number system with still greater precision and range. Thus what we need is an infinite sequence of number systems, each offering more precision (at least 20 extra bits or 6 extra digits in the fraction field) and more range (at least 3 extra bits or one extra digit in the exponent field) than its predecessor. Obviously only the first two or three of these systems would be provided in hardware, but all of them should be conveniently available at the language level. To avoid an infinite sequence of libraries, we will need a language facility for determining the precision of a formal parameter and declaring auxiliary variables of the next longer precision.

It is interesting to consider the number formats that are suggested by these requirements. Typically we would allocate one word for short numbers, two words for long numbers, three words for long long numbers, and so forth, with the bits or digits of each word after the first being divided in some fixed ratio between the exponent and fraction fields. As an example, for binary computers with 32-bit words, I am enthusiastic about current proposals[4,5] to partition the first word into an 8-bit signed exponent and a 24-bit signed fraction with an implicit normalization bit, and for long numbers to allocate 3 bits from the second word to the exponent field and the remaining 29 bits to the fraction field.

On a decimal computer with 9 digits and perhaps 1 or 2 sign bits per word (all of which could easily be encoded in 32 bits), I would suggest partitioning the first word into a 2-digit exponent and a 7-digit fraction, and allocating one digit from each subsequent word to the exponent field and the remaining 8 digits to the fraction field.

On a computer with exceptionally long words (e.g., 60 bits or 18 digits), it may be desirable to call the single-word numbers "long". In this case, a reasonable facsimile of a "short" system can be provided by performing a range check and an optional rounding on every assignment of a long value to a short variable.

5. Evaluating the Environment Parameters

As mentioned in Section 2, the model accommodates numerically anomalous machines by adjusting their environment parameters to reduce the purported precision or range. Thus the parameters reflect primarily the dynamic behavior of the host computer rather than its static number representation.

To use the model and the environment parameters effectively in the development, analysis, and documentation of mathematical software, it is obviously crucial to determine the parameter values for each target machine, to make them conveniently available to software developers from relevant programming languages, and to publish them in human-readable form for the benefit of the ultimate users. Unfortunately, there are many possible anomalies that may affect the parameters, and some of them are quite subtle.

While it may not be easy to determine the correct values of the environment parameters, it is necessary, and N. L. Schryer has developed a test program[6] to aid in the task.

This program performs arithmetic operations and comparisons on carefully chosen pairs of operands, and tests whether the results conform to the axioms of the model. Obviously such a test cannot be exhaustive, because there are far too many pairs of numbers in any useful floating-point system to test them all. Nevertheless, Schryer's program exercises the hardware quite thoroughly, and I doubt that many errors will elude it.

The possibility of discovering the parameter values automatically by numerical experimentation has been discussed in the literature[7] but we feel that this approach has inherent limitations.[8] By contrast, Schryer's program goes to far greater lengths to achieve the more limited objective of testing whether a given computer and a given set of parameter values support the model. While we feel that a detailed analytical justification of the final values is essential, the test program can help us to acquire the necessary understanding and can strengthen our confidence that we have not overlooked any relevant phenomena.

In using Schryer's program, we recommend first setting the parameters to the values that are suggested by the representation of floating-point numbers as documented by the manufacturer. If the tests reveal any difficulties, they should be studied and understood, and the parameter values adjusted accordingly. The entire process should be repeated until the tests reveal no further problems.

To describe Schryer's test program in more detail, we need some notation from [1]. First, if x is a real number with $|x| \leqslant \lambda$, then x' denotes the smallest model interval containing x. As a special case, if x is a model number, then $x' = x$. Second, if x is any expression, then $fl(x)$ denotes the computed value of x.

To test the operation of multiplication, Schryer's program chooses model numbers x and y with $|xy| \leqslant \lambda$ and attempts to verify that

$$fl(xy) \in (xy)' \qquad (21)$$

as required by the multiplication axiom. Each of the operands is formed by choosing a fraction part and an exponent from certain specified sets. The fraction part is chosen from one of the following patterns:

$$f = 0 \ or \ \pm b^{-1}$$
$$f_i = \pm (b^{-1} + b^{-2} + \cdots + b^{-i})$$
$$f_i = \pm (b - 1)(b^{-1} + b^{-i})$$
$$f_i = \pm (b - 1)(b^{-1} + b^{-2} + \cdots + b^{-i}) \qquad (22)$$

where $i = 2,...,p$. The exponent is chosen from an array provided by the user, which should include all possible values near 0, e_{min}, and e_{max}. All pairs of operands that can be formed in this way and that satisfy the inequality $|xy| \leqslant \lambda$ are tested.

As an example of the testing procedure, let

$$x_i = b^{-1} + b^{-i}$$
$$y_j = b^{-1} + b^{-j}$$
$$z_{ij} = x_i y_j$$
$$\tilde{z}_{ij} = fl(x_i y_j) \tag{23}$$

with $2 \leqslant i,j \leqslant p$. Our task is to test the assertion that $\tilde{z}_{ij} \in z_{ij}'$. Clearly

$$z_{ij} = b^{-1}(b^{-1} + b^{-i} + b^{-j} + b^{1-i-j}) \ . \tag{24}$$

If the precision, p, is large enough, then z_{ij} is a model number, and so $z_{ij}' = z_{ij}$. Otherwise, the end-points of z_{ij}' can be constructed from (24) by substituting 0 and a suitable larger power of b, respectively, for the last term. The critical value of p depends on whether the quantity in parentheses is less than unity (the usual case) or greater than unity (the special case $i = 2$, $j = 2$, and $b = 2$). Given an array B with $B_i = b^{-i}$ for $i = 1,...,p$, it is easy to construct the end-points of z_{ij}' using only operations whose operands and exact mathematical results are model numbers. If the model is supported, these constructions will be error-free, and we can then use ordinary comparisons to test whether $\tilde{z}_{ij} \in z_{ij}'$.

If $\tilde{z}_{ij}$ is outside the required interval by less than one unit in the last place, the comparison operation might equate it to the nearest end-point, thus allowing the error to escape detection. However, if there is one such error, there are likely to be many, and we would expect to miss only about half of them. If a single error of a more serious nature occurs, it is almost certain to be detected. While multiple errors might compensate each other in a particular case, we would expect to encounter other cases where the compensation would not occur. Although sample testing can never prove the absence of errors, I am convinced that Schryer's test program will prove to be extremely helpful in uncovering machine anomalies and determining correct values for the environment parameters.

6. Tools for Numerical Performance Evaluation

The modern numerical programmer uses many tools in developing, analyzing, and documenting mathematical software that will perform efficiently and reliably wherever it is needed, and will deliver results as accurate as claimed.

Like most other software tools, those that are intended for classical performance evaluation (i.e., timers and profilers) are not confined to numerical programs, but can be used far more generally. For numerical software, it is important to extend the concept of performance to include the question of accuracy and the problems of overflow and underflow. Fortunately, some work is being done on tools for roundoff analysis[9] and the generation of test data.[10]

In this paper we have focused on a more fundamental tool, by N. L. Schryer, for testing the numerical performance of a computer. Obviously this is an essential ingredient in any evaluation of the performance of numerical software running on the computer. As an interesting sidelight, we remark that the development of this tool depended on the extensive use of other, more general, software tools, as discussed in [6].

References

1. W. S. Brown, "A Realistic Model of Floating-Point Computation," pp. 343-360 in *Mathematical Software III*, ed. John R. Rice, Academic Press, New York (1977).

2. J. L. Blue, "A Portable Fortran Program to Find the Euclidean Norm of a Vector," *ACM Trans. Math. Soft.* **4**, pp. 15-23 (March 1978).

3. C. L. Lawson, *Environment Parameters and the L2 Norm in the BLAS*, unpublished, November 1977.

4. J. T. Coonen, "Specifications for a Proposed Standard for Floating Point Arithmetic," *Revised Memorandum No. UCB/ERL M78/72*, Department of Mathematics, University of California at Berkeley, Berkeley, California (December 1978).

5. Mary Payne and William Strecker, *Draft Proposal for a Floating Point Standard*, Digital Equipment Corporation, Maynard, Massachusetts, December 1978.

6. N. L. Schryer, "UNIX as an Environment for Producing Numerical Software," *ACM SIGNUM Newsletter*, (To appear.) (1979).

7. M. A. Malcolm, "Algorithms to Reveal Properties of Floating-Point Arithmetic," *Comm. ACM* **15**, pp. 949-951 (November 1972).

8. W. M. Gentleman and S. B. Marovich, "More on Algorithms That Reveal Properties of Floating-Point Arithmetic Units," *Comm. ACM* **17**, pp. 276-277 (May 1974).

9. W. Miller and D. Spooner, "Software for Roundoff Analysis, II," *ACM Trans. Math. Soft.* **4**, pp. 369-387 (December 1978).

10. W. Miller and D. Spooner, "Automatic Generation of Floating-Point Test Data," *IEEE Trans. on Software Engineering* **SE-2**, pp. 223-236 (1976).

Performance Evaluation of Numerical Software, Fosdick (ed.)
© IFIP, North-Holland Publishing Company, 1979

CORRECTNESS PROOF AND MACHINE ARITHMETIC

by

T.J. Dekker
Department of Mathematics,
University of Amsterdam
Roetersstraat 15,
1018 WB Amsterdam, Netherlands

1. Introduction

In the last decade, there is a great activity in proving correctness of programs
and in structured programming. The purpose of this paper is to give some results
in the area of proving correctness of numerical programs. We focus attention to
correctness of programs perfomed in inexact (floating-point) arithmetic satisfying certain sets of axioms. We not only consider finite precision (section 2),
but also limited exponent range (section 3). As an example we consider some algorithms for finding a zero of a function in an interval (section 4).

The examples show that useful results on the behaviour of numerical algorithms can
be obtained concerning not only the effects of finite precision, but also those
of underflow and overflow. In particular, it becomes apparent that in some
situations an underflow trap may be very undesirable. It may, for instance, interrupt a beautifully converging iteration process. Even an overflow trap may sometimes be undesirable.

2. Finite precision

In this section, we describe four different sets of axioms for the arithmetic in
a floating-point system with finite precision, but unlimited exponent range (i.e.
we here disregard underflow and overflow). This system, $\mathbb{F}$, is determind by two
parameters, the base b and the mantissa length p which both are integers larger than 1.

Definition $\mathbb{F} = \mathbb{F}(b,p)$ is the set of numbers of the form

$$x = mx \times b^{ex-p},$$

where mx and ex are integers and $|mx| < b^p$.

Each non-zero element x of $\mathbb{F}$ has an immediate predecessor, x^- , and successor,
x^+ , in $\mathbb{F}$, and a <u>relative spacing</u> by definition equal to

$$\max(x^+ - x, \; x - x^-)/|x|.$$

Obviously, the relative spacing has minimal value $b^{-p}/(1 - b^{-p})$ and maximal
value b^{1-p} .
So, the <u>resolution</u> or <u>maximal relative spacing</u> Δ between successive non-zero
numbers in $\mathbb{F}$ equals b^{1-p} , cf. Reinsch [1978].
We consider the following operations:
the relational operations $<, \leq, =, \neq, \geq, >$, the sign reversal $-$ (i.e. the monadic
minus), and the dyadic arithmetical operations $+, -, \times, /$.

We assume that the implementations of the relational operations and of the sign
reversal in this system are exact. (In practice, this assumption holds for the
sign reversal, but not always for the relational operations; we here assume exact-
ness for simplicity.)

For the implementations of the dyadic arithmetical operations, we state four different sets of axioms which are more or less common in literature (see, for instance, Wilkinson [1963], Dekker [1971] and Reinsch [1979]), and fulfilled in all
(A) or several   (B,C,D) machine implementations.

In the axiom sets A and B, there is a constant ε , called the <u>arithmetic</u>
<u>precision</u> or <u>machine precision</u>, which bounds the errors in the implemented arith-
metical operations. It has to satisfy $0 < \varepsilon < 1$; for reasons to be explained
below, we assume $0 < \varepsilon < 1/8$ and $\Delta < 1/8$.

For $\ast \in \{+, -, \times, /\}$, let $\hat{\ast}$ denote the corresponding operation implemented in
system $\mathbb{F}$. For division, we assume that the denominator is non-zero.
<u>Axiom set A (weak arithmetic)</u>.
In this axiom set, it is assumed that multiplication and division have a low relative error bounded by ε times the magnitude of the exact result, whereas addition and subtraction have an error bounded by ε times the sum of the magnitudes
of the operands. In other words, for $\ast \in \{+, -\}$,

$$x \mathbin{\hat{\ast}} y = (x\xi) \ast (y\eta), \quad |\xi - 1| \leq \varepsilon, \quad |\eta - 1| \leq \varepsilon,$$

and, for $\ast \in \{\times, /\}$,

$$x \mathbin{\hat{\ast}} y = (x \ast y)\zeta, \quad |\zeta - 1| \leq \varepsilon.$$

Wilkinson [1963] mentions these axioms for round-off with single-precision accumulator. Disregarding underflow and overflow, we can state that these axioms
hold in every machine implementation.
<u>Axiom set B (strong arithmetic)</u>.
In this axiom set, it is assumed that not only multiplication and division, but

also addition and subtraction have a low relative error bounded by ε times the magnitude of the exact result. In other words, for $* \in \{+,-,\times,/\}$,

$$x \; \hat{*} \; y = (x * y)\zeta, \; |\zeta - 1| \le \varepsilon.$$

Clearly B implies A with the same value of ε. Wilkinson [1963] mentions these axioms for round-off with double-precision accumulator. In fact, only one guard digit is needed to achieve strong arithmetic.

Axiom set C (faithful arithmetic).

In this axiom set, it is assumed that the four arithmetical operations all yield a result which is either the largest element of $\mathbb{F}$ not larger, or the smallest element of $\mathbb{F}$ not smaller than the exact result. In other words, when $x * y$ lies between two successive elements a and a^+ of $\mathbb{F}$, then $x \; \hat{*} \; y$ equals either a or a^+; otherwise ($x * y$ is in $\mathbb{F}$ and) $x \; \hat{*} \; y = x * y$.

Clearly, C implies B with $\varepsilon = \Delta$.

Accordingly, we define the arithmetic precision ε in axiom set C by $\varepsilon = \Delta$.

Axiom set D (proper rounding arithmetic).

In this axiom set, it is assumed that the four operations considered all yield a result equal to an element of $\mathbb{F}$ nearest to the exact result.

Note, that this axiom set determines the results uniquely, except when $x * y$ lies halfway between two successive elements a and a^+ of $\mathbb{F}$, in which case $x \; \hat{*} \; y$ equals either a or a^+. Clearly, D implies C and B with $\varepsilon = \frac{1}{2}\Delta$.

Accordingly, we define the arithmetic precision ε in axiom set D by $\varepsilon = \frac{1}{2}\Delta$.

Axiom sets C and D were used by Dekker [1971].

To achieve proper rounding, only two guard digits are needed (cf. Knuth [1969], section 4.2.1 exercise 5).

The most ideal arithmetic is unbiased proper rounding arithmetic. This satisfies D with an extra condition to achieve unbiased rounding when $x * y$ lies halfway between a and a^+. Usually, this extra condition is round to even, i.e. $x \; \hat{*} \; y$ equals that element a or a^+ which has an even p-digit mantissa.

To achieve this, one needs not only two guard digits, but also a "sticky" bit indicating if beyond the guard digits some digits were discarded, cf. Coonen [1978].

3. Limited exponent range

In this section, we describe four sets of axioms taking into account the effects of finite precision as well as limited exponent range and the corresponding phenomena of underflow and overflow.

The floating-point system $\mathbb{F}$ is now determined by not only the base b and the mantissa length p, but also the minimal exponent e and the maximal exponent E which both are integers satisfying (for simplicity) $e < -p$ and $E > p$.

<u>Definition</u> $\mathbb{F} = \mathbb{F}(b,p,e,E)$ is the set of numbers of the form

$$x = 0 \quad \text{or} \quad x = mx \times b^{ex-p},$$

where ex and mx are integers satisfying

$$e \leq ex \leq E, \quad b^{p-1} \leq |mx| < b^p,$$

In this system, the largest element, λ, equals $(1 - b^{-p})b^E$. The system contains, besides 0, only elements representable with normalised mantissa (i.e. $|mx| \geq b^{p-1}$). Consequently, the smallest positive element, σ, of $\mathbb{F}$ equals b^{e-1}.

Another consequence is, that the relative spacing for elements other than $\pm \lambda$, $\pm \sigma$ and 0 is still bounded by $\Delta = b^{1-p}$, cf. Reinsch [1979].

For this system, we consider the same relational and arithmetical operations as above.

We assume that underflow and overflow do not cause a trap. This is realistic, as in most implementations traps are only optional.

Moreover, our aim is to investigate not only the effects of finite precision but also those of underflow and overflow on the performance of algorithms.

We again assume that the implementations of the relational operations and the sign reversal are exact.

For the implementations of the dyadic arithmetical operations, we state four sets of axioms, A',B',C',D', which are modifications of the sets of axioms mentioned above.

In these axiom sets, there are two positive constants υ and ω, called the <u>underflow threshold</u> and the <u>overflow threshold</u> respectively, which limit the errors in the implemented arithmetical operations in case of underflow or overflow. We assume (for simplicity) that these constants satisfy $\upsilon < \varepsilon$ and $\omega > \varepsilon^{-1}$.

<u>Underflow</u>

It is assumed that underflow can only occur when the exact result of an arithmetical operation has a magnitude smaller than υ. If underflow occurs, then the computed result has a magnitude not larger than υ and either equals 0 or has the same sign as the exact result. In other words, if underflow occurs in the calculation of $x \circledast y$, then

either $0 < x * y < \upsilon$ and $0 \leq x \circledast y \leq \upsilon$

   or $x * y = x \circledast y = 0$

   or $-\upsilon < x * y < 0$ and $-\upsilon \leq x \circledast y \leq 0.$

<u>Overflow</u>

It is assumed that overflow can only occur when the exact result of an arithmetical operation has a magnitude larger than ω. If overflow occurs, then the

computed result has the same sign as the exact result and a magnitude not smaller
than ω and not larger than the magnitude of the exact result. In other words, if
overflow occurs in the calculation of $x \hat{*} y$, then

either $\omega < x * y$ and $\omega \leq x \hat{*} y \leq x * y$

   or $x * y < -\omega$ and $x * y \leq x \hat{*} y \leq -\omega$.

Normal operation

If no underflow or overflow occurs, then in axiom sets A',B',C',D' it is assumed
that the errors are bounded as in axiom sets A,B,C,D respectively.

Additional assumptions

In C' and D', it is assumed that $\upsilon = \sigma$ and $\omega = \lambda$. A consequence is, that
the arithmetic satisfying axiom set C' or D' is _faithful_, i.e. the computed
result always equals either the largest element of $\mathbb{F}$ not larger, or the smallest
element of $\mathbb{F}$ not smaller than the exact result.

Remarks

1) In case of underflow in D', we do not require rounding to a nearest element,
0 or $\pm\,\sigma$, in $\mathbb{F}$, because for this case we prefer the value σ or $-\sigma$ with
the same sign as the exact result. This choice has the advantages that the system
does not contain zero-dividers and that the relational operations can be implemented
by means of subtraction, i.e. for

$\sim \,\epsilon\,\{<,\leq,=,\neq,\geq,>\}$, we have $x \sim y$ is equivalent to $x \hat{-} y \sim 0$.

2) These axiom sets are all easy to implement. As remarked above, axiom set D
(and, therefore, also axiom set D') requires only two guard digits to achieve
proper rounding. Moreover, none of the axiom sets presented requires infinities
or"non-numbers",i.e. bit patterns not corresponding to some numerical value. (Note,
however, that we here disregard division by zero!)

3) These axiom sets are not fulfilled in all practical implementations, even if
overflow is disregarded. For instance, in the CDC Cyber systems some kind of
underflow may occur in multiplication or division, when an operand is near zero
but the exact result is not (in particular, this system contains small positive
numbers x for which $x \hat{*} \xi = 0$ for all $\xi \,\epsilon\, \mathbb{F}$). In view of this, it would be
more natural to present weaker assumptions on underflow in axiom set A'. For
simplicity, we preferred not to do that here.

4) Our results presented in section 4 would remain valid if system $\mathbb{F}$ in axiom
sets $A,B,$ A',B' would not be a floating-point system, but any other system of
rational numbers consistent with the corresponding axiom sets; in particular, $\mathbb{F}$
must be symmetric (i.e. if $x \,\epsilon\, \mathbb{F}$ than also $-x \,\epsilon\, \mathbb{F}$), contain the element 0,
and have a sufficiently small resolution. For definiteness and simplicity, we
presented all our axiom sets with floating-point systems as described above.

4. Example: finding a zero in an interval

We consider the problem of finding a zero of a continuous real-valued function, f, defined on a real interval. We assume that the values of f at the end points of the interval are neither both positive nor both negative. Hence, according to classical analysis, f must have a zero in the interval.

The required result is a pair of real values in the interval which include a zero of f and have a difference not larger than a certain tolerance. We assume that the tolerance is a real function, t, which is linear in the magnitude of its argument and has a positive minimal value.

We consider some iterative algorithms to solve this problem and focus attention on their performance in an arithmetic satisfying any of the sets of axioms given above. Accordingly, we assume that all data and results are in $\mathbb{F}$.
Moreover, we disregard effects of errors in computing values of f. Thus, we assume that f is given as a routine yielding a value in $\mathbb{F}$ for each argument value in the intersection of $\mathbb{F}$ and the interval considered.
Furthermore, we assume that the tolerance function t has the form

(4.1) $t(x) = \max(\rho \hat{\times} |x|, \alpha)$,

where ρ and α are certain positive constants.
For $x, y \in \mathbb{F}$, let $J(x,y)$ denote the closed real interval whose end points are x and y. Let <u>sign</u> denote the monadic operator yielding the value $+1, 0, -1$ for positive, zero, negative operand respectively.
In view of the foregoing, we can formulate the problem as follows:
the given data are

> x0, y0 $\in \mathbb{F}$,
> f,t $\in J(x0,y0) \cap \mathbb{F} \to \mathbb{F}$,

where f satisfies the <u>initial condition</u>

(4.2.)   <u>sign</u> $f(x0) \times$ <u>sign</u> $f(y0) \le 0$

and t has the form (4.1);
the required results are

> x,y $\in J(x0,y0) \cap \mathbb{F}$

satisfying the <u>final condition</u>

(4.3) $\begin{cases} \text{sign } f(x) \times \text{sign } f(y) \le 0, \\ |f(x)| \le |f(y)|, \\ |y \hat{-} x| \le t(x). \end{cases}$

Algorithms

To solve this problem, we consider iterative algorithms of the following type.

In each iteration step a current interval is reduced to a proper subinterval
leaving a certain assertion invariant.

Let b and c denote real variables whose values are the end points of the current interval. Then the <u>invariant assertion</u> is

$$(4.4) \quad \begin{cases} \underline{\text{sign}}\ f(b) \times \underline{\text{sign}}\ f(c) \le 0, \\ |f(b)| \le |f(c)|, \end{cases}$$

and the stopping criterion is

$$(4.5) \quad |c \mathbin{\widehat{-}} b| \le t(b).$$

In particular, we consider the bisection algorithm and a class of algorithms which combine linear and/or three-point rational interpolation with bisection.
Two algorithms of this class were presented by Bus & Dekker [1975], another related algorithm by Brent [1971].

We consider two numerically different versions of bisection. In the first version,
$Z1$, the next iterate is calculated in each step by the formula

$$m1 = (b \mathbin{\widehat{+}} c) \mathbin{\widehat{/}} 2,$$

in the second version, $Z2$, by the formula

$$m2 = b \mathbin{\widehat{+}} h,$$

where

$$h = (c \mathbin{\widehat{-}} b) \mathbin{\widehat{/}} 2.$$

Let

$$r = \underline{\text{sign}}\ (c \mathbin{\widehat{-}} b) \times t(b) \mathbin{\widehat{/}} 2,$$
$$s = b \mathbin{\widehat{+}} r.$$

In the class of other algorithms, denoted by $Z3$, each step yields as next iterate
a value which either equals an approximate value of a certain interpolation formula
if this value is (approximately) between s and $m2$, or otherwise equals s or
$m2$.

We distinguish between two kinds of steps, <u>normal steps</u> in which a certain interpolation formula is used (different formulas may, however, be used in different normal steps), and <u>abnormal steps</u> in which no interpolation formula is involved, but immediately the value $m2$ is delivered as next iterate. The abnormal steps have the purpose to ensure that the length of the current interval becomes (approximately) a factor 2 or more smaller at least every 4 or 5 steps.
In a normal step, the next iterate is obtained as follows. Firstly, one calculates approximations of numerator, p, and denominator, q, of the value of the interpolation formula chosen. (Thus, the division $p \mathbin{\widehat{/}} q$ is postponed in order to avoid

division by 0 or overflow.) Subsequently, if $p < 0$ then p and q are replaced by $-p$ and $-q$ respectively (because the next formula yields the correct
result only for non-negative p). The next iterate then equals

> if $p \le q \times r$ then s,
> otherwise if $p < q \times h$ then $b \mp p/\hat{q}$,
> otherwise m2.

In axiom sets A' and B', however, one here has to replace $q \times r$ by
$\max(q \times r, \upsilon)$. In other words, the meaning of this formula is that, apart from
rounding errors, the value s is delivered when p/q is in the closed interval
$J(0,r)$, the value $b \mp p/\hat{q}$ when p/q is between r and h, and the value m2
otherwise.

<u>Remarks</u>

1) Our results on algorithm Z3 stated in the subsequent theorems do not depend
on the choice of the interpolation formulas, but on the fact that the value
$b \mp p/\hat{q}$ is delivered only when certain conditions stated in the formula are fulfilled.

2) In the class of algorithms Z3 as defined above, some calculations are slightly simpler (and nicer from numerical viewpoint) than the corresponding calculations
in two algorithms, M and R, presented by Bus & Dekker [1975] in the form of
Algol 60 procedures.

These procedures, modified according to the "note added in proof" mentioned in
that paper, perform the calculations as described above for Z3 with the following
changes. In the formulas given above, h and m2 are replaced by
$h' = ((c \mp b) \times 0.5) \triangleq b$ and $m' = b \mp h'$ respectively; moreover, the stopping
criterion is replaced by $|h'| \le t(b)$. Obviously, these changes would (slightly)
affect the corresponding results stated in the subsequent theorems.

3) The algorithm presented by Brent [1971] in the form of Algol 60 procedure zero
fits into our class Z3 with the following changes. It has a different strategy
to decide when an abnormal (bisection) step is to be used; moreover, in the formulas given above, h is replaced by $h'' = 0.5 \times (c \triangleq b)$ and the stopping
criterion is replaced by $f(b) = 0$ or $|h''| \le t(b)$, which is approximately
equivalent to $|c \triangleq b| \le 2 \times t(b)$.

Consequently, our results stated in the subsequent theorems remain valid when the
lower bounds for ρ and α mentioned in theorem 1 are halved. Brent uses as
tolerance function $t(b) = 2 \times \varepsilon \times |b| + \alpha$, where α is supplied by the user,
and gives an error analysis assuming axiom set A.

Our theorem 2 shows, however, that the factor $2 \times \varepsilon$ may be too small and should,
for safety, be larger than 3ε.

Theorem 1

Let algorithms Z1,Z2,Z3 be performed in arithmetic satisfying any of the sets
of axioms A,B,C,D, A',B',C',D'.

Let Δ be the resolution of $\mathbb{F}$ and ε, the arithmetic precision in these axiom
sets (recall that we defined $\varepsilon = \Delta$ in axiom sets C,C' and $\varepsilon = \frac{1}{2}\Delta$ in axiom
sets D,D', and that we assume $0 < \varepsilon < 1/8$), and let υ be the underflow
threshold in axiom sets A',B',C',D'.

Let tolerance function t have the form (4.1):

$$t(x) = \max(\rho \hat{\times} |x|, \alpha),$$

where α is positive in axiom sets A,B,C,D and larger than $2\upsilon/(1-8\varepsilon)$ in
axiom sets A',B',C',D', and where ρ is larger than $1/(1-8\varepsilon)$ times a constant specified for the various algorithms and axiom sets by the following table

	A,A'	B,B'	C,C'	D,D'
Z1	6ε	4ε	$3\varepsilon(=3\Delta)$	$4\varepsilon(=2\Delta)$
Z2,Z3	6ε	2ε	$2\varepsilon(=2\Delta)$	$2\varepsilon(=\Delta)$

Let, for algorithm Z1, no overflow occur in the calculation of b $\hat{\mp}$ c.
Then the algorithms have the property that each iteration step yields a next
iterate between the end points, b and c, of the current interval.
We prove this theorem only for algorithms Z2 and Z3. The proof for algorithm
Z1 is similar to that for Z2 and is left to the reader.

Proof for algorithm Z2 in axiom sets A,B,C,D.

Let

$$w = c \hat{-} b.$$

In view of the stopping criterion, the iteration is continued only if

$$|w| > t(b) = \max(\rho \hat{\times} |b|, \alpha);$$

hence, in all axiom sets,

$$|w| > \rho \times |b| \times (1-\varepsilon) \text{ and } |w| > \alpha.$$

Without loss of generality, we may assume $w > 0$; otherwise, we reverse the sign
of b and c and consider the corresponding step for the problem of finding a
zero of function g defined by $g(x) = f(-x)$.
In axiom set A, we have

$$w = c\gamma - b\beta, \quad |\gamma - 1|, \quad |\beta - 1| \leq \varepsilon,$$
$$m2 = b \hat{\mp} w\hat{/}2 = b\zeta + w\delta\eta/2, \quad |\zeta - 1|, |\delta - 1|, |\eta - 1| \leq \varepsilon.$$

Hence,

$$m2 - b = w\delta\eta/2 + b(\zeta - 1) > |b|\{\rho(1-\varepsilon)^3/2 - \varepsilon\},$$

$$c = w/\gamma + b\beta/\gamma,$$

$$c - m2 = w(1/\gamma - \delta\eta/2) + b(\beta/\gamma - \zeta)$$
$$> (|b|/(2\gamma))\{\rho(1-\varepsilon)(2 - (1+\varepsilon)^3) - 2(3\varepsilon + \varepsilon^2)\}.$$

In axiom sets B,C,D, the same formulas hold with $\beta = \gamma$ and $\zeta = \eta$. This leaves the bound for m2 − b invariant, but changes the bound for c − m2 as follows

$$c - m2 > (|b|/(2\gamma))\{\rho(1-\varepsilon)(2 - (1+\varepsilon)^3) - 2\varepsilon(1+\varepsilon)\}.$$

Since ρ is larger than the lower bound specified in the theorem, it follows that m2 > b and, using the inequality

$$(1+\varepsilon)/[(1-\varepsilon)(2 - (1+\varepsilon)^3)]] < 1/(1-8\varepsilon),$$

also that c > m2, i.e. m2 is between b and c.   □

Proof for algorithm Z2 in axiom sets A',B',C',D'.

The proof given above remains valid if only normal operations are involved. So, we here have to consider only underflow and overflow.

Underflow

Since α has the lower bound specified in the theorem, underflow cannot occur in the calculation of w and h; so it can only occur in the calculation of m2 = b ∓ h. We then have $|b + h| < \upsilon$ and

$$m2 = b + h + \varphi, \quad |\varphi| \leq \upsilon.$$

Hence,

$$m2 - b = w\delta/2 + \varphi > \alpha(1-\varepsilon)/2 - \upsilon,$$

$$c - m2 = w/\gamma + b(\beta/\gamma - 1) - h - \varphi$$
$$= w(1/\gamma - \beta\delta/(2\gamma)) + (b+h)(\beta/\gamma - 1) - \varphi$$
$$> (1/(2\gamma))\{\alpha(2 - (1+\varepsilon)^2) - 2\upsilon(1+3\varepsilon)\}.$$

From this it follows that m2 > b and, using the inequality

$$(1+3\varepsilon)/(2 - (1+\varepsilon)^2) < 1/(1-8\varepsilon),$$

also that c > m2, i.e. m2 is between b and c.

Overflow

Overflow can only occur in the calculation of w and h, or in the calculation of a value of f. According to the axioms, w and h have the correct sign and are not larger in magnitude than the corresponding exact values c − b or w/2, from which it follows that m2 is between b and c. It is easy to see that overflow in the calculation of f is also harmless.   □

Proof for algorithm Z3 in axiom sets A,B,C,D.

In each step there are three possible values for the next iterate, viz. m2, s and $b \mathbin{\widehat{\div}} p\widehat{/}q$. We have to show that each of these values is between b and c. For the first value, $m2$, it has been shown above; for the second value, s, it can be shown similarly; for the third value it is shown as follows. According to the description of Z3 given above, the value $b \mathbin{\widehat{\div}} p\widehat{/}q$ is delivered only when

(4.6) $q \mathbin{\widehat{\times}} r < p < q \mathbin{\widehat{\times}} h.$

Hence, $q > 0$ and

(4.7) $r(1 - \varepsilon) < p/q < h(1 + \varepsilon).$

So, in a similar way as above for $m2 - b$ and $c - m2$, we obtain

$$b \mathbin{\widehat{\div}} p\widehat{/}q - b > r(1 - \varepsilon)^3 - |b|\ \varepsilon > |b|\{\rho(1 - \varepsilon)^5/2 - \varepsilon\},$$
$$c - (b \mathbin{\widehat{\div}} p\widehat{/}q) > w/\gamma + b(\beta/\gamma - 1) - |b|\varepsilon - h(1 + \varepsilon)^3$$
$$> (|b|/(2\gamma))\{\rho(1 - \varepsilon)(2 - (1 + \varepsilon)^5) - 2(3\varepsilon + \varepsilon^2)\},$$

whereas in axiom sets B,C,D the latter formula may be replaced by

$$c - (b \mathbin{\widehat{\div}} p\widehat{/}q) > (|b|/(2\gamma))\{\rho(1 - \varepsilon)(2 - (1 + \varepsilon)^5) - 2\varepsilon(1 + \varepsilon)\}.$$

Since ρ is larger than the lower bound specified in the theorem and

$$(1 + \varepsilon)/[(1 - \varepsilon)(2 - (1 + \varepsilon)^5)] < 1/(1 - 8\varepsilon),$$

it follows that $b \mathbin{\widehat{\div}} p\widehat{/}q$ is between b and c.   □

Proof for algorithm Z3 in axiom sets A',B',C',D'.

In axiom sets A',B', the value $b \mathbin{\widehat{\div}} p\widehat{/}q$ is delivered only when

$$\max(q \mathbin{\widehat{\times}} r, \upsilon) < p < q \mathbin{\widehat{\times}} h,$$

i.e. when p satisfies (4.6) and $\upsilon < p$. Hence, if underflow occurs in $q \mathbin{\widehat{\times}} r$, then

$$q \times r < \upsilon < p,$$

and, if overflow occurs in $q \mathbin{\widehat{\times}} h$, then

$$p < q \mathbin{\widehat{\times}} h \leq q \times h.$$

In axiom sets C',D', we have $\upsilon = \sigma$. Hence, relation (4.6) implies $\upsilon = \sigma \leq p$. So, relation (4.7) remains valid in all cases and the result follows in a similar way as above.   □

Theorem 2

The lower bounds for ρ and α in theorem 1 are best possible within a factor $1 + O(\varepsilon)$ in axiom sets A,B,A',B'.

Proof

We prove this theorem by giving some examples of values b and c for which w may be equal to $1 - 8\varepsilon$ times the bounds given in theorem 1, and a value $m1$ or $m2$ not between b and c may be produced within the limitations of the axioms.

Examples concerning the lower bounds for ρ

for algorithm Z1 in axiom sets A,A':

$b = 1$, $c = 1 + 4\varepsilon$, $b \not= c = 2 + 6\varepsilon$, $m1 = 1 + 4\varepsilon = c$, $w = 6\varepsilon$, and in axiom sets B,B':
the same values except $w = 4\varepsilon$; for algorithms Z2 and Z3 in axiom sets A,A':
$b = 1$, $c = 1 + 4\varepsilon$, $w = 6\varepsilon$, $h = 3\varepsilon$, $m2 = 1 + 4\varepsilon = c$, and in axiom sets B,B':
$b = 1$, $c = 1 + 2\varepsilon$, $w = 2\varepsilon$, $h = \varepsilon$, $m2 = 1 + 2\varepsilon = c$.

Examples concerning the lower bound for α

for algorithm Z1:

$b = 0$, $c = 2\upsilon$, $b \not= c = 2\upsilon(1 - \varepsilon)$, $m1 = 0 = b$, $w = 2\upsilon$; for algorithms Z2 and Z3:
$b = 0$, $c = 2\upsilon$, $w = 2\upsilon$, $h = \upsilon(1 - \varepsilon)$, $m2 = 0 = b$.   □

Remark

The author does not know if the bounds given in theorem 1 for ρ in axiom sets
C,D,C',D', and for α in axiom sets C',D' are best possible within a factor
$1 + O(\varepsilon)$.

Theorem 3

Let the conditions of theorem 1 be satisfied. Then algorithms Z1, Z2, Z3 terminate in a finite number of steps.

Proof

According to theorem 1, each iteration step yields a next iterate between the end
points of the current interval. Hence, the algorithms must terminate in a finite
number of steps, because $\mathbb{F}$ is finite in axiom sets A',B',C',D', and $\mathbb{F}$ is
discrete, except in a neighbourhood of 0, in axiom sets A,B,C,D; termination
in a neighbourhood of 0 is ensured, because the values of t have a positive
lower bound α.   □

Conjecture

Let the conditions of theorem 1 be satisfied, except that ρ and α are larger
than the bounds given in theorem 1 times a certain constant larger than 1.
Then the number of steps required for algorithms Z1, Z2, Z3 performed in any of
the sets of axioms mentioned in theorem 1 is of the same order of magnitude as the
corresponding number for the same algorithms performed in exact arithmetic.

Acknowledgment

The author is grateful to Dr Peter van Emde Boas for stimulating discussions on
this subject and to him and Mr Walter Hoffmann for reading the manuscript.

References

[1] R.P. Brent: An algorithm with guaranteed convergence for finding a zero of a
    function, Comp. J. 14 (1971) 422-425.
[2] J.C.P. Bus, T.J. Dekker: Two efficient algorithms with guaranteed convergence
    for finding a zero of a function, ACM Trans. on Math. Softw. 1 (1975) 330-345.

[3] J. Coonen: Specifications for a proposed standard for floating-point arith-
    metic, Comm. IEEE Microprocessor standards committee, June 1978.

[4] T.J. Dekker: A floating-point technique for extending the available precision,
    Num. Math. 18 (1971) 224-242.

[5] D.E. Knuth: The art of computer programming, vol. 2, Addison-Wesley (1969).

[6] C.H. Reinsch: Principles and preferences for computer arithmetic, to appear
    in SIGNUM Newsletter (1979).

[7] J.H. Wilkinson: Rounding errors in algebraic processes, Her Majesty's
    stationery office (1963).

Performance Evaluation of Numerical Software, Fosdick (ed.)
© *IFIP, North-Holland Publishing Company, 1979*

DISCUSSION OF FIRST SESSION ON GENERAL ASPECTS
OF PERFORMANCE EVALUATION

Bo Einarsson

Swedish National Defence Research Institute (FOA)
Department 2/Proving Grounds
S-147 00   Tumba,   SWEDEN

Session chairman's remarks

The talks by Professor Hull and Professor Dekker stressed that testing alone is
not enough for a complete certification. Professor Hull and his colleagues had
extended Floyd's method to numerical computation, where clean floating-point
arithmetic and precision control are useful. Symbolic manipulation is often
required. The method had with some success been applied to real programs, but it
is impossible to really prove correctness of a computer program.

Professor Dekker introduced four different sets of axioms for floating-point
arithmetic (weak, strong, faithful, or proper rounding) and applied these to the
zero-finding problem.

Dr. Brown discussed some fundamentals for software evaluation, namely a model of
the floating-point computation and a set of environment parameters. He also
recommended many of the software tools available at his laboratory.

As an example the model was applied to an accurate, efficient, and portable
algorithm to compute the mean of a vector.

A portable program to test the conformance of a computer and a set of
environment parameters to the model has been developed by N. L. Schryer, and was
briefly described.

Since the model, for simplicity, omits many details of the behaviour of real
machines, its precision and range parameters may be penalized in a way that is
unfair to certain manufacturers. However, Dr. Brown said the degree of
unfairness is generally less than one digit of precision and one decade of
range.

45

DISCUSSION

Question (W. Kahan)

The TI Business Analyst calculator nominally carries 11 significant decimal digits, but gives

$$1/3 = 0.33333\ 33333\ 0$$
$$9/27 = 0.33333\ 33333\ 3$$

so $1/3 \neq 9/27$. Is this behaviour compatible with your model, and do you think it should be?

Answer (W. S. Brown)

I believe this behaviour reflects an error in the design or construction of the calculator, but the model happens to allow it if the claimed precision is reduced from 11 digits to 10. In fact, such flexibility in the model is essential to accomodate certain reasonably well designed computer/compiler environments in which, for example, $a = b = c = d = 1/6$, but the logical expression $a + b = c + d$ is false, because $(a + b)$ is stored while $(c + d)$ is computed and retained in extended precision. Such anomalies are not overly burdensome in practice, because in any event the hazards of roundoff require great caution in the use of comparisons.

Question (J. L. Schonfelder)

I think relative spacing is not adequate as the sole measure of precision. Instead, we need to measure the maximum error generated by each of the binary operators. Other parameters that should be measured include the conversion errors for external data and for literal constants. Measurements I have done on a number of machines indicate that single precision numbers are normally converted very well, but double precision conversions are far from satisfactory. As much as two decimal digits of accuracy may be lost in the conversion process.

Answer (W. S. Brown)

I see no compelling reason to have separate error parameters for each of the operators, since all of them would be the same on an ideal machine, and all of them are likely to be affected about equally (in the sense of my floating-point model) by anomalies such as unguarded arithmetic. I view conversions in the same way as library functions; they should be as accurate as reasonably possible, and their accuracy (measured in terms of the environment parameters) should be clearly documented.

Question (Iain S. Duff)

I am a little concerned by the use of some terms in the talk of Professor Hull. I feel that it is important to differentiate between the algorithm and the implementation or program. Your description of Gaussian elimination and the use of backward error analysis was a study and analysis of an algorithm, which could be performed without a computer, and does not by itself guarantee the correctness of any program based on this algorithm. I feel that there are two distinct levels here; that of proving that the underlying algorithm is correct (e.g. stable), and then that of showing that the coded implementation correctly represents this algorithm.

Answer (T. E. Hull)

My talk was intended to be about programs. I realize that I discussed the ideas
at what you call the "algorithmic" level. However, I did this for two reasons:
one is that we have admittedly worked at that level to some extent because it
was easier to think out the ideas at that level, and the other is that I thought
it would be easier to explain those ideas at that level. However, I can assure
you that we were motivated by real programs, and we actually tried the technique
on a number of real, and fairly large, programs.

I realize now that I only made brief mention of the work we did with real
programs, and may have given the impression that we were concerned exclusively
with what you call "algorithms".

Question (John R. Rice)

Termination and convergence proofs of programs for problems involving functions
require assumptions on the functional behaviour. What type of hypotheses do you
see used in program proofs and how does one verify these hypotheses in practice?

Answer (T. E. Hull)

In mathematics certain "natural" assumptions have been identified, such as
continuity and a Lipschitz condition for ODEs. It would be nice to find
"natural" ones for computing. At present, the best we seem to be able to use are
closely related to the mathematical ones (such as a bound on the Lipschitz
constant, or a bound on a high derivative in quadrature). However, if we try to
be too general the theorems we can prove become too conservative.

Another point to keep in mind is that we need information about the accuracy
with which the function values can be computed - this factor should be taken
into account as well.

Question (John R. Rice)

Are people involved in automatic program proving concerned about the difficulty
or ease of verifying the hypotheses of the program proofs?

Answer (T. E. Hull)

Automatic proving does try to deal with the hypotheses of the program - but,
remember, the people in this field are not interested in numerical computation
and always deal with hypotheses that are relatively easy to check, e.g., that N
is an integer between 0 and 100. I don´t see how we could imagine verifying
assumptions about functions - e.g., continuity.

SESSION 2 : GENERAL ASPECTS OF PERFORMANCE
                                  EVALUATION

Performance Evaluation of Numerical Software, Fosdick (ed.)
© IFIP, North-Holland Publishing Company, 1979

PERFORMANCE PROFILES AND SOFTWARE EVALUATION*

J. N. Lyness
Applied Mathematics Division
Argonne National Laboratory
Argonne, Illinois

1.   Introduction

We discuss here the problem of comparing and evaluating different items of numerical software which carry out the same task, requiring input of a similar nature.   The author's experience in this area is limited to evaluating automatic quadrature routines and to evaluating routines for unconstrained minimization. These are Class 2 routines (which are defined below) and usually have to be compared by means of numerical experiment.   The results of such experiments are often extensive.   The evaluator, if he is not careful, may be faced with an enormous amount of numerical evidence, from which it may be a daunting task to extract information sufficiently coherent to make any convincing value judgement about the relative merits of the routines he is evaluating.

In this presentation, we describe the sort of numerical measurements which we feel should be made in such experiments.   In order to justify these suggestions, we start by introducing Class 1 and Class 2 numerical software and describing their performance profiles.   With this background we outline various basic principles in the design of numerical experiments for Class 2 routines.   Simply following these principles leads naturally to a testing procedure which produces results which are significantly easier to use in an evaluation process than some of those which are traditionally used.

While this is illustrated in the context of evaluating unconstrained minimization routines, the underlying principles are general and we hope that they can be applied with profit in many other areas of numerical software evaluation.

2.   Class 1 and Class 2 Software and their Performance Profiles

We shall start by introducing the term performance profile by means of a discussion.

Suppose that we have a routine which is intended to produce an approximation to some desired result q, this approximation being within a prescribed accuracy $\varepsilon > 0$.   The behavior of the routine depends on the input it is given.   Suppose we envisage a series of runs using this routine in which the input depends continuously on a parameter λ.   For each individual value of λ, the routine produces a result $\tilde{q}(\lambda)$ which is an approximation to the exact result $q(\lambda)$.   A plot of the quantity $e(\lambda) = \tilde{q}(\lambda) - q(\lambda)$ against λ is called an error type performance profile. If the routine actually carries out the intentions of its constructor and invariably succeeds in producing an approximation within the prescribed accuracy ε, then we would find $\varepsilon \geq e(\lambda) \geq -\varepsilon$ and a plot of the error type performance profile would lie within a band of width 2ε centered along the λ-axis.   However, routines constructed by human beings are rarely so consistently reliable.

In evaluating numerical software, another interesting property of a routine is the cost of the calculation.   We could plot a measure of this cost $c(\lambda)$ against λ.   This would be termed a cost type performance profile.   One obvious measure for

---

*
Work performed under the auspices of the U.S. Department of Energy and supported in part by the U.S. Army Research Office.

the cost is the actual machine time. Others include operations counts or if
applicable the number of incidental function evaluations required. All measures
seem to have advantages and disadvantages.

The particular types of performance profiles described above are those which
the author has found to be useful concepts in examining and describing routines.
In general the term performance profile could be applied to a plot of any functional of the output or behavior of a routine against λ where the input to the
routine depends on λ in a continuous manner.

Performance profiles can be useful in a variety of circumstances. For
example, one could choose λ in such a way that as λ is increased, the intrinsic
difficulty of the problem increases and perhaps determine the value of λ above
which the routine fails to operate properly. We shall be interested in this paper
in a different sort of profile, one in which changing λ does not seem to affect
the difficulty of the problem at all. We term these profiles equal difficulty
performance profiles.

Our principal interest in this paper is simply in what an equal difficulty
error type performance profile looks like qualitatively. There is rarely any need
to actually plot such a profile. But its qualitative features are important when
it comes to designing numerical experiments for software evaluation.

Let us now consider the nature of an error type performance profile obtained
using a library function SIN(X). We may denote by $\tilde{q}(\lambda)$ the result returned by the
routine when we use $X = \lambda$ as input, and by $q(\lambda)$ the actual value of $\sin\lambda$. We
should expect $e(\lambda) = \tilde{q}(\lambda)-q(\lambda)$ to be a small quantity, but not precisely zero.
Since the routine uses machine arithmetic, in general $e(\lambda)$ might take values of
either sign and of magnitude perhaps one or two units in $\varepsilon_M \sin\lambda$. Here ε_M is the
machine accuracy parameter, which we shall take to be about 10^{-15} for illustrative
purposes.

The visual appearance of the error type performance profile would depend on
the scale on which it is plotted. On a macroscopic scale (for example $1:10^{-4}$) one
would see only a horizontal line, coinciding with the λ-axis. On this scale one
would not notice divergence from this line of magnitude 10^{-15}. But on microscopic
scale (for example $1:10^{-15}$) the effect of the machine arithmetic would be observa-
ble. What had previously appeared to be a horiontal line would now be revealed as
a sequence of discontinuous horizontal segments (like a silhouette of an unfin-
ished spasmodically constructed brick wall). This plot would have discontinuities located between each machine expressible number λ. However, the entire
plot would lie within one or two units of $\varepsilon_M \sin\lambda$ of the λ-axis.

We shall refer to routines like SIN(X) which have error type performance
profiles of the character just described as Class 1 routines. The important feature, not shared by Class 2 routines, is that, on a macroscopic scale it appears
that the profile provided by a Class 1 routine is simply $e(\lambda) = 0$.

In a previous paper (Lyness and Kaganove, ACM Trans. on Math. Soft. 2, pp.
65-81, 1976) we discussed in considerable detail an error type performance profile produced by an automatic quadrature routine FUNCTION QUAD(A,B,EP,FUN). Here
A and B are integration limits, EP is a required tolerance and FUN the name of a
user-provided integrand function. We used a function $f(x,\lambda)$ which had a narrow
peak at $x = \lambda$. As λ varies within (A,B) the problem presented to the quadrature
routine varies essentially only in the location of this peak. Thus, at first
sight, the intrinsic difficulty of each problem seems to be the same. The performance profile was constructed by setting A=1, B=2, EP=0.1. The routine returns
an approximation $\tilde{q}(\lambda)$ to the exact result $q(\lambda)$ for which $e(\lambda) = \tilde{q}(\lambda)-q(\lambda)$ hope-
fully satisfies $|e(\lambda)| \leq$ EP.

In that paper we presented a plot of $e(\lambda)$ on a macroscopic scale. Unlike the plot for SIN(X) described above, this plot on this scale contained many discontinuities. Between these appeared sections of apparently continuous curves. While most of the plot lies within a distance EP of the λ-axis, there are significant sections of the plot which do not. This plot resembled the sky-line of the downtown area of an American city. If one were to plot $e(\lambda)$ on a microscopic scale, one would simply find that the sections which previously appeared to be continuous were in fact discontinuous in just the same way as previously described for SIN(X). Some of the reasons why the macroscopic plot of the error type performance profile for an automatic quadrature routine is "jagged" are described in detail in the previous paper. These are related to the fact that the "exact arithmetic algorithm" on which the routine is based is not reliable and that the algorithm may or may not follow quite different strategies for two neighboring values of λ. In fact, the macroscopic picture reflects the exact arithmetic algorithm and the microscopic picture includes the effect of the use of machine arithmetic in the implementation as a routine.

For the purposes of this paper we define a Class 2 routine as one for which a macroscopic plot of an equal difficulty error type performance profile is jagged.

We assert that much of the more sophisticated numerical software being presently produced consists of Class 2 routines. And in the subsequent sections we shall demonstrate that numerical experiments have to take this into account if they are to produce useful numerical evidence for software evaluation.

3.   Software Evaluation - The Battery Experiment

Software evaluation is a large complicated and expanding area of computational analysis. One significant part of it is concerned with comparing different software items which carry out the same task, using input of a similar nature. The results of such a comparison might be useful to:

(i)   a general user who has to choose a single routine for his problem;

(ii)   a librarian who has to choose a small selection for his library;

(iii)   a software package constructor who has to choose one to use in his complicated interrelated conglomerate of routines; and even

(iv)   a constructor of a single routine who wants to assign numerical values to individual somewhat arbitrary parameters required by his individual routine.

It became apparent very early on that a completely mathematical or theoretical comparison of different algorithms is usually not feasible. But it was believed for some time that one could determine which routine was "best" by numerical experiment. In fact very good progress in this direction has been made in comparing different algorithms for the simpler tasks. For example, when comparing different algorithms for SIN(X), one might proceed by choosing a random selection of values of X and obtaining performance statistics. One would report perhaps the median values of $|SIN(X) - \sin x|$ and $\tau(x)$, the machine time required. Based on this type of numerical evidence, decisions were made whose overall result is that we now have excellent library functions SIN(X).

Generalization of this approach to the more complicated Class 2 routines was not an immediate success. In the more general context significant difficulties

were encountered, not the least of which was in agreeing what the term "best"
meant. In the end experimenters were led, slowly but surely to a more sophisticated attitude to the problem. Some of the aspects of the evaluation experiments
which have been widely recognized are mentioned below.

1) For nearly every item of software written there are some problems for
which it performs very well indeed, others for which it is very poor and yet
others for which it actually fails. As it is Class 2 software, one cannot avoid
failures. All one can hope for is that the class of problems for which it does
well is large and the class for which it does poorly or fails is as small as
possible.

2) Moreover in a comparative context, the various classes of problems mentioned above are different for different routines. One rarely finds that routine
A is better for all problems than routine B. One finds instead that it is better
for some and worse for others.

One principal purpose of software evaluation then is to identify classes of
problems for which one routine is likely to be more suitable than another when
either could be used.

3) If results of this sort are to be of lasting value, they must take into
account that other routines may be written and there will be a need to extend the
comparison. Since these numerical experiments are expensive, it should be the case
that the constructor of routine C can carry out the same experiment using his routine and need not repeat the runs using the original routines. This leads to the
desirability of using a quantitative measure. That is, the results of the numerical experiment are sets of numbers, each set being associated with one routine.
The corresponding set of numbers for any other new routine can be calculated
independently at a later time.

There is yet another point which has not been fully recognized. Before going
on to it, I should like to pause and describe briefly the nature of some of the
numerical experiments in minimization with which most of this audience must be
familiar. These have been termed Battery type experiments.

The experimenter uses a list of twenty or thirty different objective functions. Associated with each is a starting value. Each of his selection of
routines is applied in turn to each of the problems. Results pertaining to the
behavior of each of the routines are recorded. Depending on the individual
preference of the experimenter, few or many results are recorded.

Using present-day computers one can generate an enormous amount of information in this way. The experimenter faces a daunting task in extracting from
these results a selection which gives a coherent idea of which routine is suitable
for what. Nevertheless, the experiment conforms to items 1, 2, and 3 of the description given above. In his documentation the experimenter has usually explained
how such an experiment can be repeated with another routine, and his results comprise sets of numbers, usually somewhat extensive, ready to be used as numerical
evidence.

It is perhaps important to distinguish two independent aspects of software
evaluation. One is the set of numerical experiments. These produce _evidence_ in
the form of numbers. This evidence is normally of a permanent nature. The other
is the _value judgement_. On the basis of the evidence, any individual may make a
judgement as to which routine should be used for what. Such judgements are subjective. Most but by no means all writers in this area are very careful to keep
these two aspects of the same problem separate.

The difficulty about the Battery approach is certainly not that it does not provide abundant numerical evidence. The difficulty is that there is a great amount which, while not self contradictory, points in all sorts of different directions. In fact there is so much available that the problem of making the value judgement required becomes well nigh impossible.

This brings us to our fourth and final point.

4.   Software Evaluation - The Performance Profile Aspect

The ultimate beneficiary of this software evaluation is the user, librarian or package constructor mentioned above. He is not really interested in single examples. What he is interested in is classes of functions. He will want to extrapolate from the results which are presented. If a result pertaining to a particular curved valley is stated, and his objective function is a different but similar curved valley, he would expect different but similar results. If he cannot expect this, the result about the curved valley is of little interest to him.

It is here that the nature of the performance profile becomes important. Unconstrained minimization is a Class 2 problem. Certainly one may measure the behavior of a routine, given a starting point and an objective function. But if one changes the starting point by a very small amount, the subsequent behavior can be quite different. So one cannot extrapolate from individual results of this sort. The fourth point, which is flouted by many Battery type numerical experiments is this.

4) The numerical results on which evaluation is based should be stable with respect to small variation in the parameters which define the problem.

If we accept this fourth point, and are still interested in software evaluation by numerical experiment, then a natural way out of the difficulty would be to carry out two or three or more runs which are nearly identical with one another. Perhaps, in the case of an unconstrained minimization problem, one could use the same curved valley, but use different starting values, reasonably close to one another. Then in some way we could average the results over these various starting values to provide a composite result for the curved valley problem.

To do this is a major step in what the author considers to be the right direction. However, the problem of how many different starting values to use and where to situate them remains. There is also a problem of how to treat what is clearly exceptional behavior such as a chance convergence in two iterations when normally fifty are required or a chance failure, which might be expected only in a small fraction of the runs.

In cases considered by the author, these particular problems have a natural resolution if one is prepared to go one step further forward and rephrase what one is doing in a mathematical context. We do this by means of an example.

Let us suppose that we have a curved valley and that we are interested in how fast the minimization algorithm makes its way down this valley. What we might do is to measure the number of objective function calls required by the routine to reduce the value of the objective function from h_1 to h_2. To do this we might provide starting values at points $x^{(0)}$ much higher up the valley located in some specified region R, and in each run we could choose a starting value using a random number generator. Thus

$$\min_{x^{(0)} \in R} f(x^{(0)}) \gg h_1 > h_2 \gg f(x_{min}) .$$

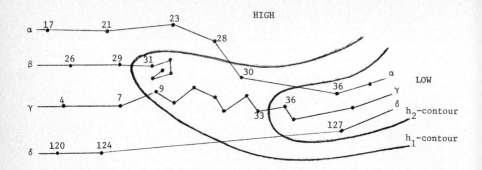

$$
\begin{array}{llll}
\alpha: & \nu_1 = 28.8 & \nu_2 = 32.0 & \nu_{12} = 3.2 \\
\beta: & \nu_1 = 29.2 & \nu_2 = \infty & \nu_{12} = \infty \\
\gamma: & \nu_1 = 8.2 & \nu_2 = 35.2 & \nu_{12} = 27.0 \\
\delta: & \nu_1 = 125.3 & \nu_2 = 125.7 & \nu_{12} = 0.4
\end{array}
$$

In the figure, four trajectories denoted by α, β, γ and δ having four starting points are illustrated schematically. The points marked are iterates and the associated numbers are the number of function values required up to and including the calculation of the particular iterate at that point.

We note that, in trajectory β, the algorithm decided mistakenly that it had reached a minimum and terminated. In trajectory δ, the algorithm had a difficult time reaching this section of the valley but found this section easy. On the other hand in trajectory γ, the same algorithm found earlier sections easy and this section difficult.

For each trajectory we have noted ν_1 and ν_2, a measure of the number of objective function values required to reach heights h_1 and h_2 respectively and defined $\nu_{12} = \nu_2 - \nu_1$. The quantity ν_{12}, which is a measure of what it costs this algorithm to negotiate this section of this valley, is the quantity in which we are interested. Since in this experiment all other parameters are fixed, we may treat the quantity ν_{12} as if it were a function of only one variable

$$
\nu_{12} = \nu_{12}(x^{(0)}) \,,
$$

i.e., it depends only on the starting value.

Any evaluation based on any of these individual values of ν_{12} could be misleading. However, what we can do is to accept the idea that we need more information and press on until we get it. After perhaps 100 or even 1000 runs we may end up with a clearer picture of what is happening.

Taking an average of the values ν_{12} is complicated by the circumstance that some values are not finite. A more rewarding approach is to consider the statistical distribution function defined by

$$
\phi_m(y) = \frac{\text{number of the first m runs for which } \nu_1 < \infty \text{ and } \nu_{12} < y}{\text{number of the first m runs for which } \nu_1 < \infty}
$$

As m becomes large, this function approaches a limiting function $\phi(y)$ illustrated in the figure

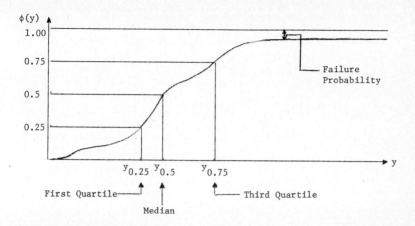

The reasons for thinking in terms of the statistical distribution function are well known. They include:

(a) The function

$$\phi(y) = \lim_{m \to \infty} \phi_m(y)$$

is a well defined functional of the region R, the objective function $f(x)$ the numbers h_1 and h_2, the precise method used to interpolate for ν_1 and ν_2 and, of course, the algorithm. It may be remeasured if there is any doubt by a different investigator or measured for some other algorithm for comparison purposes.

(b) This function (a one-dimensional plot) provides virtually all the quantities of interest. The median $y_{0.5}$ represents what a user might expect and quartiles $y_{0.25}$ and $y_{0.75}$ could be interpreted as what to expect if one is lucky or unlucky. Also of interest is the failure probability, $1-\phi(\infty)$, which is indicated unambiguously.

However, the really important aspect of such a measurement is this:

(c) If one were to make a small variation in the objective function or in the region R, the resulting statistical distribution function $\phi(y)$ would be different from but close to the previous one. With respect to continuous variation of some problem defining parameter μ, results such as $y_{0.5}$ or $1-\phi(\infty)$ are usually continuously varying functions of μ. Thus these results are stable and can be used for extrapolation.

The question of how many individual runs are carried out is resolved in the following way. It depends on how accurately one wants to measure the quantities one is interested in. And this in turn depends on the reason why these quantities are required. In the above context, one plans to compare say $y_{0.5}$ measured for one routine with $y_{0.5}$ measured for another. And one is really only interested to see if one is significantly different from the other. If the results are 25.3 and 25.7, respectively, normally one would conclude that the difference in the behavior of these routines on this problem is too small to be significant in a general evaluation process. However, if one were 25.3 and the other 35.5, this 40% difference in effective cost might be an important factor. In experiments carried out by the author, we have aimed for 2% accuracy in $y_{0.5}$. Thus we have continued to make runs until the value of $y_{0.5}$ has settled down to within 2%.

And this has determined the number of runs. Incidentally, the statement of the results should include a statement about the numerical accuracy of numbers that are reported.

To provide some sort of ending to this hypothetical experiment, I simply suggest that while the experimenter would retain records of the details of the function, he might report for comparison purposes the values of $y_{0.25}$, $y_{0.5}$, $y_{0.75}$ and $(1-\phi(\infty))$.

It is pertinent to look back to the evaluation of Class 1 software with the hindsight of our experience with Battery testing and performance profile testing. The sort of testing for the routine SIN(X) might consist of choosing values of λ randomly from some finite range and constructing statistics about the error. For example, the quantities usually reported include median values of the errors and the probability (usually small) of the relative error exceeding values like $2\epsilon_M \sin\lambda$ or $3\epsilon_M \sin\lambda$ where ϵ_M is the machine accuracy parameter. Such numbers are in fact mathematically defined abscissas on a statistical distribution function.

The battery type test generalizes the measuring procedure by attempting to vary (instead of λ) the argument FUN in the calling sequence. There is no mathematical basis for such a generalization.

The performance profile test generalizes the same procedure by choosing λ as a single parameter in FUN and varying this.

5.   Concluding Remarks

The details of the numerical experiment described in the previous section were invented by the author to provide "versimilitude in an otherwise bald and unconvincing narrative." In fact, successful experiments for curved valleys have been carried out (see Lyness, Math. Comp. 33 (1979)) but the measured quantities were not precisely as stated above. The description in this presentation has been simplified almost out of recognition in order to emphasize the general effect of applying the various principles involved.

The sort of experiment described above requires significantly more human effort on the part of the programmer than the Battery type experiment. We had to choose problem families with care, run pilot calculations which revealed further difficulties. We stored on tape a huge number of individual results so that these could be processed in different ways at leisure without having to rerun expensive series of numerical experiments. The reward in our case was that when at last we produced numerical results, the subsequent evaluation was childishly straightforward. It was obvious which routines were more suitable for the somewhat narrow class of problems we considered.

My own belief is that there are many other areas in which the evaluation of software using an approach based on the same principles could be carried out. The details, of course, may be quite different from area to area.

Performance Evaluation of Numerical Software, Fosdick (ed.)
© *IFIP, North-Holland Publishing Company, 1979*

A PERFORMANCE EVALUATION OF LINEAR ALGEBRA
SOFTWARE IN PARALLEL ARCHITECTURES

Thomas L. Jordan
Computer Science and Services Division
Los Alamos Scientific Laboratory
Los Alamos, New Mexico

This report provides performance data of "parallel" computers
on several of the problems of linear algebra using direct
methods. The computers considered include some software pipeline, hardware pipeline, SIMD, and MIMD types. Special features of each architecture are considered. We discuss such
factors as start-up time, scalar-vector break-even points,
consistency in operation count, parallel steps required, and
speed-up and efficiency of the hardware.

A reasonably broad comparison is given for LU factorization
without pivoting. A less extensive comparison is given for
LU factorization with pivoting. Also, various intracomputer
comparisons are presented to show the performance of different
implementations of a particular algorithm as well as the performance of different algorithms for solving the same problem.
The eigenvalue problem is not addressed.

INTRODUCTION

We shall provide some insight into the performance of various computers with
parallel architectures in the area of linear algebra. Our survey contains performance data on the solution of various types of linear systems by direct methods.
It is necessary to make clear at the outset that, for all practical purposes,
it is impossible to achieve meaningful comparisons in the conventional way of
running and timing a given code written typically in Fortran. The architectures
considered are in many respects special-purpose. All require careful attention
to data organization and algorithms in order to achieve the performance for
which they were designed. Furthermore, these considerations vary from architecture to architecture. For example, the facility for handling rows and
columns of matrices may differ dramatically and this affects data organization.
As an example of algorithmic considerations, explicit matrix inversion is seldom
done on conventional computers to solve Ax = b, but is the cheaper and simpler
method for some of the computers under consideration.

Only recently have we been able to map standard Fortran into the parallel syntax
of the hardware. Either extensions to Fortran have been required or hardware
oriented Fortran-like dialects have been designed. Even in those cases where we
can write entirely in Fortran, it is imprudent to do so without considering the

ultimate hardware code generated. Also, we find that the parallel code produced
from a high-level syntax is generally further from optimal than its scalar coun-
terpart. Part of this is due to a lack of sophistication in the compilers compared to their scalar counterparts. Part is due to the lack of an adequate syntax in Fortran that reflects the hardware capabilities. Thus, we cannot test
these architectures fairly with a common code. Consequently, we have chosen to
accumulate data for specific computational tasks from various computers on the
assumption that the implementation has been done with reasonable consideration
given to the particular architecture.

HARDWARE

First, we identify the computers of interest and classify them according to
pipeline--multiple instruction, single data (MISD); single instruction, multiple
data (SIMD); and multiple instruction, multiple data (MIMD). We then discuss
significant features of each architecture that are important to the problems of
linear algebra. We also consider certain theoretical properties of parallel processing. Such factors as start-up time, scalar-vector break-even points, consistency in the number of operations used, the number of parallel steps required,
speed-up and efficiency of the hardware, and average vector or array size are
discussed and quantified in a general way. Finally, we discuss performance. We
are able to make the broadest comparison in the performance of LU factorization.

We consider computers that appear capable of exploiting the parallelism of large-
scale scientific calculations and for which we can obtain data on the problems
of interest in linear algebra. Although a simple and accurate classification may
be impossible, we have categorized those computers as pipeline (both hardware and
software) and multiprocessor (both SIMD and MIMD). Such a classification provides an adequate basis for both theoretical and practical considerations.

Pipeline computers are streamlined sequential computers, and a rigorous taxonomy
might well exclude them from the class of parallel computers. However, effective
use of pipeline computers depends heavily on the design of efficient parallel
algorithms. In this sense, pipeline computers are treated as parallel computers.
The major pipeline computers are the Control Data Corporation (CDC) STAR-100,
Texas Instruments (TI) ASC, and the Cray Research Incorporated (CRI) CRAY-1.
Large sequential computers that have some pipelining capabilities include the CDC
7600, IBM 360/91, and IBM 370/195. Since we have an abundance of data from the
CDC 7600, we use this computer as generic of the large sequential computers with
pipelining capabilities. In addition there are a number of special-purpose com-
puters with pipelining or multiprocessing capability that are commonly identified

as array computers. Representative examples include the IBM 3838, CDC MAP III, and Floating Point Systems (FPS) AP-120-B. Data from the Floating Point Systems AP-120-B provides some perspective on the performance of array computers. The array computers typically work in conjunction with sequential host computers.

The major scientific SIMD computers are the Burroughs BSP, Burroughs ILLIAC IV, International Computers, Ltd., DAP, and Burroughs PEPE. PEPE is a special-purpose computer designed for antiballistic missile defense. However, recent work has been undertaken to study the effectiveness of this computer for more general applications |1|. The Goodyear STARAN computer does not appear to be appropriate for this problem because it lacks floating point arithmetic.

We are unable to find an appreciable amount of data for the problem of interest for MIMD computers. The TI ASC (four-pipe) can be considered to be in this class since the four pipes may be independently designated for execution and may be performing different instructions. We understand that the Siemens SMS computer has been programmed to execute matrix multiplication and inversion. The SMS is a laboratory model with 128 processors tied to a common bus and a host computer. Each of the 128 processors has its own memory and can operate its own independent instruction stream. However, this data is not available to us and would not be competitive since much of the arithmetic is implemented through software.

FPS AP-120-B

The AP-120-B is a small inexpensive processor designed for efficient signal analysis. The add and multiply units are pipelined. Program, constant, and data memories are independent so that bank conflicts may be avoided. Not only may two operands be fetched simultaneously from separate memories, but multiple instruction tasks may be executed within the same machine cycle. All of these features permit a very high level of performance relative to the machine cycle time and low cost for signal processing type problems, including the problems of linear algebra of interest to us.

To achieve the speeds that make this computer competitive with the other computers considered in this report, one must program in assembly language to schedule instructions optimally and manage data across the various memories and registers in the most conflict-free manner possible. Consequently, a rather extensive library of Fortran callable basic functions has been written to obtain the ultimate performance of this computer. The computer is somewhat limited in vocabulary, because it is designed to solve a limited class of problems. It is well designed as a transform array processor.

CDC 7600

The CDC 7600 has been converted to a "software vector" processor. The hardware features of multiple, independent, and segmented functional units along with an instruction stack of sufficient size to hold sizable loops (thereby freeing memory for data transfers only) permitted the development of a complete set of vector subroutines that make up the STACKLIB |2|. Loops are unfolded (multiplexed) to allow elimination of wait times associated with memory references and passage of results through the various functional units. Also, loop overhead is prorated over more processing. Use of the CDC 7600 in this way has caused it to be referred to as a "Vector-7600" |3|. Subroutine linkage causes it to have a software start-up time not unlike that of a CDC STAR-100 or a TI ASC.

CDC STAR-100

This computer has a very high processing rate (100 million 32-bit floating adds per second) in the vector mode. In view of the slow memory (1280 ns random access) that supports it, this is a rather remarkable feat. As a consequence of the slow memory, both start-up time* for vector instructions and scalar arithmetic are large. Vector instructions execute four to six times that of the CDC 7600, whereas scalar performance is about one-fourth that of the CDC 7600. At least 80% of a code must be vectorized to make it compete with the CDC 7600. One other feature of the CDC STAR-100 that forces considerable algorithmic redesign is its ability to process only contiguous vectors.

TI ASC

Significant features of this computer are that it (1) can operate on vectors from rows as well as columns of a matrix; (2) can process up to three nested DO loops as a single vector; and (3) may have one to four vector pipes. The first feature allows for simple vectorization of codes compared to the STAR-100. The second feature reduces the bad effect of its slow start-up times. The third feature provides some MIMD-like features because the pipes are scheduled through software and may be executing different arithmetic instructions concurrently.

---

*Start-up time is defined to be that part of the total time for a vector operation that is not accounted for by the result rate.

The scalar performance of the ASC is reasonably good but not up to that of the good sequential computers. Memory conflicts for noncontiguous vectors may degrade performance significantly. Also start-ups cannot be overlapped to a significant extent. Combining this with the resultant shorter vectors causes the four-pipe computer to perform worse than the one-pipe computer for problems of small order.

CRI CRAY-1

The novel and powerful features of this computer are (1) 8-vector registers consisting of 64 words of 64 bits each and (2) a feature called "chaining" that permits the overlapped use of functional units required in a given sequence of operations, such as a multiply followed by an add, that is so important to the class of problems considered here. This permits sustained rates as high as 135 megaflops.

Burroughs ILLIAC IV

The ILLIAC IV is extremely fast (100 megaflops) when all 64 processors can be used simultaneously. However, the machine is currently plagued by disks serving as the main memory for many problems. This computer is tedious to program. Consequently, its use has been confined to very special applications in which the algorithms were tailored to the computer architecture. With detailed attention paid to both problem and algorithm, the computer has been made to perform at a few times that of a CDC 7600 or an IBM 360/195. We had expected to find performance data on the problems of linear algebra since the memory can contain sizable systems. However we were unsuccessful. This reflects the special purpose use of this computer.

Burroughs PEPE

PEPE consists of up to 288 parallel processing elements, each with its own memory. Data transfer between processing elements occurs sequentially through the control processor. Experience with an 11-processor version and a simulator for the full 288 system led Blakely |1| to study the feasibility of PEPE for more general applications; for example, weather problems and LU factorization. Because data is scarce for SIMD machines, we use Blakely's findings.

ICL DAP

This computer is a bit serial array processor. It is similar to the ILLIAC IV in connectivity properties. It is different in that there are many more processors (32x32) each operating at much slower speed. A 64x64 version is being built and will be installed at Queen's University in London. The 64x64 processor system will increase full array operations by a factor of 4 over that of the 32x32 system. Each processor is a one-bit processor (bit serial) and, therefore, is very slow. The arithmetic is performed in software. The bit serial approach is effective in many applications, for example, the searching and swapping operations required in pivoting. However, it does not perform the floating point arithmetic effectively.

Burroughs BSP

The BSP is an SIMD system with 16 processing elements. The memory is designed around 17 banks to avoid bank conflicts, and this simplifies programming considerations. The maximum megaflop rate is 50, and is easily approached without requiring long vectors. In fact, vector performance is degraded more through over-simplification of expressions. For example, the vector task $z = x + y$ is dominated by memory references, whereas $y = ax + y$ provides enough arithmetic to be computation limited. The basic cycle time tends to limit the scalar performance of the BSP.

BASIC MODELING

Typically the time required for vector operations on a pipeline computer is

$$t = ms + r_v n, \tag{1}$$

where m is the number of start-ups required, s is a start-up time, r_v is the time for which each element of the vector is processed, and n is the number of elements in the vector. If r_s is the time for processing a scalar operation, and if $n > ms/(r_s - r_v)$, it is profitable to use a vector operation. Thus, the necessity for processing long vectors is obvious when the total start-up time ms is large. This limits the performance on the CDC STAR-100 and the TI ASC for the linear system problem by Gaussian elimination.

The efficiency e of a pipelined computer for processing an n-element vector might be defined by

$$e = \frac{\text{computation time}}{\text{total time}}$$

$$= \frac{r_v \cdot n}{m.s + r_v n}$$

$$= \frac{1}{1 + \left(\dfrac{s}{r_v}\right)\left(\dfrac{1}{n/m}\right)}$$

where $m = \lceil n/p \rceil$, p is the vector length per vector operation, and $\lceil \cdot \rceil$ is the ceiling function. The factor (s/r_v) shows the importance of start-up and the factor (n/m) displays the dependence upon average vector length.

The average vector length is an important consideration in algorithm design. Clearly, performance is proportional to vector length; unfortunately, temporary storage requirements are also proportional to it. In Fig. 1 we plot the average vector length for simple loops implementing matrix multiplication, general LU factorization, and tridiagonal LU factorization by cyclic reduction. The TI ASC computer can process up to three nested loops as a single vector. We have, therefore, included in Fig. 1 the average vector length for nested loops for both

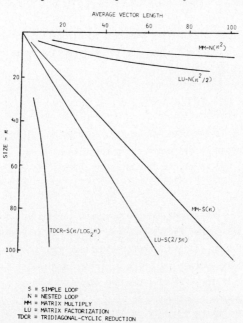

S = SIMPLE LOOP
N = NESTED LOOP
MM = MATRIX MULTIPLY
LU = MATRIX FACTORIZATION
TDCR = TRIDIAGONAL-CYCLIC REDUCTION

Fig. 1. Average vector length vs order.

matrix multiplication and LU factorization. However, temporary storage requirements are likewise $O(n^2)$ for these algorithms.

Whereas the work in SIMD and MIMD systems is typically measured by the number of
parallel steps required, work in pipeline computers is normally measured, as with
sequential computers, by the number of arithmetic operations required. Following
Lambiotte and Voigt |4|, a vector algorithm is <u>consistent</u> if the order of the
number of operations, $O(n)$, required in the vector operations is asymptotically
the same as the number of scalar operations required in a "best" sequential algorithm. Hence, algorithms suitable for a pipeline computer may be closer to that
of a sequential computer than a multiprocessor.

In the multiprocessor environment, particularly with SIMD computers, one has many
processors and must "use them or lose them." Although this is not theoretically
the case for MIMD computers, it may still be the practical case in their utilization because of the time currently required for task switching and initiation.
In the multiprocessor environment, the number of parallel steps required is
generally more important than the total number of operations required. Requirements for numerical stability as well as hardware resources usually temper the
use of algorithms with a large, total operation count but few parallel steps.

If we let s be the total scalar time required to effect an arithmetic operation
in a multiprocessor environment, then the total time to process an n element vector with p processors is given by

$$t = m \cdot s$$

where $m = \lceil n/p \rceil$ and s is the time for an array operation. This processing time
is similar to that given in (1) where r_v is taken to be zero. See Fig. 2 to compare the two models. Obviously, multiprocessor systems offer the long term solution to speed up.

The existence of many independent calculations and the limitation of the speed
of light on sequential computers are the driving forces for developing multiprocessor systems. However, since all computations are not independent, the efficiency with which these systems can solve a particular problem becomes an important consideration. If the size of a problem is a function of n, let the time to
compute this problem with p computers be $t_p(n)$. Then the speed-up gained by using p processors is given by $f = t_1(n)/t_p(n)$ and efficiency of using p processors
is $e = f/p$. Hence $e \leq 1$ and $f \leq p$.

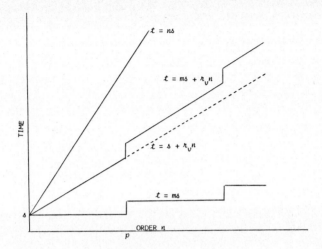

Fig. 2. Processing time required
for various types of processors.

Unfortunately, most of the calculations of linear algebra are sufficiently dependent that there are theoretical lower bounds of $O(\log_2 n)$ on the number of
steps required to solve a problem of order n. This kind of theoretical lower
bound exists for the dot product of two vectors, solving tridiagonal linear systems, matrix multiplication, and full matrix inversion and LU factorization.
This strong dependence on $\log_2 n$ derives from the necessity to sum n elements,
which can be done in $\log_2 n$ steps with the important associative fan-in algorithm.

There are stable algorithms that can be executed in $\log_2 n$ steps for all the problems mentioned except that of the general linear system. The algorithms that
can be performed in less than $O(n)$ steps use methods other than Gaussian elimination and for which stability is poor or yet an open question. Much work is being
done to bridge the gap between a theoretical lower bound of $O(\log_2 n)$, and the
current best parallel algorithms for Gaussian elimination are $O(n)$ with no pivoting and $O(n\log_2 n)$ with pivoting. Note the emergence of pivoting as the dominant
operation.

This brings us to another interesting point to be made about increased paralleli-
zation. The high order term in the work estimates of various tasks are often
cubic and can be made small in pipeline computers or nonexistent in multiprocessor systems. In the case of pipeline computers, this is an additional cost.

TABLE I

Cubic work polynomials (time in nanoseconds)

ASC 1-pipe		
	inner	$26.9n^3 + 1235n^2 + |31,850n| + 0(1)$
	outer	$54.6n^3 + 186n^2 + |19,107n| + 0(1)$
ASC 4-pipe		
	inner	$7.3n^3 + 232n^2 + |100,200n| + 091)$
	outer	$14.9n^3 + 240n^2 + | 33,890n| + 0(1)$
CRAY-1		
	inner	$|4.81n^3| + 139.2n^2 + 5,792n + 0(1)$
	outer	$|9.63n^3| + 77.1n^2 - 282n + 0(1)$
STAR-100		
	outer	$20.7n^3 + |6328n^2| + 7,166n + 0(1)$

The start-up time increases the quadratic and/or linear terms of the work polynomial. For small systems the $0(n^2)$ tasks of searching, swapping, and scaling have always been important; for small n even dominant. Consequently this situation is aggravated by the start-up time associated with all of the known pipeline computers. We can see this effect in Table I taken from the benchmark study of Calahan, Joy, and Orbits |5|.

Even for pipeline computers with good start-up times, say $s/r_v \leq 10$, the relative improvement of pipeline computers over conventional computers on systems of small order or small bandwidth is modest. It is not uncommon for many such systems to exist simultaneously; for example, multiline relaxation. In this case one may be able to vectorize the problem by solving many systems simultaneously. This has been done with tridiagonal systems by Jordan for CRAY-1 |6| and Boris on the ASC |7|.

Students of Calahan have written CRAY-1 codes to solve multiple general systems of small order as well as multiple banded systems with small bandwidths |8|. In this manner one is able to obtain good vector performance on small systems for a limited yet frequently encountered class of problems.

Despite the attention that has been paid to sequential and parallel complexity, data communication is a formidable problem in multiprocessor environments. In SIMD machines, such as the ILLIAC IV and the ICL DAP, the problem of data routing to other processors can be severe |10|. Similarly with MIMD computers, data transmission and/or memory conflicts are major difficulties.

PERFORMANCE OF LU FACTORIZATION

We have extended the work of Calahan, Joy, and Orbits |5| in which they compared the CDC STAR-100, CRI CRAY-1 and TI ASC, along with the IBM 370/195 and the AMDAHL 470/V6, for the problem of matrix LU factorization without pivoting. We have used their data and have added data for PEPE, CDC Vector-7600, ICL DAP (32x32), FPS AP-120-B, and the Burroughs BSP (estimated). Some of this data comes from the literature. Data for the Vector-7600 was measured, whereas that for the BSP was estimated from a reasonably detailed timing analysis of the two inner loops. A more complete description of how the data displayed in Fig. 3 was obtained is described below for each computer.

The CDC Vector 7600 times were obtained with a Fortran code calling two subrou-

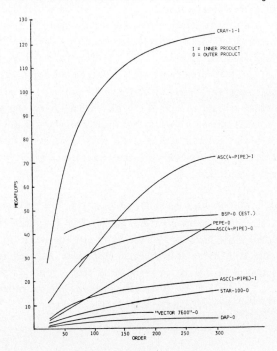

Fig. 3. LU without pivoting.

tines written in assembly language, one for scaling the pivot column and the
other for performing the vector function y = ax + y that is used repetitively in
each stage of the reduction.

The CDC STAR-100 times come from Calahan, Joy, and Orbits |5|. This code was
written in CDC Fortran with vector extensions. The code used the outer product
method since the hardware inner product instruction of the STAR-100 is much
slower. The timing formula in nanoseconds for the fitted work polynomial is
given by

$$t = 20.7n^3 + 6328n^2 + 7166n + 38949. \tag{2}$$

We see from (2) that we will not come close to the asymptote for orders that will
fit in memory. At n = 1000, about 1/3 of the total time is accounted for by the
n^2 term. This can also be seen from the behavior of the curve in Fig. 3.

Data for LU factorization times on the ASC computer are those reported by Calahan,
Joy, and Orbits |5|. The codes are primarily written in Fortran with some vector
extensions. Codes were written for both the outer product and Crout (inner product) algorithms. The latter algorithm uses the hardware's ability to form inner
products at the rate of a multiply and an add in one cycle. The results are
shown in Fig. 3 for LU factorization without pivoting. These codes were run on
both one- and four-pipe ASC machines. In the latter case we have an unusual
characteristic of the MIMD pipelines. This creates more start-up overhead so
that the one-pipe computer is faster for the outer product case when n < 12 and
faster for the inner product case when n < 35. Not only are there crossover
points for the optimal number of pipes, but there are also crossover points within each architecture of the two methods. The outer product is faster for orders,
say n < 80. Hence, an optimal code for the four-pipe computer would be very
hybridized or polyalgorithmic.

It is clear from the high order terms ($7.3n^3$ and $26.9n^3$ in Table I) of the fitted
work polynomial that an asymptotic rate of almost four is theoretically possible.
However, it is also clear from Fig. 3 that it is not achievable in practice and
the average gain made by using four pipes as opposed to one pipe might be close
to two on average even in the region where it is best. Table II shows the efficiency of the ASC (four-pipe) for LU factorization at various orders. To our
knowledge this is the only data available for MIMD architecture on this particular problem.

TABLE II

ASC efficiency of 4-pipe computers (relative inner product times)

order	25	50	75	100	150	200	250	300
efficiency	.18	.31	.44	.54	.69	.77	.82	.85

PEPE data is taken from the work of Blakely |1|. Because this computer was specifically designed for the antiballistic missile defense system, studies were undertaken to evaluate the capability of the system for more general problems. The data is taken from an abbreviated PEPE (11-PE's) and a software simulator that allows the number of processors to be extended to the full complement of 288. The timing approximation is quadratic and is given in nanoseconds by

$$t = 4418n^2 + 5705n + 6544 \quad . \tag{3}$$

With 288 processors, the megaflop rate would attain its maximum at slightly under 45. We understand from private communication with Blakely that a better algorithm has been found to improve upon this data. However, the new data is not yet available to us. The necessary data transfers between processing elements must be made by transferring the data through the control processor. The hardware is not designed to perform this task quickly.

We have attempted to estimate performance of the ICL DAP on LU factorization based upon data obtained from Flanders, Hunt, Reddaway, and Parkinson |9|. They give a matrix inversion time, including pivoting, of 29 ms (recently improved to 26 ms) for n = 32. On this computer, inversion for n $\leq$ 32 is cheaper than LU factorization and the time over this range should be linear in n. For larger n, we estimate the time for doing block LU using the time quoted for matrix multiplication (16 ms at n = 32). Because inversion by Gauss-Jordan is so close to matrix multiplication when pivoting is not performed, we have also used 16 ms as the inversion time for n = 32. The matrix add time is comparatively short and has been ignored. We then count the inversions and multiplications for n = 32, 64, ..., 256. These points provide a bound on the megaflop rate. Note in Table II the effect of the extra work done in inversion before the full megaflop rate is achieved.

The ICL DAP (64x64) is expected to provide a factor of 4 in speed up. This means that inversion and multiplication times used for our purposes for matrices of order 64 will be twice that of order 32 on the ICL DAP (32x32). Observe in Table III that n must increase four times the block size of 32 before we are

within 10% of the asymptotic rate. The same will be true of the 64x64 configuration, that is, n = 256 before we reach 10% of the asymptote. Hence, one may
seldom see the $1/3n^3$ effect in practice in that architecture.

TABLE III
DAP block LU

Order	Time	Megaflops
32	16	1.3
64	64	2.7
96	176	3.3
128	384	3.6
160	720	3.8
192	1216	3.9
224	1904	3.9
256	2818	3.9

The BSP timing formula for the triple DO loop that performs LU factorization is
given in nanoseconds by

$$t = t_{scaling} + t_{reduction}$$

$$= \sum_{k=1}^{n-1} (480\lceil m/16\rceil + 640\lceil m/16\rceil m)$$

where n is the order and m = n-k. This formula ignores the initial start-up time.
Loop time is assumed to be covered by the vector operations. This provides us
with an upper bound on performance and undoubtedly is somewhat optimistic. The
computer is memory limited when scaling, whereas the arithmetic units are the
limiting resources for reduction. This formula predicts a megaflop rate of 44.8
at n = 100, or almost 90% of the maximum rate of 50 megaflops.

We show in Fig. 4 the few more realistic comparisons we have available to us.
All the data presented here are from complete codes with partial pivoting. All
are reasonably close to optimal for each computer and exploit the parallel features of each architecture. We have repeated the graph of the CRAY-1 kernel to
demonstrate the growing importance of the $O(n^2)$ tasks required for partial pivoting. Observe that the time required for this task is nontrivial and becomes
relatively more important in the parallel architectures. As stated earlier, the
pivoting step, which is $O(n\log_2 n)$, becomes the dominant one in the limit of multiprocessor systems since arithmetic can be done in $O(n)$ steps. Here we see the
effect showing up significantly in the pipeline computers.

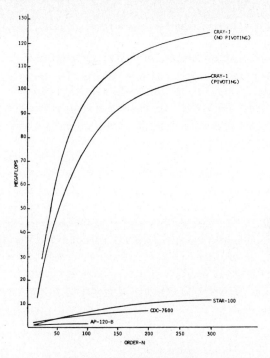

Fig. 4. LU with pivoting.

The CRI CRAY-1 codes are written entirely in assembly language. The CDC STAR-100
codes are written in Fortran using vector extensions. The CDC Vector 7600 is
written in Fortran, calling STACKLIB for vector arithmetic. The FPS AP-120-B is
written in assembly language but does complete matrix inversion. The megaflop
rate given for this computer is for the $2n^3$ floating point operations required to
invert, whereas the others are based on the $2/3n^3$ operations required in LU factorization. We do not know why FPS has not supported the conventional LU factorization. If indeed it is because matrix inversion is faster, then one would want to
decrease the megaflop rate shown by a factor of 3. We believe that the hardware
can be programmed to produce an LU code that will approach the megaflop rates indicated in Fig. 4.

Jordan and Fong |6| have characterized performance level on CRAY-1 into scalar,
vector, and supervector. Nominal values of performance for these three levels
are 2, 4, and 12 times the CDC 7600, respectively. All are assumed to be good
machine language codes. Scalar (S) performance implies no vector instructions are
used. Vector (V) performance implies vector operations are from memory to memory
as they would be if supported by the basic linear algebra subroutines (BLAS) or
STACKLIB. Supervector (SV) performance is achieved by keeping intermediate results in the vector register and thereby maximizing the use of vector registers

as a cache memory. To date, compilers can achieve at best V speeds.

Figure 5 displays the LU performance with code implementations in scalar Fortran,
vectorized Fortran, Fortran calling BLAS written in CRAY assembly language (CAL),
and a code written entirely in CAL that fully exploits the machine architecture.
The first three codes are LINPACK compatible and were written and timed by Don-
garra |11|. The last code was written by Jordan |6| and is LINPACK compatible
except for the permutations in the L matrix. The usual algorithm with partial
pivoting performs at most $1/2n^2$ interchanges. Our algorithm requires up to n^2 interchanges in order to exploit the hardware on the $O(n^3)$ operations. This is an
example of trade-off of extra work for speed, which appears to be characteristic
of the switch from scalar to parallel hardware. However, this is not the first
time in computing history that we have seen this phenomenon. When we went from
hand computing to automated computing, we saw much of this. For example, in relaxation problems we computed new corrections in an unordered fashion at those
points for which there was greatest activity. With the advent of automatic computing, we could not afford this testing and proceeded to a successive relaxation
of points.

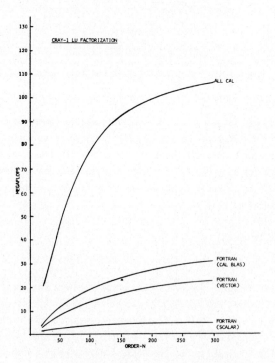

Fig. 5. CRAY-1 performance on LU factorization.

SUMMARY

We have attempted to portray the current status of "parallel" computers for solving several important problems in linear algebra. We believe the data presented
and the material referenced support the following points.
1. Although the hardware to date requires special attention, we see the hardware
   development becoming more general purpose with time. The CRAY-1 and BSP can
   produce good performance from Fortran codes written so as to be vectorizable
   with a small number of rules and quidelines to users.
2. Optimal performance will be difficult to achieve in a high-level language,
   although the BSP appears most likely to overcome this difficulty.
3. Relative start-up time and average vector or array size are important factors
   in parallel processing.
4. Increased parallelism leads to considerably increased temporary storage requirements.
5. Arithmetic is generally not the critical resource. Such items as memory performance, data transfers, and searching may well be the resources that pace a
   calculation.
6. Hybrid or polyalgorithms are more prevalent for exploiting parallel architectures.
7. Architecture is still dictating algorithm selections.
8. Portability to these architectures will require parallel considerations at
   the design stage.
9. Parallelism may be exploitable in more than one way. First, there is the
   natural vector or array expression in a single system. Often the problem may
   be expressed as multiple systems.

The reader's attention is called to a recent article by Heller |12| for an up-to-
date report on the development of parallel algorithms in linear algebra.

ACKNOWLEDGMENTS

We are indebted to W. Brainerd of the Los Alamos Scientific Laboratory for developing the BSP Asymptotic timing formula for the LU factorization without pivoting
and the recursive hardware that can be used in the forward and back substitution.

REFERENCES

|1| C. E. Blakely, "PEPE Application to BMD Systems," 1977 Conference on Parallel
    Processing, Wayne State U., Detroit, Michigan, IEEE Computer Society Catalog
    No. 77 CH1253-C (August 1977).

|2| "STACKLIB, A Vector Function Library of Optimum Stack-Loops for the CDC 7600,"
Los Alamos Scientific Laboratory internal document, January 1977.

|3| L. L. Barinka, K. W. Neves, and P. G. Tuttle, "Performance of Some Vectorized
Mathematical Software," Babcock and Wilcox Co., Nuclear Power Div. report
TM-361 (Lynchburg, Virginia, April 1976).

|4| J. J. Lambiotte and R. G. Voigt, "The Solution of Tridiagonal Linear Systems
on the CDC STAR-100 Computer," ACM Trans. on Mathematical Software, Vol. 1,
No. 4, 308-329 (1975).

|5| D. A. Calahan, W. N. Joy, and D. A. Orbits, "Preliminary Report on Results
of Matrix Benchmarks on Vector Processors," U. of Michigan report SEL 94 (May
1976).

|6| K. W. Fong, and T. L. Jordan, "Some Linear Algebraic Algorithms and Their
Performance on CRAY-1," Los Alamos Scientific Laboratory report LA-6774 (June
1977).

|7| J. Boris, "Vectorized Tridiagonal Solvers," Naval Research Laboratory report
NRL-3408, Washington, D. C. (1976).

|8| D. A. Calahan, U. Of Michigan, personal communication, 1978.

|9| P. M. Flanders, D. J. Hunt, S. F. Reddaway, and D. Parkinson, "Efficient High
Speed Computing with the Distributed Array Processor," in High Speed Computer
and Algorithm Organization, D. Kuch, D. Lawrie, and H. Sameh, Eds. (Academic
Press, Inc., New York, 1977) pp. 113-128.

|10| W. M. Gentleman, "Some Complexity Results for Matrix Computation on Parallel
Processors," Journal of the ACM 25, No. 1 (1978).

|11| J. J. Dongarra, "LINPACK Working Note #11: Some LINPACK Timings on CRAY-1,"
Los Alamos Scientific Laboratory report LA-7389-MS (August 1978).

|12| P. Heller, "A Survey of Parallel Algorithms in Numerical Linear Algebra",
SIAM Review, Vol. 20, No. 4, (October 1978).

Performance Evaluation of Numerical Software, Fosdick (ed.)
© IFIP, North-Holland Publishing Company, 1979

DETECTING ERRORS IN PROGRAMS*

Lloyd D. Fosdick
Department of Computer Science
University of Colorado
Boulder, CO 80309, U.S.A.

A review of work on the occurrence and detection of errors in
computer programs is presented. This includes: experiments
to measure the frequency and distribution of errors; the use
of simulation to determine the effect of typing mistakes;
data flow analysis for static error detection; and measures
to quantify program testing.

1. INTRODUCTION

"It is natural at first to dismiss mistakes in programming
as an inevitable but temporary evil, due to lack of
experience, and to assume that if reasonable care is taken
to prevent such mistakes occurring no other remedy is
necessary".

Stanley Gill wrote these words over twenty-five years ago [1], pointing out
that experience refuted this position: experience still refutes it, arguments
for proving programs correct notwithstanding. We are human, and being human it
is inevitable that we make mistakes which result in errors in computer
programs. With improved languages, with careful attention to program design
and programming style we can reduce the occurrence of errors. But we cannot
eliminate them, so we cannot ignore the problem of error detection or questions
concerning the distribution of errors, their types, and their symptoms.

This paper is concerned with the occurrence and detection of errors in
computer programs. It is a review intended to show the nature of the research
being conducted and the state of our knowledge to those not well acquainted
with the subject. It will show that our knowledge of the occurrence of errors
in computer programs is scanty and difficult to interpret, that error detection
in practice is weaker than it needs to be, that software testing is guided
largely by intuition and is lacking a scientific foundation, and finally that
there is no shortage of interesting problems in this area.

2. TERMINOLOGY

The terminology used to describe errors in the literature varies in meaning,
making it difficult to interpret and compare results. For example, in one
article "syntax error" means any error reported by a compiler, in another it
means a violation of a grammatical rule. While it might seem easy, anyone who
attempts to analyze the problem of error classification will find it very
difficult; indeed the notion of error itself is hard to characterize. This is
confirmed by the experience of Gerhart [2] who has recently examined this issue
in connection with an unsuccessful attempt to build a taxonomy of errors.
However we must have a common understanding of the terms used here so some
brief definitions suitable for this purpose follow.

Here the word "error" means a construction in the program which violates a
language rule, or which may cause a computation that is not intended according
to the specification of the program: an error of the first type is called a
language error, and an error of the second type is called a specification error.

---

* This work was supported in part by U.S. Army Research Contract
No. DAAG29-78-G-0046.

Language errors are divided into two types, syntax errors and semantic errors, according to the following scheme which is similar to that used by compiler writers [3]: the language rules are separated into two groups such that those in one group can be represented by a context free grammar, then a language error is called a syntax error if it is a violation of the rules of the context free grammar, otherwise it is called a semantic error. If a program contains no language errors it is called well-formed, and a well-formed program with no specification errors is called correct. A construction in a well-formed program that is extremely unusual is called an anomaly; it is often a symptom of a specification error.

It will be convenient to use the term "flow graph". This is a directed graph representing the control structure of a program, with the nodes corresponding to statements and the edges to pairs of statements executable in immediate succession. Thus a flow graph is an abstraction of the conventional flow diagram. In speaking of program execution I use a loose but convenient terminology, speaking of executing a node (i.e. a statement) or executing a path (i.e. a sequence of statements).

Two approaches to error detection, called static and dynamic, are distinguished. Static error detection is characterized by the use of techniques that do not require actual execution of the program. The most common form of static error detection occurs during parsing in the compilation of a program. Dynamic error detection, on the other hand, uses information gathered during actual execution of the program to detect errors. The most common form of dynamic error detection occurs during the testing of a program.

Unfortunately language rules and specifications are often incomplete or ambiguous, so the application of these definitions runs into difficulty; therefore in practice the distinction between syntax error and semantic error is often fuzzy, as is the distinction between well-formed programs, and correct and incorrect programs. FORTRAN programs present a particularly difficult problem in this respect. Furthermore the notion of a correct program is idealistic for most practical purposes, especially so when it comes to numerical software, since we lack the mechanisms required for insuring that a program meets its specifications unless the specifications are exceptionally simple. Thus these definitions must be taken as a statement of intent, recognizing that in practice they will be applied imperfectly.

3. OCCURRENCE OF ERRORS

Our concern here is with the frequency of errors in program text and with the types of errors that result from simple mistakes made in the preparation of program text. Data on errors has been obtained by examination of a collection of arbitrary and naturally occurring program text, such as that submitted to a computing facility within a particular period of time [4,5]; by examination of error reports in controlled programming environments [6,7]; and by experiments using deliberately chosen problems, and a controlled group of programmers [8,9,10]. Because it involves human subjects gathering this information is difficult; often students are used as subjects, casting doubt on the generality of the results.

The difficulties of dealing with human subjects can be removed, or at least isolated, by using simulation; an approach which conveniently divides the study of the occurrence of errors into two distinct components which can be studied separately -- the frequency and kinds of mistakes that humans make in the construction of programs, and the effect of these mistakes on program text. The second component can be studied with simulation, thus it can be precisely controlled and large amounts of data can be obtained at low cost.

3.1 Errors produced by humans. For programs in the stage of development when the first compilations are being made, the frequency of errors found in the text ranges from about fifty errors per thousand statements to over one hundred errors per thousand statements.

In discussing their work on DITRAN, a FORTRAN compiler, Moulton and Muller [8] reported on the errors detected in students' programs using it. For errors detected during compilation, which would include all syntax errors and most semantic errors, their data indicate an error frequency of more than forty errors per thousand statements. For errors detected during execution time which would include semantic errors not caught in compilation, their data indicated at least eight errors per thousand statements. Since this data does not appear to include specification errors it seems reasonable to conclude that the frequency of errors in these programs was in excess of fifty errors per thousand statements. These numbers are reasonable but crude because of problems in interpreting the data. They counted diagnostics and one error can cause more than one diagnostic, but they do not distinguish two compilations of the same program from compilation of two different programs. I have assumed one error per diagnostic in estimating frequencies which will tend to make the frequency too high, but failure to distinguish multiple compilations of the same program tends to reduce the frequency.

Youngs [9] has made a careful study of errors in programs and his results are consistent with the error frequencies indicated by the Moulton and Muller data. In Youngs' study the languages ALGOL, BASIC, COBOL, FORTRAN, and PL/1 were used, and programs written by "professionals" and novices were considered. Most of the programs were simple numerical routines of less than eighty statements in length. His data indicates a frequency of more than seventy errors per thousand statements. Moulton's and Muller's data applies largely to beginners, as does part of Youngs' data and one might guess that this would bias the error frequency to be high; however Youngs says "... on first runs of all programs both beginners and advanced programmers had an average of 5.6 errors in their programs ...", and he said that beginners committed 19.4 errors per program compared with 15.1 errors per program for advanced programmers. (There is an inconsistency in these figures that I have not been able to resolve). The significant difference that Youngs did observe between the two groups was in the type of errors committed: for the beginners the error distribution according to type was 12% syntax, 41% semantic, 35% "logic" (specification), 5% "clerical" (syntax or semantic), and 12% other; for advanced programmers it was 17% syntax, 21% semantic, 51% "logic" (specification), 4% "clerical" (syntax or semantic), and 11% other.

Gannon [10] made a careful study of the occurrence of errors in programs in connection with an investigation of the effect of language design decisions on programming errors. In this study two systems programming languages, TOPPS and TOPPS II, were used. The programmers did not have prior experience with either language but were "reasonably experienced programmers". His data indicated a frequency of 164 errors per thousand statements, including errors made after the first submission to the compiler.

Boies and Gould [5] have measured the frequency of errors which prevent a program from compiling or assembling. The languages used were FORTRAN, PL/1 and assembly language (IBM). All jobs submitted through the IBM TSS/360 computer system at an IBM research center during a 5 day interval were considered. They found that one in five programs contained one or more errors which prevented compilation or assembly. Moulton and Muller in the study mentioned earlier found one in three programs contained such errors. The difference might be due to the fact that the programmers in the Moulton and Muller study were less experienced, but data reported by James and Partridge [4] on errors observed in programs, which included ones written by experienced programmers, also shows that one in three programs "would have been rejected by a conventional computing system". On the other hand results provided by Kulsrud [6] with experienced programmers is somewhat at variance with these results. Kulsrud's results indicate that only about one in eight programs have errors preventing compilation. (I have inferred this from his data, but he did not claim it.) His data also suggests that half of the discovered errors were language errors, and the other half were specification errors. Other data that

seems inconsistent appears in the paper by James and Partridge [4] in which
they say they observed three errors per thousand statements. (It is likely
they are referring to simple typing mistakes only).

The frequency of residual errors in programs, errors which remain after the
program is put into regular service, is difficult to estimate. Reports of
incorrect results from programs which have been used successfully over a period
of years are not uncommon but they may not reflect program errors: they may be
due to changes in use or environment not accounted for in the specifications.
But data which is available suggest that a residual error frequency of five
errors per thousand statements would not be unusual. In Youngs' study it was
found that 27 errors out of an original 383 errors remained in the collection
of programs after they had been compiled and corrected ten times; there were
about 3500 statements altogether, implying a residual error frequency of about
eight errors per thousand statements. In Gannon's study twenty four errors in
about 4800 statements were never found implying a residual error frequency of
about 0.5 errors per thousand statements. It has been reported [11] that
IBM-TSS in its twentieth revision had "12,000 distinct new bugs" implying a
residual error frequency of more than four errors per thousand statements.
Finally, in a recent study [12] in which a program for the simulation of radar
reports was carefully examined, after this program had been used for three
years, an average of ten errors per thousand statements was found.

3.2 Errors produced by simulated mistakes. As noted earlier there are
advantages in the use of simulation for the study of the nature of errors in
program text. The basic idea is to simulate certain types of mistakes made by
programmers and then to observe the nature of the errors caused by these
mistakes. In experiments of this type we can measure the influence of language
design and programming style on the nature of errors caused by human mistakes
in programming.

Surprisingly there is almost no work of this type reported in the literature.
Some years ago Weinberg and Gressett [13] made a study of the effectiveness of
a FORTRAN compiler in detecting errors caused by simulated typing mistakes.
However the scope of their experiment was very limited and there was no
analysis of errors according to type.

Recently I have conducted experiments to measure the distribution of the
kinds of errors caused by simulated typing mistakes in FORTRAN programs [14].
The simulated typing mistakes were: substitution (e.g. DIG instead of DOG);
deletion (e.g. OIL instead of FOIL); insertion (e.g. FRIEND instead of FIEND);
transposition (e.g. SILT instead of SLIT). Two kinds of substitution mistakes
were simulated: nearest-neighbor, in which the substituted character is a
nearest-neighbor of the correct character on the keyboard; and random, in
which the substituted character is any character. In all cases the only
characters considered were the characters in the FORTRAN alphabet. Four
algorithms published in the ACM Transactions on Mathematical Software were used
as subjects: Algorithm 495, Algorithm 498, Algorithm 505, and Algorithm 513.
One thousand samples, each an algorithm with a single simulated typing mistake,
were created. This ensemble had the following composition: for each algorithm
fifty samples with each kind of mistake, thus 50 x 5 samples of an algorithm
were created. For each sample the place where the mistake occured was selected
at random, ignoring blanks and comments. Analysis of the one thousand samples
yielded the distribution displayed in Fig.1.

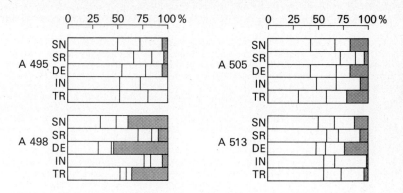

Figure 1. Distribution of errors caused by typing mistakes.

In this figure the bars are labeled SN for nearest-neighbor substitution
mistakes, SR for random substitution mistakes, DE for deletion mistakes, IN for
insertion mistakes, and TR for transposition mistakes. The bars of the figure
are divided into segments corresponding to the different kinds of errors: the
first is for syntax errors; the second is for semantic errors; the third is for
simple anomalies, such as an identifier only occuring once; and the fourth,
shaded, is for errors which are impossible or very difficult to detect at
compile time. With these results as a basis for comparison we could, by
performing similar experiments on programs written in other languages, determine
whether the use of these other languages increases the likelihood that typing
mistakes could be detected at compile time. It is evident that similar
experiments in which other kinds of mistakes are simulated could be conducted
in order to measure the influence of language design and programming style on
the ease of error detection.

4. DETECTION OF ERRORS

This subject has been divided into static and dynamic error detection. A
further distinction should be drawn between detecting an error and detecting
the error, where the latter implies detection of an error and knowledge of what
the construction should be. Detecting the error implies human interaction as
normally takes place in proofreading text but this topic is not included here
except for the short digression which follows.

Experienced programmers assume too much about the error detecting capability
of the systems they use and are inclined to be careless in proofreading the
text of their programs. The evidence presented in the last section supports
this conjecture. Unformatted program text makes careful proofreading difficult
and so an elementary but important tool for error detection is a simple text
formatter such as POLISH [15]. One aspect of formatting peculiar to numerical
software which is often ignored is the appearance of numerical constants. The
designers of mathematical tables have long recognized that digits should be
grouped and separated as in

$$1.32404\ 74631\ 77167$$
$$1.56663\ 65641\ 30231$$

to avoid transcription errors. This formatting should also be used for
constants in computer programs for the very same reason. This point is worth
emphasising because the placement of blanks or other separators, such as an
underline, within constants is not permitted in all languages and represents a
simple language design consideration which can have an important effect on
numerical software reliability.

4.1 Static error detection. Compilers normally detect all syntax errors but are weak in detecting semantic errors expecially for programs written in FORTRAN. Scowen [16] has constructed a small set of erroneous ALGOL and FORTRAN programs and attempted to compile them on various systems. Out of eleven examples in which a semantic error should have been found during compilation the IBM FORTRAN-G compiler did not detect an error in five of the examples.

In the experiment on typing mistakes [14] mentioned earlier, the mutilated programs were submitted to various compilers. All of the compilers considered (MNF, IBM FORTRAN-H, CDC FTN, WATFIV) were weak: MNF missed as many as 20% of the easily detectable errors; FORTRAN-H, FTN, and WATFIV missed as many as 40% of the easily detectable errors. MNF did detect almost all semantic errors but was weak in detecting simple anomalies which clearly signaled the presence of an error; FTN, FORTRAN-H, and WATFIV missed semantic errors and anomalies. Because of the recognized weaknesses of many compilers, and to reduce errors in moving FORTRAN programs from one system to another, a software tool called the PFORT verifier [17] was developed by a group at Bell Telephone Laboratories. It is widely used for static error detection; it is especially useful for detecting errors in the communication between subprograms and for the information it supplies on the utilization of identifiers.

Almost all static error detection systems, including the PFORT verifier, ignore path dependent anomalies or errors such as the statement sequence

```
X=1.0
X=2.0
```

or a statement sequence in which a reference is made to an uninitialized variable as in

```
SUBROUTINE XAMPL (X,Y)
K=1
IF(X .LT. Y) J=J+1
```

The detection of these requires the recognition of certain combinations of events, or their absence, on a path in the flow graph. Many anomalies and errors in this category can be detected statically without excessive cost. The problem is similar to that faced in global optimisation and, like it, can be treated by data flow analysis [18]. Osterweil and I have built a prototype system called DAVE [19] which uses data flow analysis to detect anomalies and errors in FORTRAN programs. In DAVE a program is represented by its flow graph and each node is labelled with information describing the actions taking place on variables at the node. Three actions, reference, define, and undefine, are recognized: "reference" means a value for the variable must be used, as for X in Y = X+1.0; "define" means a value is assigned to the variable, as for Y in Y = X+1.0; and "undefine" means a value for the variable becomes not known, as is the case for the loop control variable upon satisfying a DO loop in FORTRAN and all local variables upon entry to a subprogram. Error detection then depends on recognizing paths in the graph, including those which cross subprogram boundaries, which contain erroneous or anomalous sequences of actions. For example a path in which an "undefine" is followed by a "reference" without an intervening "define" is erroneous, and a path in which a "define" is followed by another "define" without an intervening "reference" is anomalous.

The critical part of data flow analysis is the search technique which is used to detect combinations of events on paths in the flow graph. DAVE uses depth-first search [20]. This has two advantages: the path containing the detected anomaly or error is obtained without extra work as a byproduct of the search and often the search can be terminated before the entire graph has been examined; it has the disadvantage that a separate search is required for each variable. The time for depth-first search is bounded by $K|V| (|E|+|N|)$ where K is a constant, $|V|$ is the number of variables, $|E|$ is the number of edges in the flow graph, and $|N|$ is the number of nodes in the flow graph. A different search algorithm described by Hecht and Ullman [21] which is iterative has the advantage that all variables can be treated simultaneously and, assuming bit-parallel operations take one time unit, has a time bound $KD(|E|+|N|)$ where D, the number of iterations, is close to unity: this scheme is not favored with the two advantages just mentioned for depth-first search but it seems likely that this scheme will give a faster system in practice than one based on depth-first search because usually $|V|>>D$. DAVE executes about 50 times slower than the FTN compiler on a CYBER 175. This is too slow for general use. We are now building a new version, designed for efficiency, which uses the Hecht and Ullman algorithm and are hoping for a factor of ten improvement in speed.

There are several problems, some of which may prove intractable, that limit the effectiveness of data flow analysis. The address of the variable subjected to an action may depend on the computation as with $A(J)$ in the statement $A(J) = C$. In some cases it may be possible, at least in principle, to determine the value of J from the program itself but in many cases J will depend on data supplied at the time of the computation: we can try to deal with this problem by looking for the possibility of an erroneous sequence of actions for some J but this approach produces false alarms. The practical solution in this case seems to be to issue an alarm if the error or anomaly will arise for every value of J. Another problem concerns flow of information across subprogram boundaries. In FORTRAN, because it does not use recursion, the tracing of data across subprogram boundaries is not too difficult, but in a language supporting recursion tracing data actions across these boundaries is quite difficult and no practical system in which this problem is treated exists. Recent theoretical work on this problem has been done by Barth [22]. A third problem is concerned with the fact that errors or anomalies on paths which are unexecutable cause false alarms unless the unexecutability of the path can be recognized. It should be noted that data flow analysis ignores the predicates associated with nodes in the flow graph and along some paths the predicates may be contradictory (e.g. $X \geq 0$ and $X < 0$) which would mean that the path would never be traversed in any execution of the program. A technique known as symbolic execution has been explored by Clarke [23] to treat this problem; Howden [24], Cheatham, Holloway, and Townley [25] and others have also been exploring this technique.

The essential idea of symbolic execution is to derive symbolic expressions to denote the values of program variables; for instance, from the statement sequence $X = 1.0$, $Y = X+A$ the expression $A+1.0$ can be derived to denote the value of Y. Loops and branches cause obvious problems. In some applications the path is specified and these problems can then be avoided; this is the case, for example, in determining whether a particular path is executable. The difficulties of doing symbol manipulation and algebraic reduction on a machine are intrinsic problems of this approach and necessarily cause it to be expensive, but the possibility of being able to make inferences about a large set of computations makes this technique worthy of further study and development.

4.2 Dynamic error detection. In this approach, which takes place during program testing, errors are detected by examination of the results produced from executions of the program assuming an oracle exists for verifying the correctness of these results. An essential difficulty for this approach is that it is restricted to considering the results of a small number of executions because of cost considerations (note that static analysis implicitly takes into consideration many, if not all, executions but the results are less detailed) so it is important to choose the executions carefully in order that any errors which are present are likely to be exposed. Hence work in this area is concerned with measures of effectiveness of executions with respect to the probability of error detection, and with the mechanics of selecting input data sets associated with these measures.

Although numerous papers, e.g. [26,27,28,29], have appeared in recent years discussing measures of test effectiveness, the state of our knowledge of this subject is very unsatisfactory. There has been no significant work done to

relate any proposed measure to a probability of error detection; the argument
for accepting a measure is completely intuitive. The simplest measure is the
fraction, F_1, of nodes in the flow graph executed. This seems like a
reasonable measure since errors at nodes not executed in testing will not be
found; therefore we might expect that a higher F_1 would correspond to a lower
chance of an undetected error. However this presupposes each node is equally
likely to contain an error, and while this might serve as a convenient first
approximation to the truth it is certainly of questionable validity. Another
measure which has been proposed, and discussed in some detail by Huang [27], is
the fraction, F_2, of edges in the flow graph which are executed. It is
evident that $F_2=1$ implies $F_1=1$ but the converse is not true, so F_2 can be
regarded as a stronger measure than F_1. Again the issue raised above about the
relative importance of nodes applies to F_2. It is evident that one can build
still stronger measures; for example, Woodward, Hedley, and Hennell [30] have
suggested a hierarchy of such measures.
    It is an interesting fact that these measures are not used in practice as
one might expect. There are at least two situations in which substantial
amounts of software are produced, software which is supposed to be reliable,
where the value of F_1 at the end of testing is not known. In one situation
human life depends on the reliability of the software. The developers of this
software rely on their knowledge of the problem handled by the software, the
special cases involved, and their experience, to create data for testing the
software. In short the reliability provided by a guarantee that $F_1=1$ seems to
be not worth the effort. It is easy to dismiss this attitude as foolish, but
it would be wiser to examine carefully the reasons for it. Some of these
reasons are: confidence based on intimate knowledge of the software that all
nodes will be executed; the difficulty of finding test data to achieve $F_1=1$;
the difficulty of testing after changes are made to the software to repair
errors discovered in testing. Particularly important is the fact that a high
level of trust is associated with correct results produced by the software on
test problems. It would be worthwhile investigating how well placed this trust
is. In a small example studied by Di Millo, Lipton, and Sayward [31] it was
found that 8% of the errors deliberately inserted in a short (31 statement)
FORTRAN program did not cause erroneous results on tests proposed by
experienced programmers.
    Two approaches to finding test data to achieve some level of testing, say
$F_2=1$, have been discussed in the literature. As might be expected, one of
these is based on random selection of test data [32,33]. For large programs
this seems impractical unless some carefully designed stratified sampling is
used and even then the costs may be too high. The other approach is based on
symbolic execution [23,24]. From a theoretical viewpoint this is the more
attractive but, as presently proposed, is too expensive to use in practice.
    Mills, Gilb, Weinberg and others [11] have discussed seeding programs with
known errors and then using the number of errors discovered in testing as a
measure of the effectiveness of a test. This is a well-known practice
followed by naturalists in studying wild-life populations but no one yet has
provided convincing evidence that errors in programs obey the same statistical
behaviour as fish in a pond. This approach may be effective but far more needs
to be known about the nature of the distribution of errors in programs before
it can be applied in a scientific manner. Recently Di Millo, Lipton and
Sayward [31] have described a method of test data selection which is related to
the error seeding idea. They measure the effectiveness of test data selection
by the proportion of errors detected out of a group that has been deliberately
inserted in the program.
    Another technique advocated [34] for dynamic error detection is the
instrumentation of the program with predicates at carefully chosen points in
the program. Typically these predicates describe expected values for
variables at these points. Like other techniques described here it does not
appear to be widely used in practice though forms of this technique have been

used since the earliest days of computing to identify the location of a
suspected error by tracing or monitoring an execution.

Finally, it should be mentioned that some efforts have been made to answer
the question of whether the correctness of a program can be inferred from a
finite set of tests [35,36]. Recently Howden [37] presented a result showing
that for a particular class of Lindenmayer grammars it was possible to obtain
an interesting correctness result based on a finite number of tests. Work of
this kind, while not having any direct application to practice, should deepen
our understanding of the testing problem and provide us with information on its
limits.

4. CONCLUSION

We must assume that scientific and technological advances will lead to far
more powerful computing machines than we have today and therefore that there
will be a demand for larger and more complex programs. If this demand is to
be met it will be necessary to improve our techniques for developing and
maintaining large programs. Error detection is an important part of this and
if we are to improve our error detection techniques, then we must know about
the distribution of errors and how the distribution is affected by programming
language, style, and human psychological factors; we must understand how to
analyze programs and to measure their reliability.

As this brief review shows, our knowledge of the occurrence of errors is
very limited. The experiments with human subjects are mainly limited to
students and so it is questionable whether or not the results could apply to a
laboratory or company producing software on a regular basis. Intuition
suggests that modular or structured programming will reduce the occurrence of
errors and improve our chances of detecting those that do occur. On the other
hand there is evidence to indicate a strong correlation between the number of
errors in a program and the number of bits required to specify it: this
result comes from a theory called "software science" which has not been
discussed here but has been reviewed recently by Fitzsimmons and Love [38].
Thus it appears that we still do not fully understand the factors which
influence the occurrence of errors in programs.

There is now available a good set of algorithms for performing data flow
analysis for non-recursive languages and it seems unlikely that there will be
much more improvement in the basic algorithms. Here the work that is necessary
for analysis of programs written in these languages is concerned with the
implementation of these algorithms, and we need to understand how language
features can affect the implementation. There remains a need for theoretical
work on the analysis of recursive programs.

It is evident that the quantification of program testing is in a very
primitive state. The measures which have been proposed have not been related
to the probability of error. Furthermore, in practice the proposed measures
are ignored and subjective evaluations are used in their place. There
appears to be much room for theoretical and applied work in this area.

5. ACKNOWLEDGEMENT

This paper was written during a visit with the Numerical Algorithms Group
Limited in Oxford, England. I thank them for their kind hospitality and I
thank Eleanor Capanni of their staff for assistance in preparing the
manuscript.

References

1. Gill, S.: The diagnosis of mistakes in programmes on the EDSAC.
    Proc. Roy. Soc. London 206A (1951), 538-554.
2. Gerhart, S.L.: Development of a methodology for classifying software
    errors. Final Technical Report (2 July 1976) Computer Science Department,
    Duke University, Durham NC 27706, U.S.A.
3. Horning, J.J.: What the compiler should tell the user. Lecture Notes in
    Computer Science, Goos, G. and Hartmamis, J. eds, Vol 21, Ch.5 (1974)
    Springer-Verlag.
4. James, E.B. and Partridge, D.P.: Tolerance to inaccuracy in computer
    programs. Comp. J. 19,3 (Aug.1976), 207-212.
5. Boies, S.J. and Gould, J.D.: Syntactic errors in computer programming.
    Human Factors 16 (1974), 253-257.
6. Kulsrud, H.E.: Some statistics on the reasons for compiler use.
    Software P.E. 4 (1974), 241-249.
7. Thayer, T.A.; Lipow, M.; and Nelson, E.C.: Software reliability study.
    TRW-SS-76-03 (March 1976) TRW, Redondo Beach, CA 90278, U.S.A.
8. Moulton, P.G. and Muller, M.E.: DITRAN - A compiler emphasizing
    diagnostics. Comm. ACM 10,1 (Jan.1967), 45-52.
9. Youngs, E.A.: Human errors in programming. International Journal of
    Man-Machine Studies 6,3 (May 1974), 361-376.
10. Gannon, J.D.: Language design to enhance programming reliability.
    Ph.D Thesis, Tech. Rept. CSRG-47, U. of Toronto (Jan.1975).
11. Gilb, T.: Software Metrics. Winthrop Publishers, Inc. (1977).
12. Benson, J.P. and Saib, S.H.: A Software Quality Assurance Experiment.
    Talk presented at NASA Workshop for Embedded Computing Systems Software,
    Hampton VA (Nov.1978).
13. Weinberg, G.M. and Gressett, G.L.: An experiment in automatic
    verification of programs. Comm. ACM 6,10 (Oct.1963), 610-613.
14. Fosdick, L.D.: The effect of typing blunders in FORTRAN programs.
    Tech. Rept.
15. Dorrenbacker, J.; Paddock, D.; Wisneski, D; and Fosdick, L.D.:
    PLOISH, A FORTRAN program to edit FORTRAN programs, Tech. Rept.
    CU-CS-050-74 (July 1974) U. of Colorado, Boulder, CO 80309.
16. Scowen, R.S.: The diagnostic facilities in ALGOL and FORTRAN compilers.
    Tech. Rept. NAC 81 (July 1977), National Physical Laboratory, Teddington,
    Middlesex, England.
17. Ryder, B.T.: The PFORT verifier. Software P.E. 4 (1974), 359-378.
18. Fosdick, L.D. and Osterweil, L.J.: Data flow analysis in software
    reliability. ACM Comp. Surveys 8,3 (Sept.1976), 305-330.
19. Osterweil, L.J. and Fosdick, L.D.: DAVE - a validation, error detection
    and documentation system for FORTRAN programs. Software P.E. 6 (1976),
    473-486.
20. Tarjan, R.E.: Depth-first search and linear graph algorithms. SIAM J.
    Computing (Sept.1972), 146-160.
21. Hecht, M.S. and Ullman, J.D.: A simple algorithm for global data flow
    analysis problems. SIAM J. Computing 4 (Dec.1975), 519-532.
22. Barth, J.M.: A practical interprocedural data-flow analysis algorithm.
    Comm. ACM 21,9 (Sept.1978), 724-735.
23. Clarke, L.: A system to generate test data and symbolically execute
    programs. IEEE Trans. on Software Engineering 2,3 (1976), 215-222.
24. Howden, W.E.: DISSECT - A symbolic evaluation and program testing system.
    IEEE Trans. on Software Engineering 4,1 (Jan.1978), 70-73.
25. Cheatham, T.E. Jr.; Holloway, G.H.; and Townley, J.A.: Symbolic evaluation
    and the analysis of programs. Tech. Rept. TR-19-78 (Nov.1978),
    Harvard U., Cambridge MA.
26. Hennell, M.A.; Woodward, M.R.; and Hedley, D.: On program analysis.
    Information Proc. Letters 5,5 (1976), 136-140.

27.   Huang, J.C.: An approach to program testing. ACM Comp. Surveys 7,3
      (Sept.1975), 113-128.
28.   Pimont, S. and Rault, J.C.: A software reliability assessment based on
      a structural and behavioral analysis of programs. Proc. 2nd Int. Conf.
      on Software Engineering (Oct.1976) 486-491. San Francisco CA. IEEE
      Cat. No. 76CH1125-4C.
29.   Brown, J.R.: Practical applications of automated software tools.
      Tech. Rept. TRW-SS-72-05 (1972), TRW, Redondo Beach, CA 90278.
30.   Woodward, M.R.; Hedley, D.; and Hennell, M.: Observations and experience
      of path analysis and testing of programs. Tech. Rept. (1978),
      U. of Liverpool, Liverpool, England.
31.   Di Millo, R.A.; Lipton, R.J.; and Sayward, F.G.: Hints on test data
      selection: help for the practicing programmer. Computer 11,4
      (April 1978), 34-41.
32.   Ramamoorthy, C.V. and Ho, S.F.: On the automated generation of program
      test data. Proc. 2nd Int. Conf. on Software Engineering (Oct.1976),
      Supplement 95-102. San Francisco, CA.
33.   Hennell, M.A.; Woodward, M.R.; and Hedley, D.: Towards more advanced
      testing techniques. Tech. Rept. (1978), U. of Liverpool, Liverpool,
      England.
34.   Stucki, L.G. and Foshee, G.L.: New assertion concepts for self-metric
      software validation. Proc. Int. Conf. on Reliable Software (April 1975),
      59-71, Los Angeles CA, U.S.A. IEEE Cat. No. 75CH0940-7CSR.
35.   Goodenough, J.B. and Gerhart, S.L.: Towards a theory of test data
      selection. Proc. Int. Conf. on Reliable Software (April 1975), 493-510,
      Los Angeles CA, U.S.A. IEEE Cat. No. 75CH0940-7CSR.
36.   Howden, W.E.: Elementary algebraic program testing techniques.
      CSTR 13 (Sept.1976), Dept. Applied Physics and Information Science, UCSD,
      San Diego, CA.
37.   Howden, W.E.: Lindenmayer grammars and symbolic testing. Information
      Processing Letters 7,1 (Jan.1978), 36-39.
38.   Fitzsimmons, Ann and Love, Tom. A review and evaluation of software
      science. ACM Comp. Surveys 10,1 (March 1978), 2-18.

Performance Evaluation of Numerical Software, Fosdick (ed.)
© *IFIP, North-Holland Publishing Company, 1979*

DISCUSSION OF SESSION ON

GENERAL ASPECTS OF PERFORMANCE EVALUATION

W. Morven Gentleman
University of Waterloo
Department of Computer Science
Waterloo, Ontario

The three papers in this session illustrate how diverse questions of performance evaluation of numerical software can be. The paper by Dr. Lyness discusses the fundamental question of what performance assessment of numerical software should mean, and how it should be measured, in those situations where no finite algorithm can be guaranteed to solve the problem to the specified accuracy. The paper by Dr. Jordan discusses the impact of newer computer architectures on what had previously been a well understood area, simple linear algebra. Lastly, the paper by Professor Fosdick includes correctness in the definition of performance, and discusses how errors can be detected in programs.

Turning first to the paper by Dr. Lyness, we are addressing an extremely difficult issue. We know that all algorithms for these problems must fail sometimes, even fail without warning. Yet it is intuitively plausible that some algorithm will be better than another over the universe of problems to which it might be applied, or at least over some identifiable subclass. "Better" can be measured various ways, such as more likely to succeed, cheaper to obtain the same accuracy, etc. The ultimate purpose of such assessment is always predictive: how would the competing algorithms fare on some new problem. Theoretical analyses rarely provide sufficient evidence, and so we resort to numerical experiment.

Historically, the approach taken was the "test battery". A collection of test problems was proposed, either samples from typical parts of the problem domain or cases known to be hard, and the algorithms were scored on how well they did on this set.

Dr. Lyness has pointed out previously how unsatisfactory such collections of isolated examples are for predictive purposes. Many algorithms in this class are quite sensitive to specific values in the data they sample, and either good luck or bad luck in the test cases may result in behaviour wildly different from that in what would reasonably be construed as nearby problems. This has lead Dr. Lyness to propose the use of "performance profiles", where rather than studying isolated problems, we parameterize a problem, and study the behaviour of the performance criterion of interest over some interval of the parameter. By greatly reducing the possibility of luck providing misleading results, this provides the stability in our assessment which is essential for prediction. Unfortunately, it also provides masses of data. Dr. Lyness has considered various ways to reduce this, and recommends a statistical statement, with medians to indicate typical performance and quantiles to indicate dispersion of results, and with the probability of failure well defined. Such statements, however, must be made in the context of distributional assumptions over what is sampled in order to have any meaning. Dr. Lyness has not really addressed this question, and while it is possible that any reasonable distributional assumptions would lead to similar statistical statements about performance, it is also quite possible that the conclusions could be radically altered.

Although performance profiles are much more informative than isolated examples, they do not avoid two of the other problems exhibited by the battery approach: (1) how does the tester know that the examples chosen truly cover the space of problems to which the algorithm might be applied, and (2) how does the user decide which of the test problems his problem is most like, so as to predict the behaviour on his problem. I hope Dr. Lyness and others will continue to work on resolving these issues.

Turning next to the paper by Dr. Jordan, there is always a certain excitement to hearing about the fastest computers in the world. What is not so attractive is to learn that in the search for speed, new architectures that facilitate exploiting parallelism must be used, and these have deep ramifications on many of our assumptions about which way of doing things outperforms which other. Fortran, which has been practically the standard language for numerical computations, has lost one of its main virtues in that it no longer is a realistic abstraction of how the machines work, and careful but straightforward compilation no longer produces acceptable code. It is very tempting to look at the painful struggles of those working with these new machines and to decide to adopt an ostrich attitude: "there are only a few such machines and I can avoid thinking about them". However, the enormous economic pressure arising from cheap but slow microcomputers may well mean that the traditional medium scale computing which most of us do will soon be done with parallel machines built from these.

Turning finally to the paper by Professor Fosdick, we have a subject of much interest and much discussion not only in the numerical software community but amongst everyone concerned with computing. So much of this discussion has been based on prejudice and myth that Professor Fosdick has done a considerable service sorting out what has a real scientific basis and summarizing it for us. It is astounding for something which is so important to us all that so little is actually known. As Professor Fosdick points out, even categorizing errors is extremely difficult. One taxonomy that is possible is defined operationally by how the error is detected - language errors which are found by the context free grammar being called syntax errors, other language detected errors being called semantic errors, and non language detected errors being called specification errors. However a taxonomy based on how errors arise would be much more interesting - eg. typing errors, misunderstanding of language definition, misunderstanding of problem specification, overlooking ramifications of some action, blunders in manipulating source file structure, etc.

The published empirical knowledge of errors, error types and error rates, is very limited, and much pertains to student programming. One very rich source of documented experience are the error logs of manufacturers and other builders of large software systems. Many of these have been studied internally - it would be very nice if publication could be encouraged.

Professor Fosdick made a very attractive suggestion in simulating error sources, to study the effects, and how well they are detected. His results are very interesting, but I regard them preliminary. They need to be extended to other languages, and to other error sources, eg. insertion, deletion, and transposition of lines as well as of characters.

Many of the tools and techniques for static or dynamic detection of errors are in similar preliminary state. Professors Fosdick and Osterweil have done a significant amount of work on developing path analyses tools for Fortran, but we have yet to see the effect in other languages. Symbolic execution is also largely undeveloped. Tools for choice of test data are just being explored. Debugging has been advocated for some time now, but there is no evidence yet that it does provide accurate predictions of the yet undiscovered bugs. Assertion testing is still very ad hoc.

In short, this area is wide open for a great deal more activity.

The papers stimulated a lively discussion.

Dr. Smith asked Professor Fosdick how much path analysis would help in detecting errors. Professor Fosdick replied that it would not help much for typing errors, but could be helpful for other error classes. Professor Osterweil augmented this reply by stating that a study had shown that one-

half of the errors in a Fortran subroutine library could be found by path
analysis.

Dr. Meyer stated that he felt some of Professor Fosdick's comments applied only to Fortran, and that programming languages which required explicit declaration of all identifiers prevent bugs arising from typing errors.   He also stated that optimizing compilers often do much of the path
analysis for their own purposes, and that they should tell the user what
anomolies they have uncovered.

Dr. Shampine asked Dr. Jordan how successful are "vectorizors",
software tools that restructure programs to increase possible vectorization. Dr. Jordan replied that Lawrence Livermore Laboratories had had to do
a total redesign of codes manually in order to obtain adequate vectorization for the Star. At Los Alamos, it had been found that raw codes don't
vectorize on the Cray. Applying the Vectorizor produced codes which could
vectorize to some extent, but which were far from optimal.

Professor Kahan made some personal observations on the subject of errors in programs. His first observation was that programming languages do
make a difference. After compiler detected errors were corrected, he had
found that with Burroughs Algol, of eight job executions, seven produced
satisfactory results.   By contrast, with either IBM or CDC Fortran, he
found that after compiler detected errors were corrected, still out of six
job executions only one produced satisfactory results. His second observation was that computational results which people claim to be right,
about one in three are actually wrong, and that this proportion does not
seem to have changed in twenty years. His third observation is that hardware could, and should, provide more aids for program checking, for example
by detecting uninitiatized variables. Of course, such hardware aids to not
help to reveal errors if the erroneous program path is not executed.
Finally, he mentioned that Professor Lipton had observed that the technique
of bebugging, where known bugs are deliberately inserted into programs, is
very useful for testing testing procedures.

Professor Hull reminded everyone that a principal objective of
bebugging was to estimate the number of bugs yet to be found in a piece of
software by observing how many of the deliberately inserted bugs were found
and how long it took to find them.

Dr. Shampine advocated the construction of error resistant software.
He cited the example of a well known differential equation solver which has
had a bug in it for several years because a low order digit in one of the
constants is wrong, however nobody ever sees an effect of this bug because
the software is error resistant.

Dr. Sherman raised the point that many variable initialization type
errors are related to subroutine calls. Although these might be caught if
all the subroutines were compiled together with the main program, current
separate compilation techniques cannot detect them because insufficient
information is available at compile time and no checking is done at binding
time.   Since separate compilation is a partical necessity in many situations, usch as when using program libraries, he wondered whether Professor
Fosdick had considered any other techniques to aid in the detection of this
type of error.

Dr. Curtis enquired what was known about detecting logical errors in
programs. He felt his worst bugs were of this kind, and nothing he had
heard would help him.

Dr. Lawson asked Professor Fosdick whether structured programming
would assist static analysis of programs. Professor Fosdick replied that
if programming languages were changed to admit only structured constructs
it would indeed help, but that for current languages there was little
benefit.

Professor Rice claimed that in his experience, character manipulating

programs were easier to produce than numerical programs. He wondered
whether there was any quantitative evidence to support this.

Professor Kahan noted that on early computers, where hardware was
often unreliable, people programmed assertion tests in their codes to ensure nothing had gone awry. He suggested that since microprocessors
sometimes are unreliable for quantum mechanical reasons, we may need to
return to this practise.

Professor Stetter directed the last question to Dr. Lyness: How do
you apply performance analysis to an area like differential equations,
where the problem space is so vast? With this question left unanswered,
the session closed.

SESSION 3 : PERFORMANCE EVALUATION
               IN LINEAR ALGEBRA

Performance Evaluation of Numerical Software, Fosdick (ed.)
© *IFIP, North-Holland Publishing Company, 1979*

PERFORMANCE EVALUATION OF LINEAR ALGEBRA SOFTWARE

I.N. Molchanov
Institute of Cybernetics
Ukrainian Academy of Sciences
Kiev, USSR

The difficulties are considered which exist in
solving the problems of linear algebra describing the applied problems and requirements imposed upon numerical methods as well as upon
computer programs entering into computer software.

INTRODUCTION

The problems of linear algebra present themselves when certain
scientific and technical problems are described. The types of com-
putational problems of linear algebra are discussed in |1| .
A sufficiently comprehensive bibliography on numerical algebraic
methods is given in |2| . A survey of methods of numerical solu-
tion of linear algebraic problems as of 1975 is offered in |3| .
The analysis of some modern conceptions of constructing libraries
of numerical software, including linear algebra, is given in |4| .

The difficulties of computational solution of problems in linear
algebra are caused, on the one hand, by the difference between the
statement of linear algebraic problems for numerical solution and
their classical statement; and, on the other hand, by the fact that
computer realization of linear algebraic algorithms are underrated.
Therefore, the assessment of the degree of confidence in computer
solutions of applied problems acquires an ever increasing importance.

We are concerned here with the solution of mathematical problems
on electronic computers. These mathematical problems have been obtained from the construction of models of problems arising in sci-
ence and engineering |5| .

When describing applied problems the systems

$$\widetilde{A}\widetilde{x} = \widetilde{b} \quad (1) \qquad \text{or} \qquad \widetilde{A}\widetilde{v} = \widetilde{\lambda}\widetilde{v} \quad (2)$$

arise rarely with exact initial data. The most typical situation
is the statement of approximate equations

$$Ax = b \quad (3) \qquad\qquad Av = \lambda v \quad (4)$$

with the error in the initial data

$$\|\widetilde{A} - A\| = \|\Delta A\| \leqslant \mathcal{E}_A \quad , \qquad \|\widetilde{b} - \widehat{b}\| = \|\Delta b\| \leqslant \mathcal{E}_B . \qquad (5)$$

Thus, the applied problem is described by a whole class of equations. As a formal solution of equation (3) we can take any which satisfies this equation with A and b satisfying the inequalities (5). This class of formally allowed equations can be wide, therefore it is important to determine the solution of this class of problems. According to this we can discuss the questions of solving exactly the systems (1), (2) given individually. Depending on the statement of the applied problem we must define for each application what is meant by the mathematical solution of the discussed class of equations. Note that the inherited error which is dependent on both the properties of the matrix and the error in the initial data, is in the mathematical solution.

The computer solution means the solution of equations obtained by executing a numerical algorithm on a computer.The estimate of the validity of a computer solution depends on the definition of the desired solution, and the estimate of the inherited error and the error generated in the computer arithmetic.

Necessary conditions for obtaining this estimate are:
- the existence of the classical solution or of the generalized solution of the mathematical problem under study;
- the possibility of finding the unique classical solution or the generalized solution;
- the stability of this found solution by the initial data |6,7|.

CLASSIFICATION OF LINEAR ALGEBRAIC PROBLEMS DESCRIBING APPLIED PROBLEMS

To obtain the valid solution of a problem it is important to specify it correctly, and to estimate the inherited error for problems of linear algebra we can use condition numbers. We must always define the mathematical solution and the process for obtaining this solution.

If the non-perturbed (with exact data) system does not have a solution then it is classified as a problem whose statement is not correct. Linear systems with square matrices which have singular systems within the limits of precision of the statement of the coefficients are also not correctly stated |8,9| . For matrices of such systems the conditions

$$\|A^{-1}_{\Delta}A\| \leqslant 1 \quad \text{or} \quad \|A^{-1}\| \cdot \|\Delta A\| \leqslant 1 \qquad (6)$$

are not satisfied.

Examine the following example. Let system (3) be represented in the form

$$\begin{aligned} 25\,x_1 - 36x_2 &= 1 \\ 16\,x_1 - 23x_2 &= -1, \end{aligned} \qquad (7)$$

with $\|\Delta A\| \leq 0.02$, $\|\Delta B\| \leq 0.01$ being known. It is easily seen that the unique exact solution of this perturbed system is $x_1 = -59$, $x_2 = -41$. However, variations of coefficients within the limits of precision of their specification can lead to the perturbed systems

$$25.01x_1 - 35.99x_2 = 1,\qquad\qquad 25.01x_1 - 36x_2 = 1,\qquad (8)$$
$$15.99x_1 - 23.01x_2 = -1,\qquad\qquad 15.99x_1 - 23.01x_2 = -1,$$

the first is incompatible, and the second has the unique solution $x_1 = -369.04315$, $x_2 = -256.41025$.

In dealing with such incorrectly stated systems a solution is found that is generalized in one way or another $|8,10,11,9|$. By a generalized solution of systems with rectangular or square matrices for which conditions (6) are not satisfied is here meant a solution of the system

$$A^T A x = A^T B \qquad (9)$$

or determination of the vector X, satisfying the condition

$$\|B - Ax\| = min \qquad (10)$$

If the rectangular matrix A isn't of full rank, i.e. the rank of the matrix A is smaller than $min\{m,n\}$, where m is the number of rows and n is the number of columns of this matrix, then by introducing additional restrictions on the solution of the problems (9),(10), for instance, a solution with minimum Euclidean norm, it is possible to find the unique solution. Special methods for finding the generalized solutions are used to solve incorrectly stated problems.

If the non-perturbed system of linear equations has unique solution within the limits of inequalities (5) and the perturbed system has a solution which is a unique one, then it falls into the category of problems whose statement is correct. If such problems are described by systems (3), (5) with square matrices, then conditions (6) are satisfied for them. Now, let us assume that for the problem (3), (5) the condition (6) is satisfied. The magnitude of the inherited error in the unique classical mathematical solution depends not only on the magnitude of an error in initial data but on the features of the matrices of the systems being dealt with. The systems

$$100x_1 + 500\ x_2 = 1700,\qquad 100x_1 + 500x_2 = 1700$$
$$15x_1 + 75.01x_2 = 255,\qquad 15x_1 + 75.01x_2 = 255.03 \qquad (11)$$

differ in the fifth significant digit. The first of them possesses a solution $x_1 = 17$, $x_2 = 0$, the second one $x_1 = 2$, $x_2 = 3$. The determinant of both systems is equal to unity.

When condition (6) is satisfied for the problem (3), (5) we have the following estimations:

$$\frac{\|x-\tilde{x}\|}{\|x\|} \leqslant \frac{\|A\|\cdot\|A^{-1}\|}{1-\|\Delta A\|\cdot\|A^{-1}\|}\left[\frac{\|\Delta A\|}{\|A\|}+\frac{\|\Delta b\|}{\|b\|}\right] \quad (12)$$

or

$$\frac{\|x-\tilde{x}\|}{\|\tilde{x}\|} \leqslant \frac{\|A\|\cdot\|A^{-1}\|}{1-\dfrac{\|\Delta b\|}{\|b\|}}\left[\frac{\|\Delta A\|}{\|A\|}+\frac{\|\Delta b\|}{\|b\|}\right] \quad (13)$$

under the natural assumption that $\dfrac{\|\Delta b\|}{\|b\|}<1$. Both these estimates

are the best estimates on the whole class of non-singular matrices
|12| .

It is obvious from estimates (12) , (13) that the stability of solutions with respect to variations in initial data depends to a large measure on the magnitude of the value

$$cond\ A = \|A\|\cdot\|A^{-1}\|,$$

called the condition number of the matrix A . The magnitude

$$m = cond\ A\left[\frac{\|\Delta A\|}{\|A\|}+\frac{\|\Delta b\|}{\|b\|}\right]$$

may be called the condition number of the system of linear algebraic equations. In real problems it makes sense to consider those
systems for which the value of m is substantially below unity,
for example, $m \leqslant 0.05$.

Thus, correctly stated problems of solving systems of linear equations can be differentiated as ill- and well-conditioned problems.

By a solution of a well-conditioned system means the unique classical solution of the problem including the estimation of inherited error. The solutions of ill-conditioned systems are unstable
with respect to small changes of the initial data, they have a stable projection on the subspace generated by eigenvectors of the
matrix A^TA which correspond to "large" eigenvalues. If the system describing the applied problem is ill-conditioned and it is
impossible to reformulate it to obtain a well-conditioned system
(for example, to begin investigating other parameter of the physical process), then, sometimes, the solution of the ill-conditioned
system means a construction of the above mentioned projection|3|.

An analogous problem of estimating the magnitude of the inherited
error arises in finding eigenvalues and eigenvectors of matrices.
Condition numbers of the matrix A with respect to eigenvalues denoted by $cond\ \lambda$ were introduced to measure a sensitivity of an
individual eigenvalue λ to variations in initial data ΔA .

In the case of a simple eigenvalue

$$cond\ \lambda = \frac{\|y\|\cdot\|x\|}{\|y^Tx\|},$$

where X and y are eigenvectors of A and A^T, respectively, corresponding to eigenvalue λ, and

$$\| X \| = \| X \|_e .$$

Variation $\Delta\lambda = \widehat{\lambda} - \lambda$ may be characterized both in terms of A and ΔA and in terms of the spectral projection P_λ as follows:
- for the simple eigenvalue λ

$$cond\ \lambda = sup\ \frac{|\Delta\lambda|}{\|\Delta A\|_E}\ ,\quad cond\ \lambda = \| P_\lambda \|_E \cdot \| P_\lambda \|_2\ ,\quad P_\lambda = \frac{xy^T}{y^T x}\ ,$$

where sup is taken over all non-zero increments of ΔA,
- in the case of the eigenvalue λ of multiplicity m

$$cond\ \lambda = sup\ \frac{|\Delta\lambda|}{\|\Delta A\|_2}\ ,\quad cond\ \lambda \leqslant \frac{\|P_\lambda\|}{m}\ ,\quad P_\lambda = X(Y^T X)^{-1} Y^T\ ,$$

where sup is taken over all perturbations of ΔA, the multiplicity of λ is preserved, the columns X and Y form the basis of invariant subspaces for λ.

If eigenvalues of the matrix A are dissimilar, then the following estimates hold $|13|$

$$|\Delta\lambda| \leqslant cond\ \lambda \| \Delta A \|\ ,\quad \frac{\|\Delta v_i\|}{\|v_i\|} = \frac{\|\widehat{v}_i - v_i\|}{\|v_i\|} \leqslant \sum_{\substack{i=1 \\ i \neq j}}^{n} \frac{cond\ \lambda}{|\lambda_i - \lambda_j|}\ \|\Delta A\|\ .$$

A perturbation ΔA can affect the multiplicity of an eigenvalue, "a cluster" of eigenvalues close to one another occurs. The smallness of $\|P_\lambda\|$ indicates this.

Summing up, it can be stated that in solving the linear systems describing the applied problems it is important to determine if a problem is stated correctly or not, and if it is well- or ill-conditioned, and one must build a method for solving the problem and means for estimating the a posteriori error in the solution.

COMPUTER REALIZATION OF NUMERICAL METHODS

The main problems of computer realization of numerical methods are to obtain a solution on a computer system and to determine the difference between the computer and mathematical solutions of the problem.

The numerical methods for solving linear algebraic problems are usually divided into direct and iterative methods. After giving the matrix A and vector 6 of the system (3) to a computer they are translated from the decimal system into the binary one yielding $\overline{A}$ and $\overline{6}$. We will call the computer problem

$$\overline{A}\overline{x} = \overline{6}\ ,\qquad\qquad (14)$$

the problem whose matrix and right-hand part are specified in the

computer in the binary system of calculation. Solutions of the mathematical and computer problem may be dissimilar.

The combined influence of round-off errors in the direct methods
may be considered as equivalent to a perturbation of the initial
data |9,10|. Therefore, the solution of the system (3) obtained
with the computer is exact for a certain perturbed system, for example,

$$(A + dA)x^{(1)} = b + db \qquad (15)$$

and approximate, dissimilar to the mathematical solution of the
system (3) . Here dA, db stand for the corresponding equivalent
perturbations. We have the following estimates

$$\frac{\|x^{(1)} - x\|}{\|x^{(1)}\|} \leqslant \frac{\|A\| \cdot \|A^{-1}\|}{1 - \frac{\|db\|}{\|b\|}} \left[\frac{\|dA\|}{\|A\|} + \frac{\|db\|}{\|b\|} \right] \qquad (16)$$

or

$$\frac{\|x^{(1)} - x\|}{\|x\|} \leqslant \frac{\|A\| \cdot \|A^{-1}\|}{1 - \|dA\| \cdot \|A^{-1}\|} \left[\frac{\|dA\|}{\|A\|} + \frac{\|db\|}{\|b\|} \right] . \qquad (17)$$

In |14,15,16| and others, the majorant and probability estimates
for certain direct methods of solving linear systems are presented.
Note, that in one parameter of these estimates the length of the
computer word mantissa is employed. The investigation of these estimates and the estimates (16) , (17) allows us to discuss whether the system is well- or ill-conditioned with respect to numerical stability of computer solution affected by round-off errors.
But here, the notion of well- or ill-conditioned is closely connected with capabilities of a particular machine. The same system may
be ill-conditioned for one machine and well-conditioned for another.

Thus, Table 1 compares the mathematical solution of the system

$$
\begin{aligned}
0.135\,x_1 +0.188\,x_2 +0.191\,x_3 +0.178\,x_4 &= 0.3516 \\
0.188\,x_1 +0.262\,x_2 +0.265\,x_3 +0.247\,x_4 &= 0.4887 \\
0.191\,x_1 +0.265\,x_2 +0.281\,x_3 +0.266\,x_4 &= 0.5105 \\
0.178\,x_1 +0.247\,x_2 +0.266\,x_3 +0.255\,x_4 &= 0.4818
\end{aligned}
\qquad (18)
$$

with computer solutions obtained by various direct methods on a variety of machines with single and double precision binary computer
words.

A successful scaling, that is, one which transforms the initial system into an equivalent one which possesses a well-conditioned mat-
rix, enchances the stability of solution |17,18| .

Analysis of computer algorithms includes estimating the required
number of arithmetic operations and computer memory requirements,

and, finally, computer time spent for the problem solution.

Information about the difference between the computer solution $x^{(1)}$, and mathematical solution x of the system (3) can be obtained through an iterative process $|14|$:

$$z^{(s)} = b - Ax^{(s)}, \quad QP\delta^{(s)} = z^{(s)}, \quad x^{(s+1)} = x^{(s)} + \delta^{(s)}, \quad s = 0,1,\ldots, x^{(0)} \equiv 0, \quad \delta^{(0)} \equiv x^{(0)}. \tag{19}$$

Corrections $\delta^{(s)}$ $(s = 1, 2 \ldots)$ are found through the application of the two matrices Q and P obtained already from decomposition of the matrix A. In (19) residuals $z^{(s)}$ must be computed with double precision. The convergence of the iterative process with respect to corrections shows the difference between the computer and mathematical solutions. We need to increase the length of the computer word when there is no convergence of iterations or it is too slow.

If the iteration process (19) converges, then the following approximate estimation of the condition number of the matrix A holds

$$cond\, A \approx \frac{1}{\varepsilon} \frac{\|\delta^{(1)}\|}{\|x^{(1)}\|},$$

where ε is the largest number for which $1.0 + \varepsilon = 1.0$ is valid in the floating-point arithmetic of the given computer $|1|$.

Making use of the approximate value of $cond\, A$ by formulas (12), (13) it is possible to estimate the inherited error of a solution.

In order to solve non-correctly stated problems we can use the regularization method by A.N. Tikhonov $|8,9,19|$ or the matrix decomposition A by singular values $|20,21|$, for example, procedure $minfit$ $|10|$. Both of these approaches allow us to obtain the approximation to the normal solution. Table 2 compares the solutions of the system

$$2x_1 - x_2 + \sqrt{2}\, x_3 = 5 + 7\sqrt{2},$$

$$3x_1 + 2x_2 - 3x_3 = -24,$$

$$3x_1 + \sqrt{2}x_2 - \frac{15}{2}x_3 = 12 - 3\sqrt{2},$$

with a singular matrix, obtained by these methods with the exact normal solution of the problem and solution which is obtained by Gauss's method with the choice of the pivot element along the whole matrix. If the matrix of the system (3) is symmetric and positive semi-definite, then the shift of spectrum $|11|$ can be used. This method permits us to obtain one of the generalized solutions of the system (3).

When systems of linear equations with matrices of a high order and certain distinguishing features (of the band type, easily generated by programs) are being solved then, sometimes, it is profitable to use the iterative methods. In the general case, the advanta-

ges offered by a direct or an iterative method must be determined
by comparing the required memory capacity, number of arithmetical
operations, length of computer word, time required to solve a given problem on a particular computer. The theory of iterative methods and some aspects of their practical realization are given in
|6,22,23| .

In individual cases, a theoretically convergent iterative process
when carried out with the computer may give a computer solution
which is not a mathematical solution of the problem. It may be caused for several reasons. Note, in place of system (3) the computer manipulates the solution of problem (14) . Features of the
computer arithmetic may be responsible for disappearence of the
exponent part, substitution of computer zero for small values, etc.
Sometimes, the condition for completing iterations is incorrect.
At each step of the iterative process rounding takes place which
perturbs the iterative process studied theoretically. Thus, an investigation into a one-step iterative explicit method carried out
by formulas

$$x^{(k+1)} = x^{(k)} + \tau z^{(k)}, \quad z^{(k)} = b - Ax^{(k)}, \quad \tau = \frac{2}{\delta + \Delta} \qquad (20)$$

and employed to solve the system (3) with symmetric positive de-
finite matrix, when estimates of the spectrum $\delta \leqslant \lambda_j \leqslant \Delta, j = 1,2,\dots,n$,
are known, from the standpoint of computer realization leads to the
estimate

$$\|x - \overline{\overline{x}}^{(k)}\| \leqslant \frac{\tau}{1-q}\left[q^k \|Ax^{(0)} - b\|\right] + \frac{1-q^k}{1-q}\tau\omega, \qquad (21)$$

where $\overline{\overline{x}}^{(k)}$ is a computer solution obtained after k-th itera-
tion, $q = \|E - \tau A\|, x^{(0)} = x_0$ is an initial approximation to the solution,
$\omega = \max\limits_{1 \leqslant i \leqslant k} \|\varepsilon_i\|$, ε is the round-off error that occurs at the i-th
step of the iterative process. The computed vector $\overline{\overline{x}}^{(k)}$ is the
k-th approximation for the iterative scheme (20) with a pertur-
bed right-hand part, with a magnitude of the equivalent perturbation computed by the formula

$$db_k = \left(\sum_{p=0}^{k}(E - \tau A)^p\right)^{-1}\beta_k,$$

where β_k is an accumulated round-off error for all k steps.

Theoretically, the condition for finishing an iteration depends
on the matrix of the system, the method of its representation, and
the length of the computer word. Thus, for the process (20) the
execution of condition

$$\max\limits_{i} \frac{|x_i^{(k+1)} - x_i^{(k)}|}{|x_i^{(k)}|} \leqslant \frac{\tau \lambda_{min}\varepsilon}{1+\varepsilon}, \quad x_i \neq 0, \; x_i^{(k)} \neq 0$$

guarantees

$$\max\limits_{i} \frac{|x_i - x_i^{(k+1)}|}{|x_i|} \leqslant \varepsilon$$

where ε is a small number given in advance, λ_{min} is a minimal eigenvalue of A .

Certain a posteriori information obtained during the iterative process may be of assistance for estimating the inherited error of the mathematical solution. For instance, if the computer solution obtained by the iterative process happens to be sufficiently close to the mathematical solution of the problem, then the approximate condition number $\text{cond } A$ can be defined by the a posteriori formula

$$\text{cond } A \approx \frac{2}{\ln \frac{\|x^{(\kappa+1)} - x^{(\kappa)}\|}{\|x^{(\kappa)} - x^{(\kappa-1)}\|}}$$

and the inherited error by formulas (12), (13).

When solving systems with symmetric and positive semidefinite matrices use could be made of both one-step and two-step iterative processes which provide an approximation to the generalized solution, i.e., to one of the solutions (9), the Gauss's left transformation being unnecessary $|24|$. Iterative methods of finding useful data for inconsistent systems of linear equations are descibed in $|25|$ and other publications.

Computer realization of methods of finding eigenvalues and eigenvectors of the problem (4) introduces an error dependent upon properties of the matrix A and methods of its solution as well as upon the computer arithmetic.

Computed eigenvalues are eigenvalues of the matrix $A + dA$, the matrix dA being not unique and defined by a chosen algorithm and the length of mantissa of the computer word. Let $A + dA$ denote one of the matrices closest to A. If the value $\frac{\|dA\|_E}{\|A\|_E}$ is small then the error $d\lambda$ in the eigenvalue is bounded

$$|d\lambda| \leqslant \text{cond } \lambda \, \|dA\|_E .$$

Requirements that some algorithms for finding eigenvalues and eigenvectors place upon computer memory and upon a number of arithmetical operations, as well as majorizing and probability estimates of the closeness between the computer and mathematical solution are given in $|15,16|$, for example.

Prior to applying a given method of finding of eigenvalues and eigenvectors, it is wise to scale the initial matrix. Subsequent to execution of the program used to solve a problem with the scaled matrix, a special procedure is necessary to eliminate the consequences of the scaling $|10|$.

Various procedures are used to estimate validity of the computer solution. Thus, a program is suggested in $|26|$ that is based on the QR-algorithm that enables the condition number to be found avoiding construction of the orthogonal matrix changing A to the upper triangular matrix T. The number ζ of accurate decimal digits in the computed eigenvalue λ is determined by

$$\zeta = \log_{10} \left(\frac{\lambda}{\beta \, \text{cond} \lambda \, \|A\|_E} \right),$$

where β is the upper estimate of the number

$$\frac{\|dA\|_E}{\|A\|_E}.$$

A particular eigenvalue problem is to be solved twice for the practical estimation of obtained results, and each time a different $\mathcal{E}$ is taken, $\mathcal{E}$ being a value covered by the condition of termination of iterative process. The number of digits which are the same in the computer solutions is the number of guaranteed characters for a given machine representation of numbers.

Let us formulate conditions under which the computer solution may be assumed to be the mathematical solution of a problem. They are:
-existence of a classical or generalized solution of a computer problem;
-singling out the unique classical or generalized solution of the computer problem during computer realization of the algorithm;
-computer stability of this singled out solution;
-good computer conditionality of the computer problem;
-correspondence of solubility algorithm to the specification of computer problem;
-correct conditions of termination for the iterative process.

CONCLUSION

In the last few years research in the estimation of validity of solutions calculated with computers is being carried out intensively in various countries.

Thus, an interval arithmetic is being intensively elaborated (see, for instance, |27|), as well as the so-called significant digit arithmetic (see, for instance, |28|). Both a priori and a posteriori data obtained in the course of a problem solution are used to estimate the validity of solutions (see,for example, |29|). A new form of software, that is, a program complex, is a means for solving applied problems with estimation of the validity of obtained solution (see, for instance, |30|). It must   carry out the specification of a problem, construction of algorithm of its solubility, calculation of the solution and estimation of its validity.

Only some of the procedures being devised now for estimation of the validity of solutions obtained with computers were described in this paper.

The difficulties which exist in solving the problems of linear algebra describing the applied problems impose a number of requirements on both the numerical methods and linear algebraic programs of computers. They must:
-yield not only a solution for some problem, but provide a means for checking the closeness between the computer and mathematical solution as well as give the estimation of inherited error of the obtained solution;
-be computer-oriented, i.e., take into account all mathematical and technological possibilities of the computer;
-economize a number of arithmetical operations as well computer me-

mory capacity in the course of computer realizing the algorithms;
-be problem-oriented, i.e., solve a given class of problems;
-be sufficiently simple in realization.

Table 1

Comparison of computer solutions of the problem (18)
obtained on a variety of machines

E C Computer

Mathematical solution	With single computer word		With double computer word
	Crout's method	Method of	LL^T decomposition
0.4	0.6297023	0.49089962	0.4000000001535001
0.5	0.3612513	0.44510078	0.4999999999072548
0.6	0.5423611	0.51717085	0.5999999999614333
0.5	0.5343801	0.51353925	0.5000000000228749

M I R   B E S M - 6

6 decimal digits   10 decimal digits
Method of LL^T decomposition

	6 decimal digits	10 decimal digits	BESM-6
0.4	-0.0378848	0.3995346979	0.400000023
0.5	0.764402	0.5002810367	0.494999442
0.6	0.710031	0.6001168362	0.599999762
0.5	0.434782	0.4999307078	0.500000138

Table 2

Comparison of computer solutions obtained
by different methods with different length
of a computer word

Normal solution	10 decimal digits		
	Gauss's method	Tikhonov's method	minfit
0.9794438929	2.725336900	0.979393820	0.979443901
-4.227168422	100.0	-4.227130465	-4.227168427
6.161331610	77.39200359	6.161126568	6.161331606
	30 decimal digits		
0.9794438929	1.078170605	0.979394108	0.979443893
-04.227168422	1.666666666	-4.227130465	-4.227168427
6.161331610	10.189281716	6.161277290	6. 161331610

References

|1|. G.E. Forsythe, C.B. Moler : Computer solution of linear algebraic systems, Englewood Cliffs, N.J. Prentice Hall, Inc. IX (1967) 148.

|2| В.Н.Фаддеева, Ю.А.Кузнецов, Г.Н.Грекова, Т.А.Долженкова: Вычислительные методы линейной алгебры, Библиографический указатель 1828-1974 гг. Новосибирск. (1976).

|3| Д.К.Фаддеев, В.Н.Фаддеева: Вычислительные методы линейной алгебры, В сб. Вычислительные методы линейной алгебры, Параллельные вычисления, Ленинград, Изд-во Лен. отделения, Наука, (1975) 1-265.

|4| О.Б.Арашунян: Некоторые современные концепции конструирования библиотек численного анализа, Вест. МГУ, сер. выч.мат. и киб. № 1 (1977) 58-72.

|5| А.Н.Тихонов: Математическая модель, Большая советская энциклопедия, т.15, Москва, Изд-во Советская энциклопедия (1974) 480.

|6| А.А.Самарский: Введение в теорию разностных схем, Москва, Наука (1971) 552.

|7| А.Н.Тихонов: Об устойчивости обратных задач. ДАН СССР, т.39, № 5 (1943)

|8| А.Н.Тихонов: Об устойчивости алгоритмов для решения вырожденных систем линейных алгебраических уравнений, ЖВМ и МФ, т. 5 № 4 (1965) 718-722.

|9| А.Н.Тихонов, В.Я.Арсенин: Методы решения некорректных задач, Москва, Наука (1974) 224.

|10| J.H.Wilkinson, C.Reinsch: Handbook for automatic computation, Linear Algebra,Springer, Berlin-Geidelberg-New York, IX (1971) 439.

|11| I.N.Molchanov, L.D.Nicolenko: On an approach to integrating boundary problems with a non-unique solution, Inf. Proc. Lett., 1(1972) 168-172.

|12| В.М.Глушков, И.Н.Молчанов и др.: Программное обеспечение МИР-1 и МИР-2, т.1,2,3, Киев, Наукова думка (1976) 974.

|13| Д.К.Фаддеев, В.Н.Фаддеева: Вычислительные методы линейной алгебры, Москва, Физматиздат (1960) 736.

|14| J.H.Wilkinson: Rounding errors in algebraic processes, London, H.M.Stat. Off., N32, VI(1963) 161.

|15| J.H.Wilkinson: The algebraic eigenvalue problem, Oxford, Clarendon Press, XVIII (1965) 662.

|16| В.В.Воеводин: Ошибки округления и устойчивость прямых методов линейной алгебры. М., Изд-во ВЦ МГУ (1969) 153.

|17| G.E.Forsythe and E.G.Strans: On best conditioned matrices, Proc. Internat. Congr. Math. (Amsterdam, 1954) 102-104.

|18| F.L.Bauer: Optimally scaled matrices, Numer. Math. N1, 5(1963) 73-87.

|19| В.А.Морозов: Оценивание точности решения некорректных задач и решение систем линейных алгебраических уравнений, ЖВМ и МФ, т. 19, № 6 (1977) 1341-1348.

|20| G.Golub, W.Kahan: Calculating the singular values and pseudo-inverse of a matrix, J.SIAM Numer. Anal., Ser. 132 (1965) 205-224.

|21| G.Golub, W.Kahan: Least squares, singular values and matrix approximations, Aplikace Mathematiky, 13(1968) 44-51.

|22| Г.И.Марчук: Методы вычислительной математики, Москва, Наука (1977) 456.

|23| A.A.Самарский, Е.С.Николаев: Методы решения сеточных уравне-
ний, Москва, Наука (I978) 559.

|24| И.Н.Молчанов, Л.Д.Николенко, М.Ф.Яковлев: О решении одного
класса систем линейных алгебраических уравнений с вырожден-
ными матрицами, В кн. Вычисл. методы лин. алгебры Новоси-
бирск(I977) 97-I09.

|25| G.I.Marchuk, Yu.A.Kuznetsov: Stationary iterative methods
for the solution of systems of linear equations with singular matrices, Gatlinburg, VI Symposium of Numer. Algebra,
Manuscripts (München, 1974) 9.

|26| S.P.Chan, R.Feldman, B.N.Parlet : Algorithm 517 a program
for computing eigenvectors, ACM Trans. Math. Software, N2,
3 (1977) 168-203.

|27| R.E.Moore: Interval analysis, Englewood Cliffs, N.J. Prentice
Hall, Inc. XI (1966) 145.

|28| N.C.Metropolis, R.L.Ashenhurst: Significant digit computer
arithmetic, IRE Trans. Electron. Comput., N7, EC 7 (1958)
265-267.

|29| F.L.Bauer: Genauigkeitsfragen bei der Lösung linearer Glei-
chungssysteme, Z. angew. Math. und Mech., N7, 46 (1966)
409-421.

|30| I.N.Moltschanow : Uber Programmpaket zur Lösung wissenschaft-
lich-technischer Aufgaben, Wiss. Z. Techn. Hochsch. O. von
Guericke, Magdeburg, N2, 21 (1977) 275-285.

Performance Evaluation of Numerical Software, Fosdick (ed.)
© *IFIP, North-Holland Publishing Company, 1979*

CAN AUTOMATED THEOREM PROVERS BE USED TO
EVALUATE LINEAR ALGEBRA SOFTWARE?*

Brian T. Smith
Applied Mathematics Division
Argonne National Laboratory
Argonne, Illinois

Yes, we believe they can. But a practical use of such a tool
dictates that the linear algebra software be coded in a
special form.

In the past, the evaluation of linear algebra software has
been done after the software is written, usually using an
exhaustive approach and software analysis tools. Many have
suggested that a better approach would be a verification
analysis of the software as it is being prepared. However,
with our current media for expressing numerical linear
algebra software, this approach does not seem feasible.

This paper reports on some work to discover means to make
this approach at least feasible and possibly practical. The
emphasis herein is on the evaluation aspect of the proposed
approach. For the Cholesky factorization algorithm for a
positive definite symmetric matrix, we display an abstract
form, and then using program verification techniques, show
how this form can be evaluated using an automated theorem
prover.

INTRODUCTION

The answer to this question is that an automated theorem prover can be used as
a tool to evaluate some parts of numerical linear algebra software. However, it
seems necessary to write the software, not in FORTRAN, but in a form more suitable for automated analysis.

In this paper, we propose an approach to the preparation and partial automated
evaluation of software for linear algebra. To illustrate the approach, a
Cholesky factorization algorithm is formulated in an abstract form and this form
is then evaluated with respect to some specification of the program. The evaluation consists of the derivation and proof of the verification conditions that
must be satisfied in order to meet the specification. Certain of the verification conditions have been proven by the automated resolution-based theorem [9].

Work performed under the auspices of the U.S. Department of Energy and supported
in part by the U.S. Army Research Office.

The first section of this paper justifies this approach for the evaluation of
linear algebra software, particularly indicating why an abstract formulation is a
reasonable way to address the evaluation problem. The second section describes a
program for Cholesky factorization. In the third section, we formulate the specifications for this algorithm, and derive some of the necessary verification
conditions that must be proven in order to satisfy the specifications for the
Cholesky factorization. Next, we describe how these verification conditions can
be provided to an automated theorem prover in an automated fashion. Finally, we
comment on the applicability of this approach to evaluating linear algebra
software.

THE RATIONALE FOR THE APPROACH

Previously, linear algebra software has been evaluated by a combination of several
techniques: one is exhaustive testing; a second is special evaluation tools such
as DAVE [10] for static analysis, and BRANL [6] and WATFIV [4] for dynamic
analysis using specific test problems; and a third technique is the manual
analysis of correctness such as that described in [7]. Each of these approaches
evaluates different aspects of the software; taken together, with careful design
of the test cases and analysis of the program, they are often sufficient to become convinced that the software is correct.

The difficulty with exhaustive testing and the special evaluation tools using
test cases is that these techniques rely on a great deal of care and ingenuity in
the design of the tests; this type of evaluation is a sampling process where
many samples are required to ensure the software is thoroughly evaluated. The
difficulty with the manual analysis of correctness is that it is laborious and
error prone. Every time the software is changed to improve some aspect of it,
the tedious job of re-examining the software with these techniques must be repeated; it seldom is. One solution is to automate the human analysis of the
complete correctness of the software, but such analysis seems beyond the capabilities of automated verification systems for numerical software. The usual
forms of numerical software are so complicated, often entangling the algorithm
with the representation of the data, that automated analysis seems infeasible.

Rather than to evaluate software in its current form, our approach is to consider
constructing software from an abstract form, and then to evaluate the software
from the abstract form. The abstract form is chosen so as to facilitate two
processes: 1) automated evaluation of the software using automated theorem
provers; 2) the construction of correct programs within, say FORTRAN, by transformations applied to a succession of more concrete forms of the program.

Provided that the transformations are proven correct, the concrete programs derived in this manner are as correct as the original program; cf. [1,2].

The purpose of this paper is to show how the abstract form of some numerical software can be evaluated in an automated way, verifying that the software meets its prescribed specifications. The abstract form seems general enough to apply to a wide collection of algorithms in linear algebra. Although our goal is automated analysis, only certain aspects of the evaluation process have been automated to date. We use a modification of the Cholesky factorization algorithm used in [5] to illustrate the techniques.

The abstract form must be detailed enough that a set of transformations can derive the appropriate concrete form. So far the required transformations for Cholesky factorization have not been prepared but the paper by Boyle [2] describes related transformations for the back-substitution process used in the solution of upper triangle systems of equations. As yet, automated verification of the correctness of any of these transformations has not been attempted.

The use of an abstract form for the evaluation of software has been demonstrated recently in [3] for adaptive quadrature; however, their abstract form, called a logic program, is very closely related to first order predicate calculus.

AN ABSTRACT PROGRAM FOR CHOLESKY FACTORIZATION

Given below is an abstract form of the square root free Cholesky factorization with diagonal pivoting for a symmetric positive definite matrix A of order n. The algorithm is a modification of the algorithm used in the subroutine SCHDC of LINPACK [5]. The algorithm uses diagonal pivoting (or similarity pivoting) to yield a factorization of the form $PLDL^T P^T$ of A where the matrices P, L and D are respectively products of permutation, unit lower triangular and diagonal matrices. The abstract form of the algorithm is written in Algol-like language in which the matrix operations (products in this case) are encoded in several procedures, some of which are specified by procedures bodies and others are specified by assertions, describing their behavior. The abstract form follows:

```
matrix A [1:n, 1:n];
diagonal matrix Da, D[1:n];
permutation matrix Pa, P[1:n];
unit-lower-triangular matrix La, L[1:n];

Comment:  I is the identity matrix of order n;
Pa := I;   La := I;   Da := I;
```

```
for j := 1 step 1 until n do;
        form_Pj;
        accumulate_Pj_into_Pa;
        apply_Pj_to A;
        apply_Pj_to_La;

        form_Lj;
        accumulate_Lj_into_La;
        apply_Lj_to_A;

        accumulate_Lj_into_La;
        apply_Lj_to_A;

        form_Dj;
        accumulate_Dj_into_Da;
        apply_Dj_to_A;
end;
```

To clarify the abstract form, we shall briefly describe the function of each procedure. The procedure form_Pj determines a permutation matrix P(j) which defines
a permutation similarity transformation to move the largest diagonal element of A
beyond the (j-1)th diagonal position to the jth diagonal position; the procedure
form_Lj determines an elementary matrix $L(j) = I + \underline{\ell}e_j^T$ whose inverse, when post
multiplied by the current A eliminates the elements of the jth column of the
current A below the diagonal. The procedure form_Dj determines the diagonal matrix D(j) which is the identity matrix of order n except for the jth element
which is equal to the jth diagonal element of current A. The "accumulate" procedures accumulate the matrices P(j), L(j), D(j) into the accumulated product
matrices Pa, La, Da, respectively. That is, they execute the following
statements:

accumulate_Pj : Pa := Pa*P(j);
accumulate_Lj : La := La*L(j);
accumulate_Dj : Da := Da*D(j);
where * operator is matrix multiplication.

Finally, the "apply" procedures transform the current A with the matrices P(j),
L(j) and D(j) and transform La with P(j). Specifically, they execute the following statements:

apply_Pj_to_A : A := P(j)*A*Tr(P(j));
apply_Lj_to_A : A := Inv(L(j))*A*Tr(Inv(L(j)))
apply_Dj_to_A : A := Inv(D(j))*A

apply_Pj_to_La : La := P(j)*La*Tr(P(j))
where the functions Tr(·) and Inv(·) transpose and invert respectively their
arguments.

ANALYSIS OF THE ABSTRACT FORM

To evaluate this program, we use the inductive assertion method described in [8].
The verification conditions that must be proven are derived; no proofs will be
given as they are straightforward. At the time of writing this paper, only some
of these conditions were submitted to the automated resolution based theorem
prover [9] and successfully proved.

The inductive assertion method requires an assertion ϕ specifying the assumed
properties of the input entities, an assertion ψ specifying the expected properties of the output entities, and an assertion Δ that represents the invariant
properties of program entities within each loop of the program. For Cholesky
factorization, the appropriate input assertion is:

$$\phi: \quad n > 0 \quad \text{and} \quad sym(A) \quad \text{and} \quad \lambda min(A) > 0$$

where $sym(A)$ means A is a symmetric matrix, and $\lambda min(A)$ is the smallest eigenvalue of A. The appropriate output assertion is:

$$\psi: \quad Input(A) = Pa*La*Da*Tr(La)*Tr(Pa)$$

where the matrices Pa,La,Da are the final values of the accumulated factors, and
input(A) is the input or initial value of A to the decomposition program. A
suitable invariant $\Delta(j)$, valid at the start of the jth iteration of the loop, is

$\Delta(j)$: 1) The current matrix A is the direct sum of an identity matrix of
order j-1 and a symmetric positive definite matrix of order n-j+1,
and

2) Input(A) is equal to the product Pa*La*Da*A*Tr(La)*Tr(Pa) where
the matrices Pa,La,Da and A are the current values of these program
variables.

In order to evaluate this program, the inductive assertion method requires the
formulation and proof of 4 verification conditions:

1. The input assertion ϕ along with the initialization statements for the
entities La, Da, and Pa implies the invariant assertion $\Delta(1)$;

2. The invariant assertion $\Delta(j)$ along with the loop body statements and $j \leq n$ implies the invariant assertion $\Delta(j+1)$;

3. The invariant assertion $\Delta(j)$ and $j > n$ implies the output assertion ψ; and

4. The loop terminates.

Proofs of the first three verification conditions imply the partial correctness of the abstract program; proof of the last condition provides the termination condition necessary for total correctness.

Since we are interested in just illustrating the technique for this paper, we state in detail the more difficult verification condition, namely 2. For this paper the verification conditions and the program segments were translated by hand into a form suitable for input to the resolution-based theorem prover [9]; to automate this translation, we will be using the TAMPR system [1,2] on more FORTRAN-like abstract forms.

To readily express these conditions, we need some notation; let

Ai   be the initial value of the matrix A at the beginning of the loop;
Af   be the final value of the matrix A at the end of the loop;
$Pa(j)$ be the accumulated product $P(1)*P(2)*...*P(j-1)$;
$La(j)$ be the accumulated product $P(j-1)*P(j-2)*...*P(1)*P(1)*L(1)*P(2)*L(2) *...*L(j-2)*P(j-2)*L(j-1)$
$Da(j)$ be the accumulated product $D(1)*D(2)*...*D(j-1)$;
I_j be the identity matrix of order j;
matrix $(j,q,\underline{c},\underline{r},S)$ be the matrix

$$\begin{pmatrix} I_j & 0 & 0 \\ 0 & q & \underline{r}^T \\ 0 & \underline{c} & S \end{pmatrix}$$

where j is a non-negative integer, q is a scalar, $\underline{r}$ and $\underline{c}$ are column vectors of length n-j-1, and S is a square matrix of order n-j-1.

Using this notation, verification condition 2 becomes

```
(A = Pa(j)*La(j)*Da(j)*Ai*Tr(La(j))*Tr(Pa(j)) &
(∃q,c,S)(Ai = matrix(j-1,q,c,c,S) & sym(S) & q > 0 & λmin(S) > 0) &
Da(j+1) = Da(j)*D(j) & Pa(j+1) = Pa(j)*P(j) &
La(j+1) = P(j)*La(j)*Tr(P(j))*L(j) & Af = Inv(D(j))*Inv(L(j))*P(j)*Ai*Tr(P(j))*
Tr(Inv(L(j))))
=>
```

$(A = Pa(j+1)*La(j+1)*Da(j+1)*Af*Tr(La(j+1))*Tr(Pa(j+1))$ &
$(\exists q',\underline{c}',S')(Af = Matrix(j,q',\underline{c}',\underline{c}',S)$ & $sym(S')$ & $q' > 0$ & $\lambda min(S') > 0)$

Now let us consider how to obtain an automated proof of this verification condition. There are at least two approaches. One way would be to specify the matrix operations used above such as product, inverse, and transpose in terms of the way elements of matrices are manipulated and combined. Such an approach is tantamount to deriving a concrete realization of the algorithm and deriving a proof by analyzing the behavior of individual elements in the matrices. The proofs would then require algebraic manipulations of complicated expressions, involving many subscripted variables. A second way, which avoids these complicated formulae, would be to specify the general properties of these matrix operations as well as specifying the special properties of the program entities $D(\cdot)$, $L(\cdot)$, $P(\cdot)$, $Da(\cdot)$, $La(\cdot)$ and $Pa(\cdot)$. This latter approach is more practical for automated theorem provers, for the set of possible inferences that can be derived from the verification conditions and axioms is drastically reduced. Also, using the latter approach the proof will be independent of the particular formulation of the matrix operations, that is, the same proof will hold whether, for instance, matrix product is an expansion by rows, or columns, or inner products.

For these reasons, the second approach is taken to prove these verification conditions. To illustrate this approach, we give the needed axioms for the proof of the verification condition (2):

A1: Commutativity of Da's and P's.
  for $k \geq j$, $Da(j)*P(k) = P(k)*Da(j)$

A2: Commutativity of Da's and L's.
  for $k \geq j$, $Da(j)*L(k) = L(k)*Da(j)$

A3: Orthogonality of P's and Pa's and symmetry of P's.
  $(\forall k)(Tr(P(k)) = Inv(P(k))$ & $Tr(P(k)) = P(k)$ & $Tr(Pa(k)) = Inv(Pa(k)))$

A4: Nonsingularity axioms for D's, L's, P's.
  $(\forall k)(nonsing(D(k))$ & $nonsing(L(k))$ & $nonsing(P(k))$ &
   $nonsing(Da(k))$ & $nonsing(La(k))$ & $nonsing(Pa(k)))$

A5: Symmetry axiom for the function matrix $(\cdot,\cdot,\cdot,\cdot,\cdot)$
  $(\forall q,\underline{c},\underline{r},S)$ & $(\forall k \geq 0)$
   $(sym(matrix(k,q,\underline{c},\underline{r},S)) <=> sym(S)$ & $\underline{c} = \underline{r}$

A6: Positive definiteness of the submatrix operation
  $(\forall q,\underline{c},S)$ & $(\forall k \geq 0)$
   $(\lambda min(matrix(k,q,\underline{c},\underline{c},S)) > 0$ & $sym(S) \Rightarrow q > 0$ & $\lambda min(S) > 0$

A7: $L(j)^{-1}$ is chosen so that it eliminates $\underline{c}$ in matrix $(j-1,q,\underline{c},\underline{c},S)$
   $(\forall j>0)(\exists S')(q > 0$ & $sym(S)$ & $\lambda min(S) > 0 =>$
     $Inv(L(j)) * matrix(j-1,q,\underline{c},\underline{c},S) * Inv(Tr(L(j))) =$
     $matrix(j-1,q,\underline{0},\underline{0},S')$ & $sym(S')$ & $\lambda min(S') > 0)$

A8: $D(j)^{-1}$ is chosen so that it scales the diagonal of $matrix(j-1,q,\underline{0},\underline{0},S)$ to
 unity.
   $(\forall j>0)(q \neq 0 => Inv(D(j))*matrix(j-1,q,\underline{0},\underline{0},S) = matrix(j-1,\underline{0},\underline{0},S))$

A9: $P(j)$ performs a permutation similarity transformation on $matrix(j-1,q,\underline{c},\underline{c},S)$
   $(\forall j>0)(\exists q',c',S')(q > 0$ & $sym(S)$ & $\lambda min(S) > 0 =>$
     $P(j)*matrix(j-1,q,\underline{c},\underline{c},S)*Tr(P(j)) = matrix(j-1,q',\underline{c}',\underline{c}',S')$
     & $q' > 0$ & $sym(S')$ & $\lambda min(S') > 0)$

A10: Matrix deflation
   $(\forall j>0)(\exists q',c',S')(sym(S)$ & $\lambda min(S) > 0 =>$
     $matrix(j-1,1,\underline{0},\underline{0},S) = matrix(j,q',\underline{c}',\underline{c}',S')$
     & $sym(S')$ & $q' > 0$ & $\lambda min(S') > 0)$

A11: Axioms for non-commutative group of non-singular matrices under multiplication
   $nonsing(X) => (Inv(X)*X = I$ & $X*Inv(X) = I$ & $I*X = X$ & $X*I = I)$

CAN THIS EVALUATION PROCESS BE AUTOMATED

So far, we have proposed the representation of a linear algebra program that facilitates the automated evaluation of the program. The question now is whether the
approach is general enough to be useful for the evaluation of other linear algebra
software and at the same time restrictive enough to permit machine proofs by an
automated theorem prover.

We claim that the process is indeed general enough. The invariants for many
other factorization processes are quite similar, using much of the same notation
given above. The properties and axioms of the matrix operation given above are
typical for many algorithms, and more can readily be added when necessary. However, termination conditions for the iterative algorithms such as eigenvalue
computations are more complicated, and in some cases, we may not be able to prove
termination; indeed, for some algorithms, no termination proofs are known.

Our second concern, whether the automated theorem provers are up to the task of
proving the verification conditions, is more serious. There are several difficulties; one is that first order predicate calculus may not be powerful enough to
specify the required assertions for the algorithm, in a natural and convenient

way. Secondly, automated theorem provers seldom obtain proofs the first time they are given a problem. Often several attempts are required, usually selecting alternative strategies to be used by the prover after examining the progress on previous attempts. Thus, only in this sense, does one obtain automated proofs of the verification conditions.

Initially, these difficulties appear to represent serious limitations on the use of automated theorem provers for evaluating linear algebra algorithms; however, we believe this is not so. Our interest in automated verification systems is that it provides a reliable evaluation technique which is independent of the existing tools. It evaluates algorithms over a class of problems rather than individual cases. In situations where the natural or complete assertions cannot be stated, less complete assertions can often be formulated which permit limited evaluations. For instance, in eigenvalue computations, we may be able to assert and prove that the trace of the matrix is preserved throughout the computation; although not the ideal assertion for this computation, it does permit a limited evaluation that can expose many errors in an eigenvalue program. In terms of reliability, proofs, when obtained, are machine generated, are less likely to contain errors, and can be readily repeated when necessary.

Another issue related to the automated question is the need for programs which from the abstract form containing appropriate assertions can derive the verification conditions in a form suitable for input to the theorem prover. Although we do not have such programs for the abstract form illustrated in the previous sections, we have developed, using the TAMPR system [1], a collection of transformations which given certain restricted FORTRAN programs containing assertions, are able to derive the verification conditions in a form suitable for our resolution-based theorem prover. It is believed that a more FORTRAN-like abstract form than the above abstract form can be used to obtain the appropriate verification conditions automatically with only minor modifications to our transformations.

THE EVALUATION OF SOFTWARE

The approach proposed here is not very helpful in evaluating software in some concrete form, say FORTRAN or ALGOL. Indeed, with the techniques and tools available to us, we find it difficult to see how such an automated verification system can be developed for evaluating software in its concrete form. This difficulty exists mainly for the following reason. Any particular concrete form is just one of many forms derivable from a more abstract form that contains just the parts of the algorithm needed to assert its correctness. Which aspects of a given concrete form are pertinent to the program specification of interest and which correspond to other properties of the program are very difficult to

distinguish. Often these other properties correspond to certain optimization
strategies such as minimal use of memory, efficient access of memory in a paged
environment, or minimal rounding errors.

For example, the FORTRAN version of the Cholesky factorization algorithm in
LINPACK operates on just half the matrix, stores the accumulated elementary
matrices in the same half of the matrix, stores the accumulated diagonal on the
diagonal of input matrix and encodes the accumulated permutation matrix in a
vector. The assertions specifying the behavior of this software must be in terms
of these encodings and now become very difficult to specify and prove
automatically.

The verification of the abstract form can still be considered as evaluating software but only that software which is constructed or derived from the abstract form.
Provided the construction techniques are shown to preserve the various properties
of the abstract program needed in its verification proof, the derived program
will be as correct as the abstract form.

The constructive approach to preparing and evaluating software in effect dis-
tinguishes between those aspects of the algorithm that permit the software to
meet its specifications and those aspects that permit efficient use of computer
resources. That is, the abstract form limits itself to specifying the algorithm,
and the transformations that derive the concrete form are encodings of the optimization techniques needed to produce efficient software. This is an ideal
partitioning of software as it permits the evaluation of either aspect independently of the other.

The evaluation of software must take into account rounding errors. We believe
that they can be treated, much like any other property of the algorithm; cf.
[11, p. 231]. For instance, in the abstract form of Cholesky factorization given
above, the only operation that involves rounding errors is the formation of $L(j)$
and the computation of $L(j)^{-1}AL(j)^{-T}$. To specify these rounding errors, we would
introduce into the abstract form the operation $f\ell(L(j)^{-1}AL(j)^{-T})$ instead of
$L(j)^{-1}AL(j)^{-T}$, and provide the additional axiom

A12: $(\forall j > 0)(\exists E)(f\ell(Inv(L(j))*A*Inv(Tr(L(j)))) = Inv(L(j))*A*Inv(Tr(L(j)))+E$
            $\& \; norm(L(j)*E*Tr(L(j))) \leq 7*j*norm(A)*\varepsilon)$

where ε is the usual relative precision machine constant. The invariant for the
loop would change to

$(\forall j > 0)(\exists F)(F+A = Pa(j)*La(j)*Da(j)*Ai*Tr(La(j))*Tr(Pa(j))$
            $\& \; norm(F) < (7/2)*(n*(n+1)-(n+1-j)*(n+2-j))*\varepsilon*norm(A))$

Of course, additional axioms for specifying norm($\cdot$) are required. Also, axiom A7, which characterizes the elimination properties of $L(j)^{-1}$, must be modified in the presence of rounding errors as follows.

$$
\begin{aligned}
\text{A7:} \quad &(\forall j > 0)(\exists S')(q' > 0 \ \& \ \text{sym}(S) \ \& \\
&\quad \lambda\text{min}(\text{matrix}(j-1,q,\underline{c},\underline{c},S)) > \\
&\quad\quad (7/2)*(n+1-j)*(n+2-j)*\epsilon*\text{norm}(\text{matrix}(j-1,q,\underline{c},\underline{c},S)) \\
&\quad => \text{sym}(S') \ \& \ \lambda_{\text{min}}(S') > (7/2)*(n-j)*(n+1-j)*\epsilon*\text{norm}(S') \ \& \\
&\quad\quad f\ell(\text{Inv}(L(j))*\text{matrix}(j-1,q,\underline{c},\underline{c},S)*\text{Inv}(\text{Tr}(L(j)))) = \\
&\quad\quad\quad \text{matrix}(j-1,q,\underline{0},\underline{0},S')
\end{aligned}
$$

Here, the hypothesis and conclusion concerning $\lambda\text{min}(\cdot)$ are necessary to ensure that the diagonal elements remain positive. A stronger input assertion ϕ is now required, namely $\lambda\text{min}(A) > \frac{7}{2} n(n+1) \ \epsilon \ \|A\|$.

SUMMARY

As it may be clear from the above discussion, the difficult analysis, either mathematical or numerical, is encapsulated in the axioms. Rather than producing these difficult proofs, the automated prover is manipulating formulae, essentially verifying that the program statements are consistent with the given set of axioms and lemmas. Only in this limited sense is the correctness of the program being verified; rather the software is being evaluated relative to these axioms. A complete correctness proof would require the proofs of each of the problem dependent axioms, as well as proofs that the program is well formed (e.g. no illegal subscripts, no arithmetic exceptions, no undefined variables, etc.).

Complicated analyses for the automated tools have been avoided in another area, namely in the use of the abstract form. The key to the advantage of the abstract form is that it specifies the algorithm only, thereby distinguishing between algorithmic constructs and other constructs (e.g. those designed to improve efficiency in time or memory). In the context of deriving concrete software from abstract forms, it is reasonable to take advantage of this distinction. However, in the context of analyzing the concrete forms of software directly, the automated analyzer must unravel the interactions between these two kinds of constructs. One way of helping the analyzer may be to provide directives that associate program constructs with the algorithmic constructs. For example, in a concrete form of Cholesky factorization, one may direct that certain columns of A are encodings (without the redundant diagonal elements) of elementary elimination matrices. With this directive, certain columns of A can be treated by the analyzer as behaving like elementary elimination matrices. However, this approach seems intractable.

To conclude on an optimistic note, our studies so far indicate that is feasible to evaluate certain forms of linear algebra software with an automated theorem prover.

REFERENCES

1. Boyle, J. M., Dritz, K. W., Arushanian, O. B., and Kuchevskiy, Y. V. (1977). "Program generation and transformation -- tools for mathematical software development," Proc. IFIP 77, North-Holland, 303-308.

2. Boyle, J. M. (1978). "Extending reliability: transformational tailoring of abstract mathematical software," Proc. Conf. on the Programming Environment for the Development of Numerical Software, JPL & ACM-SIGNUM meeting, Pasadena, CA, JPL Rep. No. 78-92, p. 27-30.

3. Clark, K., McKeewan, W. M., and Sickel, S. (1978). "Logic programming applied to numerical integration," Tech. Rep. No. 78-8-004, Information Sciences, Univ. of California, Santa Cruz.

4. Cress, P. H., Dirkson, P. H., McPhee, K. J., Ward, S. J., and Weseman, M. A., "WATIV implementation and user's guide," Univ. of Waterloo Report, Waterloo, Ontario, Canada.

5. Dongarra, J. J., Bunch, J. R., Moler, C. B., and Stewart, G. W., LINPACK Users' Guide, Society for Industrial and Applied Mathematics, to appear 1979.

6. Fosdick, L. D., "A FORTRAN program to identify basic blocks in FORTRAN programs," Dept. of Computer Science Report No. CM-CS-040-74, Univ. of Colorado at Boulder.

7. Hull, T. E., Enright, W. H., and Sedgwick, A. E. (1972). "The correctness of numerical algorithms, Proc. ACM Conf. on Proving Assertions about Programs,, SIGPLAN Notices, Vol. 7, No. 1.

8. Manna, Z. (1974). Mathematical theory of computation, McGraw-Hill.

9. McCharen, J. D., Overbeek, R. A., and Wos L. (1976). "Problems and experiments for and with automated theorem-proving programs," IEEE Trans. on Computers, Vol. C-25, No. 8, 773-782.

10. Osterweil, L. J., and Fosdick, L. D. (1976). "DAVE - A Valiation Error Detection and Documentation System for FORTRAN Programs," Software Practice and Experience, Vol. 6, No. 4, 473-486.

11. Wilkinson, J. H. (1965). The Algebraic Eigenvalue Problem, Clarendon Press, Oxford.

Performance Evaluation of Numerical Software, Fosdick (ed.)
© *IFIP, North-Holland Publishing Company, 1979*

PERFORMANCE EVALUATION OF CODES
FOR SPARSE MATRIX PROBLEMS

I.S. Duff and J.K. Reid
Computer Science and Systems Division,
Atomic Energy Research Establishment,
Harwell, Didcot, Oxfordshire

There is a strict limit to the extent that analytical methods can be used
to assess the effectiveness of techniques employed in sparse matrix codes.
Any full evaluation therefore demands the running of a realistic set of
test problems. The collections that we have assembled at Harwell are
described, although they are still in their infancy, and we show a number of
examples of their use both during algorithm development and for assessment
of codes written elsewhere.

1. INTRODUCTION

Sparse problems arise in a wide variety of application areas, including management
science, power systems analysis, surveying, circuit theory and structural analysis.
In fact any problem that involves a very large number of variables must have some
simplifying feature if it is to be tractable for computer solution, and this
feature is almost always sparsity. Each known relationship can usually be expressed
as an equation involving a small number of variables.

To exploit this sparsity it is necessary to use code that avoids storing the zeros
and operating with them, or at least mostly does so. We concentrate in this paper
on direct methods because here the case for general-purpose software is
overwhelming since full advantage can be taken of sparsity only with the aid of
sophisticated data structures and quite complicated code. Furthermore little
general-purpose iterative software yet exists.

Our basic message is that a full performance evaluation demands testing on
realistic problems. We describe our test-beds, still in their infancy, and give
some details of how we have used them both to help in the design of our own codes
and to assess other codes and ideas exploited by these codes. We restrict
attention to the solution of linear equations because this is where our main
experience lies and because it is a basic step used widely in the eigenvalue
problem and in the solution of non-linear problems. For clarity it is necessary
to edit the numerical results by presenting just a few cases.   In each instance
we aim to include at least one "typical" case and an example of any form of extreme
behaviour.

2. TEST BEDS FOR SPARSE MATRICES

The most fundamental comment we wish to make is that sparse matrices occurring in
practice always have special features. Telling an expert that a matrix has
arisen from circuit theory or structural mechanics immediately gives him a
considerable feel for its features.   Therefore when we say that a code is "general-
purpose" we really mean that it is suitable for any matrix likely to occur in
practice.   One that performs badly for a particular class of cases is not truly
general-purpose, although it may still be a useful code.   For example the 1971
Harwell code MA18 is very slow in cases where the average number of non-zeros in
the rows of the factors is high, including problems that arise from finite-
difference calculations on a three-dimensional mesh or a fine two-dimensional mesh,
but was nevertheless successfully used in a variety of other cases.

Supported in part by U.S. Army Research Office Contract DAAG29-78-M0152.

2.1 Collections of test cases

It follows that our first task in constructing a test-bed must be the collection of a representative set of problems. Our principal collection at Harwell now contains 36 matrices, some in the form of sparsity patterns alone and others with actual numerical values, and we have supplied some fifteen copies to other researchers around the world. Brief details of all the matrices are shown in Table 2.1. Our intention is to add to the collection gradually over the years and we welcome any contributions. We expect, too, to make deletions; for instance we intend soon to remove some of the LP bases (matrices 12 to 27).

Many of these matrices are unsymmetric so we have constructed a set of symmetric matrices by replacing the upper triangular parts of our unsymmetric matrices by the transposes of their lower triangular parts. This gives a far from satisfactory test bed for symmetric codes and it is our firm intention to add more genuine symmetric matrices over the next few years. Already (see section 5) we have found ourselves supplementing this test set.

In linear programming the need for a collection of realistic test cases was recognised in 1960 by the Linear Programming Committee of SHARE. The collection they established was used very effectively by Wolfe and Cutler (1963) but these problems are small by modern standards. Saunders (1972) collected four problems in MPS-360 format and used them to assess the performance of his code. He kindly provided us with a copy of these problems and we have used them extensively for our own testing. We have supplemented these four with one obtained from P. Gill (NPL) and another obtained from S. Powell (then at Atlas Laboratory). The six problems are summarized in Table 2.2. They span a wide range of practical problems and have been invaluable in assessing the performance of our codes, but again it is our intention to enlarge the collection gradually and we would welcome any contribution.

A further set of test problems is necessary to assess the performance of codes for finite-element problems that use the frontal method (Irons, 1960) and its variants, since we require the problem to be in unassembled form. Here we are indebted to A.J. Donovan of CEGB for providing the patterns of four problems from a variety of applications, all of which were presented to the CEGB stress analysis package BERSAFE. These are summarized in Table 2.3. Again we wish to supplement this set with further problems. Indeed our tests have not been confined to these cases although these examples have provided particularly illuminating results, some of which we mention in section 5.

2.2 Use of a collection of test problems

Given a collection of problems, as described in the last section, we may use it to construct a test bed for evaluating pieces of software. Writing code to call the software for each problem will immediately give a qualitative answer to the important question

    a) is the software easy to use?

and whether it solves all the problems gives a tentative answer to the question

    b) is the software reliable and robust?

Quantitative assessments may be made for each example of

    c) the computing time required

    d) the storage required

Matrix	Order	No. of non-zeros	Description
1	147	2449)	Matrices A,B from finite-element eigenvalue problem
2	147	2441)	$Ax=\lambda Bx$, supplied by T. Johansson of Lunds Datacentral, Lund, Sweden.
3	1176	18552	Pattern of large electrical network problem, supplied by A.M. Erisman of Boeing Computer Services, Seattle, U.S.A.
4	113	655	Pattern of a matrix arising in a statistical application, supplied by W.M. Gentleman, Waterloo, Canada.
5	32	126	Pattern of matrix advertising 1971 IBM conference on sparse matrices.
6	54	291	Pattern of matrix arising when solving a stiff set of biochemical ordinary differential equations, supplied by A.R. Curtis of Harwell, England.
7	57	281	Pattern of Jacobian matrix associated with an emitter-follower-current switch circuit (Willoughby, 1971).
8	199	701	Pattern of a stress-analysis matrix (Willoughby, 1971).
9	292	2208)	Patterns of normal matrices associated with least squares
10	85	523)	adjustment of survey data, supplied by V. Ashkenazi, Nottingham University, England.
11	130	1282	Jacobian matrix of a set of ordinary differential equations associated with a laser problem, supplied by A.R. Curtis, Harwell, England.
12-14	663	1687-1712	Basis matrices obtained at various stages of the application of the simplex method to three linear programming problems, supplied by M.A. Saunders, DSIR, New Zealand (see Saunders, 1972).
15-18	363	2454-3279	
19-27	822	3276-4841	
28	219x85	438	Patterns of overdetermined sets of equations associated with least squares adjustment of survey data, supplied by V. Ashkenazi,Nottingham University, England. Cases 28/29 correspond to cases 10/9, respectively.
29	958x292	1916	
30	331x104	662	
31	608x188	1216	
32	313x176	1557	Survey pattern from the Sudan, supplied by A. Abbas-Elhag, 1975 (then at University of Newcastle upon Tyne, now at Oxford).
33-36	541	4285	Four matrices having the same pattern but varying conditioning, which arose at different stages of the use of FACSIMILE (a stiff ODE package) to solve an atmospheric pollution problem involving chemical kinetics and two-dimensional transport, supplied by A.R. Curtis, Harwell, England.

Table 2.1   Harwell's collection of sparse matrices.

and   e) the accuracy achieved

as well as a qualitative assessment of

    f) the diagnostic and monitoring information returned.

Problem	Rows	Columns	Non-zeros	Description
Blend	74	114	560	Blending problem originating from Bruce Murtagh and supplied by P. Gill (NPL).
Powell	548	1076	7131	Econometric modelling problem supplied by S. Powell (Atlas Laboratory).
Stair	362	544	4013	Dynamic multi-sector model with staircase structure (6 main blocks, each about 50x100) originating from A.S. Manne and K.W. Kohlhagen (Stanford) and supplied by M.A. Saunders.
Shell	662	1653	5033	A network problem originating from the Shell Development Company, California and supplied by M.A. Saunders via R.R. Meyer. It is very sparse and we found that all bases were permutations of triangular matrices.
GUB	929	3333	14158	A generalized upper bounding problem with 890 GUB sets and 39 coupling constraints, originating from the Crown Zellerbach Corporation and obtained from M.A. Saunders via M.G. Kazatkin.
BP	821	1876	11719	Oil industry operating model problem called problem A by Forrest and Tomlin (1972) and obtained from M.A. Saunders. It is a particularly difficult problem to solve.

Table 2.2   Linear programming test problems.

No. of variables	No. of elements	Maximum element size	Description
2919	128	60	Three-dimensional cylinder with flange
3024	551	16	Two-dimensional reactor core section
3306	791	12	Framework problem essentially in two dimensions
2802	108	60	Turbine blade.

Table 2.3   Finite-element problems supplied by A.J. Donovan.

The relative importance of these six aspects is somewhat subjective. Usually tests have given prominence to c), the computing time, but our own opinion is that far more important are a), the ease of use, and b), the reliability. The user, busy with work in his own field, will not want to spend hours mastering a complicated interface to which it is difficult to adapt his data. He will be doubly frustrated if he then finds that not all his problems are successfully solved.

Such a test bed is also useful during the design of a piece of software. Some decisions can be made easily without reference to actual testing. For example it is very desirable to avoid any loop whose body is bound to be executed n^2 times for a matrix of order n because if it is very sparse then far less than n^2 actual arithmetic operations will be needed.   Other decisions, however, are not so straight-forward and it is extremely helpful to have a test bed available to assess the effectiveness of alternatives. We give several examples later in this paper.

We end this section by mentioning the usefulness of a profiling code that counts the number of times source statements are executed. We have principally been using the Harwell subroutine OE02, which is written in Fortran 66 and adds extra counting statements at all branch points.   The output from OE02 compiles into a program that typically executes only about 30% slower than the original.   It gives fine detail about the heavily used parts of the code and indeed several tables in this paper are based on its output. Similar facilities are available in the WATFIV compiler for IBM machines, but the execution speed of the compiled code is some ten times slower.

3.   THE DESIGN OF THE HARWELL SPARSE MATRIX PACKAGE MA28

We made extensive use of our test bed of sparse unsymmetric sets of linear equations during the design of the Harwell sparse matrix package MA28 (Duff, 1977). Effectively we monitored the performance of early versions to ensure that they behave as well as intended in respect of reliability, speed, storage and accuracy. We used the test bed to check the new code against the earlier code MA18, because we wanted to ensure that in no respect could the new code be said to be inferior. Here we describe some of these tests.

In this and later sections we use the following terms:

ANALYZE - determine the pivot sequence and fill-in pattern from the original pattern, without reference to numerical values.

ANALYZE-FACTOR - determine the pivot sequence, fill-in pattern and matrix factors, given the original matrix including real values.

FACTOR - determine the factors of a matrix whose pattern has already been treated by ANALYZE or ANALYZE-FACTOR.

SOLVE   - solve a system given the factors produced by FACTOR or ANALYZE-FACTOR.

ONE-OFF - solve a system on its own without storage of the information needed
          for more rapid solution of later similar systems.

3.1 Storage management

The storage pattern used by the earlier code MA18 while choosing the pivots and
determining the fill-ins is a linked list; alongside each non-zero is stored an
integer pointer to the next non-zero in its row and an integer pointer to the next
non-zero in its column.   For MA28 we decided instead to switch to storing the
non-zeros of each row in contiguous locations of a real array with the
corresponding column indices in a parallel integer array; also required is the
pattern of the active part of the matrix held similarly as row indices in
contiguous locations of another integer array.   This mode of storage has the
advantages of avoiding a search to determine the indices of a non-zero, reducing
the restrictions implied by the use of 16-bit integers, reducing page faults in a
paged virtual memory and using less indirect addressing.

However a disadvantage is that some "elbow-room" is needed so that a fresh copy
of a row can be made if fill-ins result in there no longer being sufficient room
in the row's current position. Because the storage vacated is thereafter mostly
left unused, there is also a need for a facility to "compress" the storage whenever
there is insufficient contiguous free space.

We began checking this decision by looking at MA18 profiler counts to assess the
cost of not having row and column indices ready to hand.   Three specimen results,
shown in Table 3.1, amply demonstrate how expensive this can be.

Order of matrix	199	130	822
Number of non-zeros	701	1282	4841
Number of elimination operations	2535	2629	6266
Number of array look-ups to find column and row indices	68019	274385	586970

Table 3.1   Profiler counts on runs of MA18

Next we checked the effect on run-time of having rather little elbow-room so that
many compresses were needed.   In Table 3.2 some sample times are shown for MA18

Order of matrix	147	113	199	292	822
No. of non-zeros	2449	655	701	2208	4841
MA18	5410	230	410	2920	3090
MA28 (min.store)	5260	160	380	1540	2170
MA28 (min + 2n)	3000	110	260	1040	1790
MA28 (plenty of storage)	1540	110	240	830	1760

Table 3.2   Comparative timings of ANALYZE-FACTOR phases of MA18 and
             MA28 with and without "space-squeeze" (IBM 370/168 m.secs).

and for MA28 with varying amounts of storage.   We were encouraged by these
results to think that modest elbow-room (say 2n) would usually suffice, but
sufficiently concerned about the overheads of compression to count their number
for the user to monitor if he wishes.   These results incidently confirmed that
MA28 was able to offer a worthwhile overall speed advantage.

3.2 Block triangularization

If a sparse matrix can be permuted to block triangular form, then substantial advantages can accrue when solving the corresponding set of equations because only the diagonal blocks need be factorized and all fill-in is confined there. Over the last few years efficient algorithms have been developed for this task and we decided that MA28 should automatically call code implementing them. These are fast but nevertheless we were worried about the overheads in the many practical problems that are irreducible. However we were reassured by tests, the results of some of which are shown in Table 3.3. The first and third cases shown are irreducible, but the overhead is bearable. In the second case the overhead

Order of matrix	147	199	292	822
Number of non-zeros	2449	701	2208	4841
BLOCK	70	30	100	250
ANALYZE-FACTOR	1470	210	960	480
FACTOR	280	50	210	180
SOLVE	20	10	23	40

Table 3.3   A comparison of some timings (in milliseconds on an IBM 370/168)

happens to be exactly recovered in faster execution of ANALYZE-FACTOR. The last case has a very substantial block structure and without its recognition the ANALYZE-FACTOR time rises to 1520 m.secs.

3.3 User interface

MA18 was much criticized for its awkward user interface particularly for the FACTOR entry and we were keen for MA28 to be as convenient as possible to use. We chose therefore to permit input of non-zeros in any order, with each having its row and column number stored in parallel integer arrays. We imagined that the sorting overhead would be slight but in fact have been quite disappointed with results, a representative sample of which is shown in Table 3.4. While we are happy with the ANALYZE-FACTOR interface where the overheads are reasonably slight and well worthwhile for the added convenience, we are less pleased with the

Order of matrix	147	199	130	822
Number of non-zeros	2441	701	1282	4841
ANALYZE-FACTOR interface	74	30	36	153
ANALYZE-FACTOR itself	1443	223	277	1600
FACTOR interface	80	30	30	120
FACTOR itself	210	30	30	60

Table 3.4   Time taken in MA28 interface (370/168 m.secs)

overhead on FACTOR. It does, however, permit great generality, including the input of a matrix whose pattern is different from that of the original but included in the fill-in pattern, and is very economical in storage (for instance space for the original matrix as well as the factors is not needed). It is our intention eventually to provide a less general, but faster, interface. Meanwhile we have provided the user with the option of avoiding the interface if he is prepared to sort his non-zeros by rows in final pivotal order.

3.4 Choice of the pivot threshold parameter u

To reduce numerical instability we avoid pivots whose modulus is less than a user-set parameter u times the largest element in the pivot row. We ran some numerical tests in order to be able to give some user guidance on suitable values for u. Some results are shown in Table 3.5.   It is evident that the stability is

| Matrix order | 147 | | 822 | | 541 | |
| Non-zeros in A | 2441 | | 4790 | | 4285 | |
u	Non-zeros in factors	Error norm	Non-zeros in factors	Error norm	Non-zeros in factors	Error norm
1(-10)	4881	1(2)	6474	1(-8)	18703	3(23)
1(-4)	5028	3(-9)	6474	1(-8)	16259	1(2)
1(-2)	5867	4(-10)	6495	2(-10)	15666	1(-5)
0.1	5095	2(-12)	6653	1(-11)	13623	4(-9)
0.25	6449	3(-12)	6910	4(-12)	13417	6(-10)
0.5	6381	2(-12)	7231	1(-12)	14214	7(-10)
1.0	6772	2(-12)	8716	6(-12)	16986	3(-9)

Table 3.5   Effect of varying the relative pivot threshold u

roughly montonic with u but the number of fill-ins is not necessarily so.   We chose to recommend u=0.1 as a figure likely to be satisfactory both from a fill-in and stability point of view.

3.5 Monitoring the stability of the FACTOR entry

No numerical control is left in the FACTOR entry, since this uses the pivot sequence previously determined by ANALYZE-FACTOR.   Instability will manifest itself in the form of large values for the non-zeros manipulated during the factorization.   To check all these numbers involve. an overhead on the innermost loop of the code, and to avoid this Erisman and Reid (1974) proposed an a posteriori bound based on applying Hölder's inequality to the relation

$$a_{ij}^{(k)} = a_{ij} - \sum_{m=1}^{k} \ell_{im} u_{mj}, \qquad k<i,\ j\leq n \qquad (3.1)$$

between the elements of A and its triangular factors L and U.

To assess whether this procedure gives a worthwhile saving over inner-loop monitoring, we used our test bed and results are illustrated in Table 3.6.   This shows that there is indeed a worthwhile saving.   The only penalty that the user must pay is that sometimes the bound can lead him to believe that the instability is worse than it really is.

Order of matrix	147	199	292
Number of non-zeros	2449	701	2208
FACTOR	210	30	140
FACTOR, plus calculation of growth in inner loop	290	40	180
A posteriori bound	13	3	13

Table 3.6   Times (IBM 370/168 m.secs) for stability
                    monitoring

4. COMPARISON OF CODES FOR UNSYMMETRIC SETS OF LINEAR EQUATIONS

We now discuss some of our experience in using our test bed to compare codes for the direct solution of sparse unsymmetric sets of linear equations. These experiences are described in more detail by Duff (1979). We concentrate on the effects of features quite different from those in MA28.

4.1 Compiled and interpretative codes

For extremely fast FACTOR and SOLVE execution times, Gustavson et al (1970) proposed the automatic generation of loop-free code specifically tailored to the problem in hand.   We have experimented with Gustavson's code GNSOIN which generates a sequence of Fortran statements such as

$$A(1001) = A(1001) - A(146) * A(172) - A(139) * A(710) \qquad (4.1)$$

and requires a Fortran compiler to produce executable code.

The intervention of a compiler may be avoided by generating machine code directly, but then portability is lost. An alternative which retains portability is to generate lists of integers which a later code interprets; for instance (4.1) might be stored as 7 integers consisting of the five subscripts themselves , a code for this particular operation and the number of multiplications involved. Such a code is TRGB, written by Bending and Hutchison (1974).

Representative results of the execution of MA28, GNSOIN and TRGB are shown in Table 4.1.   There is no question that the speed of execution of the GNSOIN compiled code for FACTOR and SOLVE is very impressive.   It is hard to see how it could be significantly improved on a scalar machine such as the IBM 370/168. However it does make very substantial storage demands and most of its speed advantage would be lost if its code could not be held in main memory.   The operator list approach of TRGB does not give this very fast speed and indeed it appears to be broadly comparable with the looping indexed approach of MA28.   Its speed would be much nearer to that of compiled code if the basic operations were more significant, for instance for complex matrices.   Note that both the compiled code and operator list approaches permit variability typing (Hachtel, 1972) and advantage to be taken of special values such as unity.

Matrix order No. of non-zeros in A		147 2441	199 701	822 4790	541 4285
No. of non-zeros in factors	GNSOIN TRGB MA28	5062 5073 5095	1421 1483 1478	6491 7712 4790	14734 16771 13660
Storage in bytes	GNSOIN code TRGB operator list MA28 overhead	706289* 469704 19954	72550 40800 5760	296461 198816 32466	1816708* 1709408 44460
ANALYZE time (370/168 m.secs.)	GNSOIN TRGB MA28	10753+ 3148 1261	1505+ 466 224	8542+ 5951 1532	29101+ 19855 4043
FACTOR time (370/168 m.secs.)	GNSOIN TRGB MA28	93* 292 271	6.4 25 51	21.3 123 272	235* 1079 578
SOLVE time (370/168 m.secs.)	GNSOIN TRGB MA28	9.3* - 16.6	2.7 - 8.1	12.2 - 35.0	27* - 47

*Estimates   +Excluding compilation
Table 4.1   Performance of compiled and interpretative codes

4.2 Full code

Sparse code is much more complicated than full code and the question of just how much is gained is often asked. A test bed can give qualitative answers to this question and some specimen figures are shown in Table 4.2. MA21 is a Harwell code that performs Crout reduction and F01BTF/F04AYF are routines developed by Reid for NAG and use a blocked column approach for efficiency in a paged environment. It is clear that in all the cases shown there is a substantial gain from the exploitation of sparsity, particularly for FACTOR. Incidently only the last case was large enough for page thrashing to be a danger on our machine, and it showed the superiority of the block column approach (31065 page movements as against 240920).

Matrix order No. of non-zeros		147 2441	199 701	130 1282	541 4285
ANALYZE-FACTOR time (370/168 m.secs.)	MA28 MA21 F01BTF	1261 2440 2050	224 6547 5033	285 1690 1513	12537 160740 100473
FACTOR time (370/168 m.secs.)	MA28 MA21 F01BTF	271 2440 2050	51 6547 5033	70 1690 1513	503 160740 100473
SOLVE time (370/168 m.secs.)	MA28 MA21 F04AYF	17 50 43	8 100 77	6 37 40	44 850 930

Table 4.2 Comparison with full codes

4.3 Switching to full code

It is clear that sparse codes are only worthwhile for "sufficiently sparse" matrices and it is logical to argue that such a criterion should be applied to the reduced matrices obtained in the course of sparse elimination. This involves switching to full code once some threshold is passed. Such a switch is included in the IBM program-product SLMATH and some specimen results are shown in Table 4.3. Setting the threshold to 101% ensures that we never switch to full code and it is interesting that there can be a worthwhile gain from switching at 100% (i.e. with no loss of sparsity at all). It is interesting, too, that we usually continue to gain speed until the threshold is less than 40%. The FACTOR times are virtually unchanged down to 60% and the SOLVE times are virtually unchanged down to 40%.

Order No. of non-zeros		147 1441	199 701	822 4790	541 4285
Percentage density	101 100 80 60 40 20	2990 2611 2342 2171 2089 2862	441 413 400 388 379 376	2976 2926 2932 2944 2947 2993	15428 15155 14182 13357 12342 16049

Table 4.3 ANALYZE-FACTOR time (370/168 m.secs.) for SLMATH with varying threshold for switch to full code

5. CODES FOR SOLVING SYMMETRIC SETS OF LINEAR EQUATIONS

There have been some rather exciting developments lately in connection with the solution of sparse symmetric sets of linear equations, particularly those arising from finite-element and finite-difference calculations.   In the finite-element case a symbolic analysis may be performed by working entirely with index lists of variables associated with elements and successive amalgamations into bigger and bigger "super-elements". With each amalgamation is associated the elimination of any variables internal to the resulting super-element (in a generalized form of static condensation). Attractive features of this approach are that enormously much less storage is needed for each index list than for the associated real array, much less work is done in manipulating them than in doing the corresponding real operations and no additional storage is needed for the merged lists than for the original ones. An essentially equivalent approach is being used by George and Liu (1979).

We have developed a code to implement the pivotal strategy of minimum degree in this way. An interesting effect on its development of the presence of our test bed was in connection with the computation of the degrees themselves.   The degree of a variable v is the total number of variables in all the elements $e_1, e_2, \ldots$ to which v belongs.   If another variable w belongs to just two of these elements, say e_1 and e_2, then v certainly has degree at least as large as that of w and therefore need not be taken as a minimum degree candidate. We built such a test into our code in the hope of reducing the effort involved in degree calculation.   However we found in practice that a far more effective way of reducing this effort was to terminate the accumulation of the degree of a variable whenever greater than a current threshold, resetting the threshold and recalculating degreeswhenever that threshold is passed.

Another code that we have developed is designed for very large problems that require auxiliary storage. Rather than use minimum degree ordering we have been choosing to assemble together a pair of super-elements that give least fill-in to our overall real storage, given that zeros within each stiffness matrix are held explicitly. Its ANALYZE phase works just with the index lists and so makes much more modest use of storage and computer time than does a conventional code.   Performance figures for the CEGB test cases (Table 2.3) are shown in Table 5.1.   Notice in particular how small the ANALYZE times are in comparison with the FACTOR times, particularly in the first and fourth cases where the original elements are rather large.   This is in stark contrast to our experience with conventional sparse codes where ANALYZE usually takes several times longer than FACTOR.

No. of variables		2919	3024	3306	2802
No. of elements		128	551	791	108
Maximum element size		60	16	12	60
Non-zeros in factor U		356925	115502	100377	236085
Millions of mults for FACTOR		28.1	3.0	4.3	11.5
Time	ANALYZE	1.7	4.2	4.6	1.0
(370/168	FACTOR	70.7	13.1	25.7	32.3
secs.)	SOLVE	3.1	1.2	1.3	2.1

Table 5.1   Performance of out-of-core code on CEGB examples

We have also run our newer minimum degree code on the same examples.   In cases 1,2 and 4 the ANALYZE time reduced to between half and a third of those in Table 5.1, with the numbers of multiplications and non-zeros increasing slightly. In case 3 the ANALYZE time reduced by a factor of 10, the number of multiplications reduced to 1.1 millions and the number of non-zeros reduced to 68589. This single dramatic result leads us to the conclusion that we should replace our pivotal strategy in the out-of-core package if we want that to be a good general-purpose code.   It illustrates the importance of test beds containing a sufficiently wide variety of problems.

Another interesting development in connection with symmetric problems is the use of incomplete factorizations to accelerate the method of conjugate gradients. Munksgaard (1979) has written a code, shortly to be introduced in the Harwell library as MA31, which allows the user control all the way from permitting no fill-in through dropping fill-ins below a relative tolerance to a direct method with conjugate gradients acting as iterative refinement. We have checked its performance by running it on our symmetric test examples, supplemented by some problems arising from the solution of partial differential equations because we believe that this will be a very important application area for the code. Comparisons were run against simple conjugate gradients (scaled so that the diagonal

Order		1176	147	292	1024	1000	1054	1009
No. of non-zeros		8688	1147	958	1984	2700	5943	2928
Conjugate gradients		2.17	0.70	0.36	1.94	0.94	> 23	5.00
Preconditioned	No fill	3.53	0.95	0.32	1.52	1.23	7.73	2.45
conjugate	Some fill	2.47	0.64	0.30	1.33	1.71	7.10	2.48
gradients	All fill	2.02	0.40	0.39	2.45	11.66	14.05	4.52
Yale code		2.34	0.31	0.29	2.01	6.67	9.66	3.18

Table 5.2   Time (370/168 secs.) for ONE-OFF solutions

has elements all equal to unity) and the Yale sparse matrix package, which has a good reputation for symmetric positive-definite problems.   These results (see Table 5.2) demonstrate that the algorithm is reasonably effective as a direct code and as an iterative code can be much better than simple conjugate gradients.

6.   LINEAR PROGRAMMING

The value of a test bed for linear programming has already been established. For instance Wolfe and Cutler (1963) made extensive use of a test bed to

    a) compare the effectiveness of different starting bases
    b) compare different infeasibility objective functions
    c) compare different pivotal strategies
    d) assess the effect of sub-optimization
    e) compare the standard tableau with holding the inverse basis explicitly
        and in product form
    f) compare the overall performance of different versions,

and Saunders (1972) used his test bed of four problems to

    a) produce solution statistics for his Cholesky method
    b) compare his Cholesky method with the usual product form of the inverse
    c) monitor the growth of non-zeros in the two forms.

We will here content ourselves with a description of a little of our own experience.

6.1   The stability of the Forrest-Tomlin algorithm

The product form proposed by Forrest and Tomlin (1972) has been widely used but has been criticized by numerical analysts because no guarantee of its stability is

available.   If instability is a very rare occurrence then we can afford to re-
factorize the current basis whenever it arises. The results of running the
Forrest-Tomlin algorithm and code of our own which implements the Bartels-Golub
algorithm (Bartels, 1971) are shown in Table 6.1. Instability manifests itself
in the appearance of large matrix elements and sure enough larger elements do
appear. To check the effect we looked at the in-basis reduced costs, which would
be zero without roundoff and the largest is shown in Table 6.1. Our conclusion
is that instability of the Forrest-Tomlin algorithm is not a serious worry.

		Shell	GUB	Stair	BP
Maximum matrix	B/G	1	20	3(4)	4(2)
element	F/T	1	1(3)	3(7)	5(4)
Maximum in-basis	B/G	<1(-15)	3(-10)	6(-8)	1(-7)
reduced cost	F/T	<1(-15)	1(-9)	6(-6)	4(-8)

Table 6.1   Comparison of the stability of the Forrest-Tomlin
algorithm with a sparse variant of the Bartels-Golub
algorithm

6.2   Implementation of sparse Bartels-Golub factorization

Reid (1976) has implemented code for a sparse variant of the Bartels-Golub
factorization.   It holds the basis in a representation consisting of a product of
elementary matrices and a permutation of a triangular matrix.   On a change of
basis one column of the triangular matrix changes and the form is usually restored
by multiplying by further elementary matrices.   However our code will restore the
form by permutations alone if this is possible. Of course we were worried about
the cost of finding such permutations and so designed a data structure that would
permit them to be found easily.   A real surprise came when running our tests and
discovering that the code for finding the permutation was so successful that by
far the most heavily executed loop in practice was the very simple loop that takes
the incoming column, held as a full vector, and runs through it to pick out its
non-zeros.

This algorithm has advantages over the Forrest-Tomlin algorithm both for
stability and fill-in. To assess just how great these advantages are in practice
requires test runs.   It was particularly successful on the "Shell" problem,
where no eliminations at all were ever performed because all bases could be
permuted to triangular form.   However the Forrest-Tomlin algorithm also
performed well here.   The overall comparison is summarized in Table 6.2 and we
have to admit that the advantages of our approach are not great.

		Powell	Shell	GUB	Stair	BP
Average interval	B/G	111	∞	120	46	64
between factorizations	F/T	45	∞	79	51	64
Total run time	B/G	17.7	15.5	174	46.9	289
(IBM 370/168 secs.)	F/T	19.3	17.2	202	45.8	297

Table 6.2   Comparison of sparse Bartels-Golub code with its
Forrest-Tomlin variant

6.3 Steepest-edge simplex algorithm

In principle any column whose reduced cost is negative may be pivoted into the
basis at an iteration and the standard algorithm (see, for example, Dantzig, 1963)
takes that with most negative reduced cost.   This means that we proceed
along an edge of the polytope of feasible solutions that points downhill, but not
necessarily choosing the edge that has steepest descent in the space of all the
variables. Choosing the steepest edge was found by Kuhn and Quandt (1963) (who
call this the "all-variable gradient" method) to give good performance with respect
to numbers of iterations in their tests.   Various authors, notably Harris (1973),
have essentially approximated such a choice.   Goldfarb (see Goldfarb and
Reid, 1977), however, showed that it is possible to use recurrences to implement
the steepest-edge algorithm at an extra cost in storage and operations that is not
prohibitive.   To judge the overall success of such an idea again demands
practical tests.   We therefore ran tests on our code and modifications that
implemented the original Dantzig algorithm and Harris' algorithm, each written with
efficiency in mind so that realistic timings could be made. Some of our results
are summarized in Table 6.3.

		Blend	Powell	Stair	Shell	GUB	BP
Number of rows, m		74	548	362	662	929	821
Number of cols, n		114	1076	544	1653	3333	1876
Number of non-zeros		560	7131	4013	5033	14158	11719
Phase		2	1	1	1	1	2
	Steepest edge	48	226	210	48	475	1182
Iterations	Harris	65	198	313	77	672	1819
	Dantzig	70	285	244	78	854	>3976
Time (secs.)	Steepest edge Set up run	0.02	1.1	0.4	2.4	7.5	-
		0.90	16.6	26	2.9	57	263
	Harris	1.04	11.7	33	3.8	78	305
	Dantzig	1.06	16.9	26	3.8	96	>596
Time per iteration	Steepest edge	.019	.073	.12	.060	.12	.22
	Harris	.016	.059	.11	.049	.12	.17
	Dantzig	.015	.059	.11	.049	.11	.15
Array storage (k-bytes)	Steepest edge	20	135	190	126	268	297
	Harris	20	135	190	126	268	289
	Dantzig	19	127	185	114	248	275

Table 6.3   Comparison of different simplex column pivoting strategies
on IBM 370/168

These results certainly demonstrate the practicality of the steepest-edge algorithm
and show that it can show significant advantages over the original Dantzig
algorithm but it does not show any dramatic improvement over Harris' algorithm
although it might be preferred on the grounds of elegance.

References

Bartels, R.H. (1971). A stabilization of the simplex method. Num. Math. 16, 414-434.

Bending, M.J. and Hutchison, H.P. (1974). TRGB routines. Report from Dept. of Chemical Engineering, Cambridge, U.K.

Dantzig, G.B. (1963). Linear programming and extensions. Princeton University Press.

Duff, I.S. (1977). MA28 - a set of FORTRAN subroutines for sparse unsymmetric linear equations. Harwell report AERE-R.8730, HMSO.

Duff, I.S. (1979). Practical comparisons of codes for the solution of sparse linear systems. In Duff and Stewart (1979).

Duff, I.S. and Stewart, G.W. (Eds.)(1979). Proceedings of sparse matrix symposium at Knoxville, Tennessee, Nov. 2-3, 1978. SIAM publications.

Erisman, A.M. and Reid, J.K. (1974). Monitoring the stability of the triangular factorization of a sparse matrix. Num. Math. 22, 183-186.

Forrest, J.J.H. and Tomlin, J.A. (1972). Updating triangular factors of the basis to maintain sparsity in the product form simplex method. Mathematical Programming, 2, 263-278.

George, J.A. and Liu, J.W.H. (1979). A quotient graph model for symmetric factorization. In Duff and Stewart (1979).

Goldfarb, D. and Reid, J.K. (1977). A practicable steepest-edge simplex algorithm. Mathematical Programming 12, 361-371.

Gustavson, F.G., Liniger, W.M. and Willoughby, R.A. (1970). Symbolic generation of an optimal Crout algorithm for sparse systems of linear equations. J. ACM. 17, 87-109.

Hachtel, G.D. (1972). Vector and matrix variability type in sparse matrix algorithms. In Rose and Willoughby (1972).

Harris, P.M.J. (1973). Pivot selection methods in the Devex LP code. Mathematical Programming 5, 1-28.

Irons, B.M. (1970). A frontal solution program for finite element analysis. Int. J. Num. Meth. Engrg. 2, 5-32.

Kuhn, H.W. and Quandt, R.E. (1963). An experimental study of the simplex method. Proc. of Symposium in Applied Maths. Vol.XV. Ed. Metropolis et al, A.M.S.

Munksgaard, N. (1979). Solving sparse symmetric sets of linear equations by preconditioned conjugate gradients. To appear as Harwell report.

Reid, J.K. (Ed.)(1971). Large sparse sets of linear equations. Academic Press, London.

Reid, J.K. (1976). Fortran subroutines for handling linear programming bases. Harwell report AERE-R.8269, HMSO.

Rose, D.J. and Willoughby, R.A. (Eds.)(1972). Sparse matrices and their applications. Plenum Press.

Saunders, M.A. (1972). Large-scale linear programming using the Cholesky factorization. Stanford report STAN-CS-72-252.

Willoughby, R.A. (1971). Sparse matrix algorithms and their relation to problem classes and computer architecture. In Reid (1971), 255-277.

Wolfe, P. and Cutler, L. (1963). Experiments in linear programming. In "Recent advances in mathematical programming". Ed. Graves and Wolfe, McGraw-Hill.

Performance Evaluation of Numerical Software, Fosdick (ed.)
© IFIP, North-Holland Publishing Company, 1979

DISCUSSION OF SESSION ON PERFORMANCE
EVALUATION IN LINEAR ALGEBRA

Chairman J.H. Wilkinson
Discussants T.J. Dekker and W. Hoffmann

Discussants remarks

Prof. I. Molchanov presented a survey of error analysis for the problem of solving
systems of linear equations and for the eigenvalue problem, and showed some
computational results.

Dr. B.T. Smith reported on the use of automated theorem provers to evaluate linear
algebra software. As an example, he considered the implementation in LINPACK of
Cholesky decomposition with main-diagonal pivoting.

Dr. J.K. Reid presented a paper by him and Dr. I.S. Duff on performance evaluation
of codes for sparse matrix problems. He emphasized that a full evaluation demands
the running of a realistic set of test problems, and gave an impression of the
collections of test problems with various sparseness structures assembled by the
authors.

DISCUSSION

Question (J. Rice)

Automated program verifiers usually require code to be in a nice, structured form.
Maximum efficiency and/or robustness of numerical software usually require very
specialized and sometimes obscure, even incredible, code. How does your transformational approach resolve this conflict?

Answer (B.T. Smith)

It is just the transformational approach that can resolve this conflict, for it
permits us to design a process for preparing software which is a sequence of
elementary steps, each simple enough to analyse in a straight forward fashion.
Let us take for example the issue of software robustness. One of the elementary
steps is the specification of the algorithm in a nice structured form. Part of
that specification may be a non-obvious reformulation of straightforward code permitting some computation to be performed in a safe manner. It is just such re-
formulations that are prone to error; however, because of the nice structured form
for the algorithmic specification, we can analyse the code for either equivalence
to the straightforward formulation in the presence of no rounding errors, or more
preferably, verify the safeness of the computation in the presence of rounding
errors.
As for efficiency, the transformational approach seems most appropriate, particularly if the efficiency of the basic algorithm is distinguished from the efficiency of the code that implements it. Given that the basic algorithm attains
the desired level of efficiency, the mapping from abstract form to concrete form
can be considered as an optimization process, which preserves the meaning of the
code while utilizing computationally efficient data structures and eliminating
unnecessary computation. Each optimization transformation is then designed to be
general but simple enough that automated analyzers can verify the correctness of
the transformation. The transformations are designed to be general enough that
they can be applied to a wide collection of algorithms in some discipline, in this
case, computational linear algebra software.

Question (T.E. Hull)

I would like you to clarify a point: is it your intention (or belief, at least)

that your techniques should be able to handle realistic situations involving
roundoff aspects as well as the infinite-precision aspects of your algorithms?

Answer (B.T. Smith)

It is my goal to develop a useful tool for validating numerical software, including the treatment of realistic situations involving rounding errors, from automated theorem provers. A few experiments with current automated provers make it
manifest that in their current forms, such provers are not suited to handling the
class of problems that arise from the verification of numeric programs, and thus
we are all sceptical of the usefulness of such devices for the problem at hand.
However, I am hopeful that investigations into what is needed in the proofs about
numeric programs and into how automated provers can be tuned to the problems of
interest will provide the necessary understanding to develop useful tools for
evaluating numeric software.

Remark (W. Kahan)

Except for the simplest programs, there is no way to "prove" that a program is
"correct" in any sense that has practical value. This is so because the program
does not exist in isolation, but rather in a context (machine and system limitations) and as a component (together with documentation that specifies what the
program does) of an environment that gives the lie to many certainties. A better
way to view program validation is to entertain a collection of auxiliary computations intended to corroborate (but not prove) a program's correctness. Batteries
of tests provide one kind of corroboration, its quality depending upon the ingenuity and insight that went into the selection of test data and scrutiny of
results. Running those tests is one kind of auxiliary calculation by way of corroboration. Another kind is an error-analysis, be it a priori or a posteriori.
The error-analysis (of the program, not just the algorithm) is an essential
component of program validation; but the analysis could be wrong for various
reasons (e.g. hardware caprice undreamt of by the analyst's philosophy), so it
too is mere corroboration, not proof in the legal sense. Another corroborative
calculation establishes that the program used on a machine is in fact obtainable
from some simpler and more transparent statement of intent. Compilers, if you
trust their correctness, perform part of that calculation. Theorem-provers supply
yet another corroborative calculation; they combine statements of intent (programs)
with statements of belief (assertions or predicates) in an attempt to reveal
those inconsistencies vulnerable to routine exposure.

   The value of a corroborative calculation depends not only upon the skill with
which it is carried out but also upon the extent to which it covers risks not already covered by previous testimony. Automated theorem provers attack the risk
that some trivial slip in algebra may have invalidated a program or its error-
analysis. Since that risk is appreciable, so is the value of the corroboration.
We need not be disappointed that, despite their use, the program is not yet proved
correct.

Reply (B.T. Smith)

I agree entirely with this remark. Although the theorem-prover seems only to discover trivial algebraic errors, the point is that such trivial errors are often
masked or hidden in the structure of a program, particularly, in the case where
several segments of the program form paths to be executed that correspond to special cases. It is precisely these special cases that contain the errors _and_ are
difficult to uncover by other means of performance evaluation.

Question (C. Reinsch)

It seems to be natural to distinguish between the numerical algorithm on the one
hand and its realization as a piece of code in a chosen programming language on
the other hand. One would, therefore, expect an automatic program verifier to deal
with the question of whether a given code, run on a given machine, would corres-

pond to the execution of that algorithm including the influence of rounding errors. One would, however, _not_ expect in general a program verifier to re-establish known facts about the underlying algorithm. For example, it is well known that certain submatrices stay positive definite in the Cholesky decomposition (with proper assumptions even under the influence of rounding errors). Why must this fact be re-proven in order to show the correctness of a certain program?

Answer (B.T. Smith)

We are using known behavior of the algorithm, that must necessarily be reflected in the software, to evaluate the software. This process of evaluating the software provides corroboration that the software is performing as expected under certain models. In the model of exact arithmetic, we know in the Cholesky factorization subprogram that the algorithm must yield factors that when reformed, equal the original matrix. Hence, we use that as a criterion for the evaluation: if this criterion is not satisfied, the software is incorrect, even in the presence of rounding errors in this case. Further, if it is possible to show that in the presence of rounding errrors the usual backward error criterion is satisfied, we have corroborated the performance of the software in another way.

Comment (J. Boyle)

The last question can be answered in part by clarifying the situation described in the talks. Usually, when we think of implementing a program, we distinguish between two levels, viz. the level of mathematical algorithm and the level of executable program. The distance between these two levels is relatively great, and proving that the executable program implements the mathematical algorithm is quite difficult. The proof can be simplified by introducing a third level between these two, viz. the level of abstract program and carrying out the verification in two parts. The present talk deals with the first of these proofs. Part of the confusion arises because in linear algebra, the abstract program is very close to the mathematical algorithm; thus if one assumes the result of the error-analysis, there is very little to prove. In this case it may be interesting to try to confirm some of the theorems proved in the error-analysis by using an automated theorem prover. In addition, it may be easier to formulate the specification of the abstract program in terms of first principles than in terms of the mathematical algorithm. It now appears that the second part of the proof can be carried out by showing that the executable program is derived from the abstract program by a sequence of correctness-preserving program transformations. Of course, such transformations can be applied to any correct abstract program, so that when the abstract program is very close to the mathematical algorithm we might accept its correctness based on the error analysis and complete only the transformational part of the proof.

Question (J. Boyle)

It is well-known that many difficult problems can be solved by breaking them into smaller, hopefully easier to solve, problems. It would therefore be nice if a kind of transitivity existed whereby, if an algorithm had been proved numerically stable, and a program had been proved to correctly implement that algorithm in exact arithmetic, one could conclude the program were numerically correct. To what extent does anyone think this is possible?

Reply (B. Meyer)

In the Floyd-Hoare-Dijkstra system for program-proving, the axioms are predicate transformers, which connect (by symbolic transformation) the predicates which are true after and before execution of a given program statement. The predicates themselves are expressed in a logical system for which there are several possible interpretations and associated proof rules. One may for example correspond to the exact number system, while another is more complex and involves rounding errors. Obviously, proofs will be easier in the first case, and one will be able to prove stronger properties; as far as use of program proving techniques is concerned, however, the basic mechanism is the same. If you had inexact arithmetic in mind when writing the initial "exact" proof, then it should be easier to modify it into a proof which holds in the inexact arithmetic case.

Question (A.H. Sherman)

Perhaps I'm missing the point, but it seems to me that we are talking about whether a program correctly implements (with exact arithmetic) a fully-specified mathematical algorithm, not whether a program solves a particular mathematical problem. Regardless of the problem area, this seems to be a well-posed question, even if we don't know that the mathematical algorithm itself is either correct or appropriate.

Answer (B.T. Smith)

A fully-specified mathematical algorithm has certain properties, which must be preserved by the program that implements it. In the case of Cholesky decomposition, one property of the usual algorithm is that it solves a certain matrix factorization problem. Thus, verification that the program solves that factorization problem provides evidence that the program is at least consistent with the mathematical algorithm. Since we have not verified that the characterizing properties of the algorithm are preserved by the program, and that the program does no more than the algorithm, we have not shown that the program correctly implements the algorithm. In that sense, we are not verifying a mathematical algorithm but rather verifying that a program meets some specification, thereby obtaining evidence that it performs as intended. Such evidence of intended performance seems satisfactory in areas such as matrix factorization because the verifiable properties are close to the intended purpose of the software. In other cases, such as quadrature, verifiable properties such as correct performance when the integrand is a polynomial of limited degree are substantially different from the usual use of integrating functions with singularities. Indeed, much code is incorporated in such programs to adapt to unusual behavior of the integrand. Here, we need ways of characterizing algorithms so that more useful properties can be asserted about the program. Thus, I agree that the correct implementation of a fully-specified algorithm is the problem that needs to be addressed directly. However, the task of formulating characterizing properties of algorithms seems most difficult at this time.

Question (R.B. Schnabel)

If one wants to verify the algorithm (not just the coding of that algorithm) for what Dr. Lyness has called class II (iterative) algorithms, it seems to be far more difficult to determine what the assertion about the algorithm should be. This is especially so if we want an assertion saying what the algorithm will do on all problems, as opposed to the fact that it will work on a certain very restricted subclass of problems.

Answer (B.T. Smith)

I agree.

Question (T.J. Dekker)

Why did you implement Cholesky with pivoting? In my opinion, pivoting improves numerical accuracy for positive definite symmetric matrices only marginally and has no obvious advantages.

Answer (B.T. Smith)

In demonstrating the proposed approach to the evaluation of software, I chose the SCHDC subprogram from LINPACK which uses diagonal pivoting. One reason diagonal pivoting is implemented in SCHDC is that in practice, the ratio of the largest to the smallest diagonal element (that is, the first and last) is a good estimate of the condition number of the matrix. Also, in practice, the zero or nearly zero diagonal elements indicate the dependent columns of the symmetric matrix. This property has the additional advantage that when this algorithm is used to factor projection matrices of rank-deficient projections, the columns of the original matrix corresponding to the zero or nearly zero diagonal elements form an orthogonal basis for the null space.

Question (A.H. Sherman)

There are several codes for solving sparse linear systems, and all were designed based on different assumptions about "typical" problems. It seems unreasonable to compare these codes based solely on the problem set used in the design of one of them. Wouldn't it be fairer and more indicative of relative quality to include problems which were used in the designs of each of the codes tested?

Answer (J.K. Reid)

I agree. Certainly we would welcome any extra contribution to our set of tests.

Remark (A.H. Sherman)

In regard to discussion on using small numbers of examples to justify inclusion of features in software (Dr. Reid's talk): It seems that some people missed the point that MA28 includes a number of features that may offer significant benefits in some cases, but that can be "turned off" (excluded) by simple operations like setting a switch. The hard part is adding such features when they are needed, and Duff and Reid have already done that. As long as there is an easy way to exclude such features when they appear detrimental, there is no loss in having them, even on skimpy evidence of benefit.

Reply (J.K. Reid)

Thank you for your comments supporting the design of MA28. In fact, the examples I quoted are just a selection of those actually ran and represent a cross-section of results obtained.

SESSION 4 : RELIABILITY AND WARRANTY
OF NUMERICAL SOFTWARE

Performance Evaluation of Numerical Software, Fosdick (ed.)
© IFIP, North-Holland Publishing Company, 1979

LEGAL ASPECTS
OF NUMERICAL SOFTWARE*

by

Bryan Niblett
Department of Computer Science
University College of Swansea
Swansea, United Kingdom

Various legal topics associated with the development and
marketing of numerical software are discussed. These
include the nature of property rights in numerical software, the problems of liability, and the related question
of software certification.

This paper considers some specific legal issues which arise from the development
and marketing of numerical software. We define the phrase 'numerical software' to
include not only the computer programs which implement numerical algorithms but
also the whole range of documentation which goes (or which should go) with the
program: such items as the detailed specifications, the descriptive materials, the
flow charts, and the instruction manuals. The particular legal matters we discuss
are the nature of property rights in numerical software, problems of liability,
and the associated topic of software certification. Computer programs are often
designed to be portable so that they may readily be transported from one computing
environment to another. In contrast, laws are not portable but are characteristic
of each country's legal system. This paper is concerned primarily with English
law though in practice much of what it says will have relevance in other
jurisdictions.

INTELLECTUAL PROPERTY

The nature and extent of property rights in computer programs is of growing
commercial interest because a healthy software market requires that the vendor of
a computer package be able to offer a purchaser some form of proprietary interest.
We use the terms 'vendor' and 'purchaser' but it is advisable for the developer of
a software product to retain ownership of his intellectual property and license
its use by others. So the preferred practice these days is for a software contract to consist of a grant by the owner (the licensor) to the user (the licensee)
of rights he would not otherwise have, in particular the right to copy the program for use on a designated computer.

In recent years there has been much discussion as to whether computer programs,
or more precisely an invention forming part of a program, can be patented. A
patent is a monopoly and gives its owner the exclusive right to use the invention
within the jurisdiction for a limited term of years. Such a monopoly is of course
a valuable form of personal property. The question whether program inventions can
be patented has had an uncertain answer in most countries for many years. It is
well established that a mathematical formula or an algorithm cannot be patented
per se. But can it be granted patent protection when embodied in a computer
program which is used to drive a machine? In the United Kingdom the Patents Act
1977 (which came into force on the 1st June 1978) specifically provides that a

---

*Supported in part by the Research Office of the US Army.

program for a computer is not an invention for the purposes of the Act and a
similar declaration is included in the Convention on the Grant of European Patents
1973 so it might be thought that these provisions put an end to the discussion.
In practice this is far from being so, because the Act and the Convention also
provide that a program for a computer is excluded from patentability only to the
extent that the patent application relates to the program as such. It may well be
that a patent claim directed to apparatus modified in a novel way by a computer
program, and more particularly to a programmed read-only microelectronic chip of
novel design, can be patented. The new Act may therefore continue to provide a
means of protection so that, in the United Kingdom at least, the controversial
question of patent protection for programs is by no means settled yet.

One form of protection of computer programs that is being increasingly used is the
law of confidential information, or what in the USA is more often called the law
of trade secrets. Essentially there are three requirements if numerical software
is to be protected by the action for breach of confidence. Firstly, the software
must contain some information of a confidential nature which is not in the public
domain. Secondly, the information must have been communicated to the defendant
in circumstances which impose, either expressly or by implication, an obligation
of confidence. The most obvious example is where the confidential obligation is
included as a term in a contract but this is by no means the only one. The third
requirement is that the defendant make or is about to make unauthorised disclosure
of the confidential information. There are two distinct disadvantages of the
action for breach of confidence as a means of protecting numerical software. The
first and more general one is that it provides little if any protection against
third parties who have no notice of the confidential nature of the information.
The second is that it depends upon the information being maintained confidential.
This latter requirement is unattractive in the case of numerical software because
of the importance of verifying the software and encouraging a variety of independent assessors to evaluate it. Thorough testing by the scientific community may
be incompatible with the requirements of confidentiality.

The most effective method at present of asserting property rights in numerical
software is by means of the copyright laws. In contrast to patents, copyrights
do not provide a monopoly, they protect the particular form in which an idea or
concept is expressed. The normal term of protection is in most cases the life of
the author of the work and a period of fifty years thereafter. It is generally
accepted in the UK that as long as the other requirements of the 1956 Copyright
Act are met then copyright subsists in a computer program as a literary work and
also in the associated written documentation. Copyright subsists in a flowchart,
or other diagram describing the program, as an artistic work. The requirements
that have to be satisfied are that the numerical software be original, that is to
say not copied or adapted from some other work, it must be written down in some
form of notation, and it has to be made by an author who is a qualified person at
the time of making it - but this condition is easy to satisfy for the definition
of qualified person is drawn very widely.

No special action is necessary to establish copyright in a literary work in the
UK. Registration of the work is not required and copyright subsists in unpublished works as well as in those made available to the public. There is no
expense and no delay - copyright comes into existence as the computer program is
first written down - an example of on-line creation of intellectual property
rights. The owner of the copyright (who is likely to be the author of the work or
his employer) has the exclusive right to reproduce the software, to publish it,
and to adapt it, and adaptation would include translating a program from one
computer language to another. The law provides substantial remedies to those
whose copyright is infringed. An injunction may be obtained to restrain further
infringement, the plaintiff is entitled to damages for the infringement, and also
to delivery up to him of all infringing copies of the work since in law these
copies are his. These remedies are clearly powerful ones.

An aspect of program protection that is often overlooked is the measure of control
provided by trade marks. A trade mark is a symbol or device which is attached to
goods to distinguish them from similar goods or to identify them with a particular
source. In the UK at present a trade mark cannot be registered for use with a
process of numerical calculation, but can be obtained in respect of the program
itself as scientific apparatus, and for goods, such as paper print-outs, produced
by the program. It is becoming common to register names of programs as trade
marks in both these categories and such registration can be used to help regulate
the quality and use of the program.

PROBLEMS OF LIABILITY

One of the legal aspects of numerical software that suppliers are particularly
concerned about is the problem of liability. Numerical software is often used in
circumstances where unobserved and undetected errors can cause great loss.
Suppliers may innocently misrepresent the quality of their product. How can they
protect themselves from substantial claims for damages? The natural response of
a lawyer advising the vendor or licensor of a numerical software package on
problems of liability is to draft a general exemption clause, supplemented by a
limitation of liability in respect of specific warranties to a sum such as the
total consideration to be paid under the contract. No lawyer would expect to
exempt his client from liability for patent or copyright infringement since the
purchaser or licensee of numerical software, like the tenant of a house, expects
to get quiet possession of the property for which he is paying. Similarly there
can be no exemption from liability for a fundamental breach of the contract such
that the purchaser or lessee is deprived of substantially the whole of the benefit
he is entitled to expect from the contract, or from breach of a fundamental term
which goes to the heart of the agreement. But there is no reason why suppliers of
software should not do their best to exempt themselves from liability for innocent
misrepresentation, or from inadvertent breach of warranties in the contract. The
development of novel numerical software is a risky business: the present state of
the art is such that the best numerical analyst, the most able and well-trained
programmer, cannot be sure his software does not contain unintentional errors. In
these circumstances it may be perfectly proper for the commercial risk to be
borne by the purchaser rather than the vendor.

The extent to which civil liability for breach of contract or for negligence can
be avoided by contractual terms or by notice is now governed in the United
Kingdom by the Unfair Contract Terms Act 1977 which came into force on the 1st
February 1978. The provisions of this Act may serve as a helpful guide to
exclusion clauses in contracts generally, that is to say in jurisdictions other
than the UK. The Act excludes from its provisions "any contract so far as it
relates to the creation or transfer of a right or interest in any patent, trade
mark, copyright ..... or other intellectual property ..... ". Although the
creation or transfer of property rights are excluded, the better view seems to be
that other terms in a licence agreement, for example terms which exclude or limit
liability, do come within the provisions of the Act. The legislation is concerned
mainly with terms or notices which exclude or restrict liability for things done
or to be done by a person in the course of a business and for this purpose
"business" includes a profession. Thus anyone who attempts to exclude or restrict
liability otherwise than as a business liability - for example in the gratuitous
supply of software to another - is more likely to be successful than when the
transaction is clearly directed to a profit. The Act draws a further distinction
between exclusion clauses in an agreement where the plaintiff "deals as consumer"
and one in which he does not. In a consumer agreement where the numerical software is of a type ordinarily supplied for private use or consumption, then the
supplier cannot exclude liability for breach of such implied undertakings as
conformity with description and fitness for a particular purpose. We see that
suppliers of numerical software have to be particularly careful if they make software available by way of business to consumers for private use - for example as

hobby software. Similar considerations apply if the agreement is made on the defendant's written standard terms of business, that is to say if the agreement is embodied in what is commonly called a standard form contract.

The great majority of numerical software packages are of course supplied in the course of business rather than in consumer deals. In these circumstances the supplier may exclude or restrict liability but only if the terms satisfy certain tests of reasonableness. The UK Act has been described as a piece of statutory litmus paper which has to be applied to a contract term or notice to determine whether it is reasonable. What are these tests of reasonableness or, as the corresponding US law puts it, these conscionable terms? The most general test is whether the term is a fair and reasonable one to have been included having regard to the circumstances which were or ought to have been in the contemplation of the parties when the contract was made. If it was known for example that the software was experimental, of a kind never before written, this may make an exclusion clause reasonable. It is always more reasonable to restrict liability than to exclude it altogether, and in determining whether it is reasonable to restrict liability to a specified sum regard may be had to the resources of the supplier and how far he is able to cover himself by way of insurance. The relative strength of the bargaining positions of the parties may also be taken into account, whether there were alternative means of obtaining similar software, and whether the software is designed or adapted to the special order of the customer.

These tests provided by the Act serve as pointers to guide a software supplier as to how he may most effectively limit his liability. Firstly, he should continue to include in the licence agreement carefully-drafted clauses which exclude or restrict his liability for they may be held to be reasonable. In the usual type of commercial contracts (though not in consumer deals) there is no penalty for including such terms even though they may turn out to be of no effect. Secondly, the supplier should where possible avoid standard form contracts. It is prefer-able to negotiate each contract separately and to draft it as a bespoke agreement. Thirdly, not only should the <u>contract</u> be a bespoke one but the <u>software</u> too. It is safer (and often more profitable) to adapt each licensed version of the soft-ware to the special requirements of the licensee. Fourthly, the supplier should seek to cover his warranties by professional indemnity insurance. If he cannot get the necessary cover, or get it on reasonable terms, reference should be made to this in the recital to the agreement. Fifthly, the supplier should include in the agreement ample warnings as to the performance of the software, for example a reference to the limited range over which it has been tested. Sixthly, extra special care should be taken with the specifications of the software; the supplier should naturally err on the side of caution when describing its performance. Seventhly, test data should be supplied coupled with a warning that the program has been verified only with this data. Finally, it is advisable to include an arbitration clause in the agreement. The Act provides that a term in the contract incorporated or approved by an arbitrator is to be taken as satisfying the requirement of reasonableness. In the event of a dispute an arbitrator with tech-nical experience is better able to judge whether a term is reasonable than the Court.

CERTIFICATION OF SOFTWARE

Closely associated with the question of software warranties is the formal cert-ification of software - the authoritative endorsement of a package by an independent body which affirms that certain prescribed standards have been met. A useful device that could be used in this connection is the certification mark. Whereas an ordinary trade mark is used to distinguish goods of one trader from another, a certification mark is used to distinguish goods that have reached certain standards. The prescribed standards may be for example the material from which the goods are made, or their geographical origin, their method of manufac-ture - or their quality. The proprietor of a certification mark is not allowed himself to trade in the goods for which the mark is registered but is usually a

trade association or standards body which makes the mark available to traders
whose goods reach the agreed standards. Well-known examples of certification
marks include the Kite mark of the British Standards Association, the orb used on
hand-woven Harris tweed and the 'Stilton' mark applied to cheese made by the
Stilton Cheese Makers Association.

The time may come when a certification mark is appropriate for numerical software
packages to distinguish those that reach prescribed standards of reliability.
Because of the logical complexity of all but the simplest programs it seems
likely that it will be several years - perhaps a decade? - before the correctness
of a substantial numerical software package can be guaranteed. Nonetheless there
are two features of a package that could be certified now. One is the documentation provided with the software; does it reach defined standards of clarity and
comprehensiveness? The other is portability: is the package able to run on a
variety of machines to yield identical results? It may well be that licensees are
prepared to pay a little more for numerical software that carries a certification
mark which provides an independent assurance that defined standards of documentation and portability have been reached. This is one small way in which the
devices of the law may assist in improving the performance and evaluation of
numerical software.

Performance Evaluation of Numerical Software, Fosdick (ed.)
© *IFIP, North-Holland Publishing Company, 1979*

RELIABLE-WARRANTABLE CODE*

E. L. Battiste
ELBA, Inc.
Raleigh, N. C.

A review of attitudes toward reliable and warrantable code is given. Current
software warranties are not acceptable to the scientific user. For future
acceptability, a delimitation of the scope of warrantable software is necessary,
and procedures for the production of reliable code must be honed. An effort
on reliable, portable testing mechanisms should be included within the scope of
this working group. Inability to test numerical software is one major obstacle
which hinders the production of reliable, and thus warrantable, software. If the
obstacle is not overcome, then hardware technology evolution may cause another
unsatisfactory result as a "fait accompli", for basic software production and
availability.

I. Attitude evolution

This section discusses attitudes on warranty and reliability as they have
changed since 1955. Since the major issue of this paper is to motivate the
delimitation of the scope of the reliability issue, these attitudes relate to
a limited software area. Such limitation will allow the discussion of an
incremental solution to the difficult problem of software reliability. Hence,
the possibility of warranty will come forth.

   A. A current software warranty statement by a major manufacturer is
paraphrased in figure 1.

"each licensed program....designated...as warranted will conform to its
....specification....if properly used in a specified operating
environment....Program services....will be provided. There is no
warranty that the functions will meet the customer's requirements or
operate in the combination selected for use, or that the operation of
the program will be error free or that all program defects will be
corrected. This warranty replaces all expressed or implied warranties
including the implied warranties of merchantability and fitness for a
particular purpose"

FIGURE 1

This statement does include some natural components of a reasonable warranty.
"Licensed Program", "Specification", "Specified Operating Environment", and
"Program Services" are necessary ingredients. The statement falls far short of
being a reasonable definition of warranty, but in today's climate in computing,
any software manufacturer would beware of stronger statements.

---

*Partial funding was provided by National Science Foundation Grant MCS7807932,
and by the U. S. Army Research Office

Why are reasonable warranties not possible?

1. In a limited software area, that of libraries, one assumes that
warranty would apply to each kernel and to the collection. But, such an
assumption must be considered seriously. Possible combinations of kernels
abound in such number that rigid warranty might be very dangerous. And, the
complexity problem, in growing collections, makes it possible to have warrantable
collections at time t, but not at time t + 1. Schwartz (1) describes this
problem well, and his description is given in part in figure 2.

"There is certainly a side to programming, namely the invention of
algorithms meeting efficiency constraints whose satisfiability is non-
obvious, which is as much a science as is mathematics. The fast
Fourier transform is no less an invention than the Pythagorean Theorem.
But, should the other side of programming, namely its integrative side,
i.e., the growing collection of techniques used to organize large
systems of algorithms into coherently functioning wholes, be considered
as an infant science also, or must it remain an art?

I argue that this part of programming is a science also, albeit a
science only in its infancy. To see that it is, one must observe that
the crucial obstacle to the integration of systems of programs
providing very advanced function, which will generally be large systems
of programs, is met when their complexity rises above the very finite
threshold beyond which the mind can no longer grasp them totally"

FIGURE 2

2. Very little is known about software reliability. Testing a code over
its realm of applicability is almost never done. Closure of tests over that
realm is not possible at acceptable cost. Although software engineering
principles are much discussed, a large part of released programs (including those
licensed) are released directly from the pilot project results. It is possible
today to apply principles such as those developed by Henry Ford to code production,
but that is seldom done.

3. For major manufacturers, there is a natural fear that rigorous
warranty statements applied in "uncharted seas" will cause a delighted response
from the legal profession. Such activity would seriously retard software
advances, during a period when the funds acquired from good software ventures are
small (one also sees lack of rigor in non-software documents in computing, as
pointed out by Reinsch (2). One notes that the term "uncharted seas" applies to
the whole future of computer usage, so that a company in the position of being
"suitable for suits" should be careful with descriptions of facilities. In the
Reinsch example, such care is unnecessary, however).

4. Most software companies are small, and begin with limited resources.
If reasonable components of code warranty are included with their products, they
might be subject to external pressures which would cause bankruptcy. And, the
provision of warranties by small software companies may provide customers with
an invalid sense of security. Goals can change, possibly perturbed by financial
inadequacy or modification of corporate ownership.

B. A reasonable warranty

One should consider only warranty of the machine level version of a code, and of
its associated documents. For such a warrant, the components should be

1. An environment set should be specified.

2. The tests applied should be specified (and perhaps released), and results (such as root mean square error and maximum relative error) given.

3. The user should be given information on how the testing was "closed" over the code domain.

4. The document should be included in the warranty.

5. Efficiency (time, space?) specifications within ranges of arguments should be given.

6. Program services should be provided, and would act to protect the warrantor and customer.

One notes that 2. and 3. above can be provided today only in a very limited software area.

C. A limitation on scope

In considering the weakness of the above "reasonable warranty", note that feasibility of any effort in this area is important. The near future will see movement of software to firmware or hardware. As this occurs, the economic consequences of our current lack of ability to construct reliable code may exact extreme pressure. We feel that such pressure should be anticipated now, and if such anticipation is not fruitful, the state of hard or firm ware may be still unacceptable for use in 1990. Its revision may be economically more difficult. Consider the experience of Kuki and Kahan, described in (3), with the early architecture problems of the 360 series machines.

There is, however, an area for work, and properties of

1. feasibility,
2. basic, broad need and applicability, and
3. manufacturer need for use in firmware development

exist in this area. The area is the testing of kernels of mathematical-statistical computation, including elementary functions, and a subset of that area is described below. This will allow discussion of the evolution of concepts on reliability.

Consider that $f(\underline{x};\underline{p})$, where $\underline{x} \in D\underline{x}$, $\underline{p} \in D\underline{p}$, and $f \in R$, is to be approximated by a code operating in environment $\omega_i \in \Omega$, where Ω is the current set of possible (desirable) environments.

We desire $\hat{f}$, an appropriate approximation,

$\overset{\gamma}{\hat{f}}_{\omega_i}$, an approximating code for ω_i,

and perhaps $\overset{\gamma}{\hat{f}}$, an approximating code for Ω.

Let us use this limited situation for a review of what has occurred in reliability in recent years.

1. In 1955, almost everyone felt that the availability of an adequate $\hat{f}$ and a coder would provide an acceptable $f - \hat{f}$ result (and, note that a sputnik rose in 1957).

2. By 1965, the community had gained some interest in code reliability, but expanding usage caused the total set of reliable codes, relative to the number of users, to decrease (this would seem to imply that even now reliability of

basic codes is most important, and that time spent on general reliability of software may be wasteful in the short range. System proliferation is continuing, and confounds the issue).

3. In the 1960's, it was generally accepted that the quantity of interest was $f - \hat{f}_{\omega_i}$, and that $f - \hat{f}$ was not of interest.

4. By 1972, the cost of software efforts less the worth of those efforts to the marketplace was clearly large, and was decreasing quite slowly. Only then did the community begin to consider $f - \hat{f}$. Today many codes (few special functions) minimize this quantity over some Ω. This is the major change in reliability in this limited software area. Prior to 1970, almost all scientific computer users were only interested in their current environment.

5. Until recently, software manufacturers were reticent to base warranty statements, or reliable code, on other software, which could be unreliable and unsupported. Work by Bell Laboratories and the American National Standards Institute, and portability studies were necessary ingredients to the partial solution of problems inherent in considering $f - \hat{f}$ (that is, Ω). Then, the basic machine level reliability concept began to include ω_i as a combined language-computer environment.

It is still dangerous to build software which requires other basic software as an ingredient. Even elementary function codes supplied by some manufacturers must be tested and sometimes circumvented by special coding. Such testing is usually not considered in financial planning, and such special coding does not lead to portable code.

In addition to reliability, software manufacturers have other major difficulties with presenting codes operating over the total environment set Ω. Designer-implementer teams are intent on both the production of reliable codes and portability. Individual judgment about the relative worth of these two components cannot easily be checked. Also, most codes operating over Ω show performance degradation for some subset of Ω. This degradation can be detected by users, most of whom only work in one ω_i.

II. Development status

Figure 3 gives one categorization of the current status of kernel development.

KERNEL STATUS

adequate for evolution---inadequate			
special functions; generators; cdf's	dense eigensystems; dense linear systems; ODEs; time series; tests for generators; approximations; LP; nonparametrics factor analysis; quadrature	PDEs; ANOVA; nonlinear; sparse linear systems; regression; categorical data	MVA(e.g., cluster analysis); sampling

FIGURE 3

In most categories, kernels are defined poorly, and the union of kernels does not cover the application area. In only those categories at the left has any successful study been made to provide reliable code such that reasonable warranty might someday be applied. Neither test results nor evidence of closure can be

documented in any precise fashion. Many categories will require years of research
effort, but some can be attacked today.

A. Certification

A review of an old Share Subroutine Certification Statement would show that even
in the days of early interest in reliability, good plans were developed. The
structure of the document contains all of the ingredients for reasonable
warranty given above.

What did Argonne National Laboratories (AMD) add in their more recent
certification efforts, so as to cause such successful results? We believe that
Pool, Cowell, Smith, and the many other important supporters of the project
would say that adequate long range funding and commitment, control, and the
gathering of the best talents were the main reasons for success. The Share
effort was too loosely controlled, and depended on the unfunded efforts of
interested (and talented) people whose work happened to be in the interest area
at the time.

What does such certification mean to the solution of the problem under discussion?
With all the benefits attained (good code, quite transportable code, well
certified code) the one ingredient which allows exploitation by software
manufacturers for new environments is available tests and test results. Argonne
personnel might only be unsatisfied with the emphasis placed on the problem
of "transportation to manufacturers", which, while attempted, could be improved.

Figure 4 gives an outline of the scope of the Argonne Linpack certification and
production effort in linear system solvers.

LINPACK

kernels were prepared for the cross products of the following sets

(matrix form) X (matrix data type) X (computation)

In more detail, these become

(general, general band, positive definite, positive definite band,
  symmetric indefinite, hermitian indefinite, triangular, general
  tridiagonal, positive definite tridiagonal)
                              X
(single precision, double precision, complex, double precision
  complex)
                              X
(factor, factor and estimate condition, solve, determinant-inverse-
  inertia)

FIGURE 4

Here, the two most important warranty ingredients (2. and 3.) are not handled as
simply as they can be handled in special functions. But, tests and results,
closed as well as is currently possible, are available.

We propose that any effort begin with such bases, and that directed moves be
taken to expand the effort. What is needed by manufacturers, given a category, is

  1. A kernel breakdown of the category,
  2. A specification for each kernel,
  3. At least, tests and results for each kernel, in some environment;
      more desirable would be benchmark results in the "most versatile"
      environment, and

4. Theoretical statements on expected results.

None of these are possible over the interest area. For point 4, for example,
the algorithms used in the code might be very dependent on the specific
environment. But considering the increased interest in the total environment
set, work in many of the categories will be fruitful.

B. A current testing effort

"Fruitful" can be used to describe the recent elementary function testing efforts
of Cody (4). Supported by the prior work of Malcolm (5), and Gentleman and
Marovich (6), Cody has produced two types of tests which expedite the simple
production of elementary function codes (the production of these codes is also
outlined). The tests allow the exercise of a code with the (almost) complete
elimination of errors resulting from the test code itself. Such work has real
value for software manufacturers. In the test codes, the boundaries of the
testing problem (the domain of x and the range of f) are considered, as well as
the subdomains related to the various algorithms employed.

It is important to note that the error to be minimized in a coded function
approximation is

$$f - \tilde{f} + \tilde{f}_c = (f - \hat{f}) + (\hat{f} - \tilde{f}) + \tilde{f}_c$$

where $\tilde{f}_c$ is the test code error. Seldom considered to be non-zero, this quantity
is often significantly large. $f - \hat{f}$ is the numerical mathematician's algorithm
problem; $\hat{f} - \tilde{f}$ is the problem of correct translation of the algorithm into a
reliable code. Cody properly balances these three error components, and drives
f_c near to zero.

III. Summary and restatement of need

No other computational area has more short and long range potential aid to the
scientific computing community than does such testing work. Upward evolution
of results of good effort will take place even with the extreme dynamics of the
embedding environment.

Because of the nature of the programming process, which requires the amalgamation
of talents which are unavailable to the group (often, preferably, of size one)
involved, we have accepted languages which are felt to be inadequate, and
computers which are design deficient. For basic computational tools, we have the
opportunity to avoid this situation. If we hesitate in the exploitation of our
abilities in fundamental calculation definition and testing, we may soon see
acceptance of less than adequate tools for that task, and these tools may be in
a form which makes revision difficult, economically. Note the floating point
arithmetic chip; chips or firmware with elementary function capability are nearly
at hand; certainly software will become much harder. The hard parts will be those
most heavily used, and those which have the widest potential using audience. It
seems clear that "software to hardware" movement will be beneficial. For example,
decimal to binary converters on chips would relieve the problem of compilers
converting with different results at compile and object time. But, we are not
yet in a position to have fundamental computational kernels moved to hardware.

At the initial meeting of IFIPs Working Group 2.5, Dr. Reid noted that testing was
important to consider. For the eventual solution of reliability-warranty problems
it is mandatory to make incremental steps such as those mentioned here. Software
moves to firmware will demand warranty eventually, and numerical analysts should
make their impact on such moves. Such efforts may seem too long in range, since
little is known in so many basic categories. A stable, short range approach,
with attendance to long range hopes, can be taken. What other moves would be so

well embedded in future needs? A fledgling industry will solve problems in a
haphazard fashion if action is not taken.

REFERENCES

1. Schwartz, J. T. "What constitutes progress in programming", CACM 18(11),
    November (1975), 663-664.

2. Reinsch, C. H. "Principles and preferences for computer arithmetic", draft
    for IFIP WG 2.5 presented in Toronto, May (1978).

3. Battiste, E. L. "Computer science and statistics", Proc. ACM, December (1978)
    in publication.

4. Cody, W. J., and W. M. Waite. "Software Manual Working Note #1", TM-321,
    AMD, Argonne National Laboratory, October (1977).

5. Malcolm, M. A. "Algorithms to reveal properties of floating point
    arithmetic", CACM 15(11), November (1972), 949-951.

6. Gentleman, M., and S. Marovich. "More on algorithms that reveal properties
    of floating point arithmetic units", CACM 17(5), May (1974), 276-277.

Performance Evaluation of Numerical Software, Fosdick (ed.)
© IFIP, North-Holland Publishing Company, 1979

LEGAL REMEDIES FOR MISREPRESENTATION OF SOFTWARE*

Colin Tapper
Magdalen College,
Oxford University

Programmes can be obtained in four different ways. The user can, if he has sufficiently competent staff, write them himself. Programmes produced in this way are like children. When they are good, they are very very good; but when they are bad, they are awful. Alternatively labour may be brought in to do the programming, but this leads to much the same danger. Two further alternatives, which are more relevant in the case of numerical algorithms, both involve dealing with a software house. The software house may be asked either to supply a proprietary package to accomplish the desired task, or it may be asked to design and implement a bespoke system.

If the customer merely wishes a well-defined function to be performed, and if he is sure that a proprietary package will, without any significant adaptation or modification, be able to perform it, then he may wish to buy the package. This can sometimes be arranged, but it remains more common in such circumstances for the software house to seek to lease the use of the package to the customer. This makes it much easier for the developer to retain his proprietary interest in the package. For one thing he need then supply only "object code", or sets of relatively opaque instructions which do not reveal the design philosophy of the package.[1] If this solution to the problem is adopted, and the customer may have no option, it is advisable to specify the computer environment in which the package must work and its performance specifications in some detail, and then to tie the completion of the contract, or at least the provisions for payment, to their satisfaction.

The final alternative is for the customer to contract for a set of programmes to be written by a software house so as to perform a set of operations specified explicitly by the customer. If the customer is unable to do this, he should first contract for the supply of such a specification, preferably from a firm not intended to be the ultimate supplier of the package.[2] It is vitally important in contracts for the supply of bespoke systems that the customer secure a fixed price contract based upon the successful completion of the specified operations. The danger is that the software house may seek to inveigle the customer into contracting on a time basis. One reason for this from the supplier's point of view is that it will often be very difficult for him to estimate in advance the cost of preparing such a set of programmes, especially when the specification is somewhat vague. Thus here again the most important consideration is that the specifications of the task should be made as explicit as possible. This protects the customer since he can then set up performance tests, prove breach relatively easily, and subject to what will appear later, make explicit provision for the assessment of damages. For these reasons government contracts nearly always contain elaborate specifications for the performance of software.[3] On the other

---

1 See Com-share Inc. v. Computer Complex Inc. 3 C.L.S.R. 462 (E.D. Ma., 1971) where such a package was held to be a proprietary trade secret despite installation at the sites of a number of customers.
2 See McArdle v. Board of Estimate for the City of Mount Vernon 347 N.Y.S. 2d 349 (1973) for an example of the dangers involved if this precaution is ignored.
3 See e.g. the Comptroller-General's letter to Altek Corporation of 21 April 1972 4 C.L.S.R. 1268 (1972)

* This work was supported in part by U.S. Army Research Contract No. DAAG29-78-G-0046.

hand the supplier is also protected in the case of contracts which include
such specifications against arbitrary decisions.[1] Both parties are further
protected from conflicting interpretations of ambiguous specifications. Where
software is to be prepared under this last option, the customer should seek to
secure not only the provision of programmes at source code level but also full
supporting documentation in the shape of comments and annotations on any printed
version of the programme, flow charts at different levels of abstraction, full
specifications of the package, and detailed manuals describing it complete with
full operating instructions. It is necessary to contract explicitly for the
provision of such documentation since it has been held that in the absence of
evidence of an established trade practice to the contrary in relation to a
particular application no such term will be implied.[2] The customer should also
seek to contract for as much trial of the system in advance as possible, being
careful to specify financial responsibility for the provision of such time, and
for the installation, maintenance and up-dating of the package. Finally, in such
cases the contract should deal with proprietary rights in the package. The
customer may want to secure exclusive rights, especially if he is keen to recoup
some of the expense by selling rights in it to others, but he may be concerned
only to prevent the software house from disclosing the package to his immediate
competitors. In all cases however it may prove difficult to distinguish between
the precise working out of the programme which may be susceptible of some degree
of protection, and its general philosophy which certainly will not.

    This paper will consider liability both in the law of contract and in that of
tort. It is however important to remember that the criminal law can be brought to
bear in the most serious cases of misrepresentation, where for example a valuable
contract is secured by promises which the party making them knows from the outset
can never be fulfilled. The precise extension of criminal law is a matter of
some doubt, and in the United Kingdom, at least, has not yet been tested in
relation to computer programmes. So far as liability in contract and in tort is
concerned the claims will be purely compensatory. In the English law of tort
punitive damages are now rarely awarded,[3] though 't is just conceivable that a
claim in respect of a computer contract could be brought within the range of such
categories as are still allowed. In the law of contract also penaly clauses as
such are unenforceable at common law.

    It is also necessary to consider here numerous changes in the common law which
have been brought about by statute. To some extent these have had the effect of
merging the two areas, a tendency which has also become evident in the common law
both in the United States and in the United Kingdom. Nonetheless the two areas
will continue to be distinguished from each other. In the first place they still
to some extent arise in different situations. Thus if no contract is ever
concluded the claim must be in tort. The converse is not true. Even though there
may be a binding contract this will not necessarily compel a claim in contract,
even though the breach is of a fundamental and explicit term. This has long
been recognised in the United States, but has emerged relatively recently

---

1 See e.g. the protest of California Computer Products Inc. that no benchmarking
   test was specified for software in a request for proposals issued by the Navy
   4 C.L.S.R. 1449 (1972)
2 Law research Services Inc. v. General Automation Inc. 5 C.L.S.R. 223 (CA2, 1974).
   In less specialised applications, or on more meritorious ones, the decision
   could easily go the other way.
3 Rookes v. Barnard [1964] A.C. 465

in the United Kingdom. Indeed it was not until the leading case of <u>Hedley Byrne</u>
<u>& Co.</u> v. <u>Heller & Partners</u>[1] that English law clearly recognised any liability in
tort for causing economic loss by negligent mis-statement. Even then it was still
argued, and even held,[2] that such liability was ousted, at least in some cases, by
the existence of a contractual term covering the same ground. This seems now to
have been put to rest by the latest decisions,[3] though even this year it took a
judge at first instance twenty six pages of closely reasoned argument to convince
himself that this was indeed the case.[4] It may be questioned why, if it is now
established that there is a cause of action in tort in every case, it is still
necessary to consider the law of contract, and the question of damages for breach.
The answer is that in some cases, and software contracts could easily be among
them, contractual claims offer advantages not available by suing in tort.

There have been important differences in this area between the law of the
United States and that of the United Kingdom. As remarked above English law
distinguished more sharply between contractual and tortious causes of action. In
the United States defective performance of a contract was much more likely to give
rise to a cause of action in tort. Not only this but American law, which first
imposed liability upon manufacturers to third party users by way of negligence,[5]
and then extended it by a species of warranty doctrine into strict liability,[6] now
imposes strict liability without any need to rely upon warranty.[7] This
proposition so far achieves its maximum extension in cases involving the sale of
goods to consumers of them, and has been held not to apply to contracts for the
supply of professional services where negligence is still the test.[8] It remains
to be seen how long this will continue to apply with the tide running so strongly
in the other direction. It is not inconceivable that software packages could
bridge the gap between consumer services and professional advice.

In American jurisdictions there is a bewildering array of remedies in tort and
contract. Sometimes these are based upon an express representation, sometimes not;
some rules require fraud or negligence, others are based on a theory of strict
liability. The situation can be discerned by comparison of recent cases in the
area.[9] In these two fundamentally similar suits remedies were arrived at by very
different routes. In Rhode Island where fraudulent misrepresentation actually
requires a modicum of fraud it was necessary to discover ancillary oral contracts
to avoid the disclaimers in the written contracts, while in Minnesota it was
possible to evade the disclaimers by reliance upon the benevolent definition of
fraudulent misrepresentation.

It is instructive to compare the situation in the United Kingdom where there
has been extensive statutory intervention, principally in the enactment of the
Misrepresentation Act 1967.[10] Unfortunately the situation has been little

---

1 [1964] A.C. 465
2 <u>Bagot</u> v. <u>Stevens Scanlon & Co.</u> [1966] 1 Q.B. 197
3 <u>Esso Petroleum Co. Ltd.</u> v. <u>Mardon</u> [1975] Q.B. 819
4 <u>Midland Bank Trust Co.</u> v. <u>Hett, Stubbs & Kemp</u> [1978] 3 All E.R. 571
5 <u>MacPherson</u> v. <u>Buick Motor Co.</u> 111 N.E. 1050 (N.Y., 1916)
6 <u>Baxter</u> v. <u>Ford Motor Co.</u> 35 P. 2d 1090 (wa., 1934)
7 <u>Henningsen</u> v. <u>Bloomfield Motors Inc.</u> 161 A. 2d 69 (N.J., 1960)
8 <u>Raritan Trading Co.</u> v. <u>Aero Commander Inc.</u> 458 F. 2d 1113 (CA3, 1972)
9 Cp. <u>International Business Machines Ltd</u> v. <u>Catamore</u> 548 F. 2d 1065 (CA1, 1976)
   with <u>Clements Auto</u> v. <u>S.B.C.</u> 444 F. 2d 169 (CA8, 1971)
10 Parts of this Act have been further amended by the Unfair Contract Terms Act
   1977.

clarified[1], and has indeed been authoritatively described as "almost incredibly complex".[2] There is a gamut of five different causes of action from which an action for damages might result, some sounding in tort and others in contract according to the circumstances of the misrepresentation.[2] As in the United States the most drastic remedies attach to fraudulent misrepresentation in the true sense. Such a misrepresentation will lead to an award of damages in tort and to the opportunity to rescind any contract entered into upon the strength of it. The conditions are much the same as in the United States except that the English rules make it clear that recklessness or not caring whether the statement is true or false will lead to liability. Negligence is not enough, though since the decision in Hedley Byrne & Co. v. Heller & Partners[3] in 1963 a cause of action has lain in England for negligent misrepresentation in certain conditions. The most important of these is that a special relationship must be established between the parties, and perhaps that the representer must hold himself out to give advice.[4] This will usually cover the case of the incompetent consultant or the careless salesman. However when the representer enters into contractual relations with the representee there will generally no longer be any need to rely upon the common law remedy for negligent misrepresentation. The victim can instead rely upon the statutory remedy provided by section 2(1) of the Misrepresentation Act 1967. Here he need show only that he has entered into a contract with the representer after the misrepresentation was made. This operates to shift the burden of proof so that in such cases the representer can escape liability only by showing that he believed on reasonable grounds at the time the contract was made that the representation was true. In addition to such apparently tortious remedies there is the situation where the representation becomes a term of the contract in which cases there will be the ordinary contractual action for its breach. As is so often the case the principle is clear enough but its application in practice often raises extremely difficult questions of construction as the court struggles to derive a clear set of terms from a welter of papers and recollections of what transpired. The courts are perhaps more inclined to regard a representation as a term when the representer is in the stronger bargaining position as an experienced software house would be in extolling its wares to a lay client. A variant of the term of the contract approach applies in a situation in which the representation operates as a collateral warranty, or as a consideration for entering into the main contract.[5] This has never been very satisfactory from a theoretical point of view, especially in a two party situation, but it used to be adopted to circumvent the difficulty under the old law that damages were not recoverable for an innocent misrepresentation unless it could be construed as a term of the main contract. It has lost most[6] of its force now that the Misrepresentation Act gives a remedy in damages. Under the old law the most that could be achieved was rescission, with, in a few cases only, an indemnity. The statutory remedy still relies upon the availability of rescission as a test for the availability of damages by allowing them only when the representee "would be entitled by reason of the misrepresentation, to rescind the contract". This somewhat obscure provision was clearly intended to provide a better remedy where rescission was inappropriate.

----

1 Treitel and Atiyah "Misrepresentation Act 1967", 30 Modern Law Review 369
   (1967)
2 Leaving out of account such other compensatory remedies as the indemnity which
   may exceptionally accompany rescission.
3 [1964] A.C. 465
4 Mutual Life & Citizens Assurance Co. v. Evatt [1971] A.C. 793
5 De Lassalle v. Guildford [1901] 2 K.B. 215
6 But not quite all since some cases still fall outside the Misrepresentation Act
   for various reasons.

The question then arises of how far a representer can exclude liability in
any of these cases by the insertion into the contract of an express disclaimer.
There can be no serious doubt that no disclaimer was ever possible in the case
of a fraudulent misrepresentation,[1] though there is surprisingly little direct
English authority. In other cases the situation is governed by three recent
statutes, the Misrepresentation Act 1967, the Supply of Goods (Implied Terms) Act
1973 and the Unfair Contract Terms Act 1977. Quite apart from these Acts there
has long been a principle that exemption clauses will be construed strictly. A
special application of the principle arises here when liability may be imposed
under the contract either strictly or for negligence, as is normally the case. In
such circumstances when either ground is available the exemption clause will be
construed as ousting liability only on the strict basis, thus leaving liability
for negligent breach unaffected.[2] Even if there can only be contractual liability
for negligent breach the exemption clause may still not be totally immune from
attack since in some cases courts have occasionally interpreted such clauses as
a simple warning that there is to be no strict liability under the contract.[3]
Furthermore in contractual claims no exemption clause will be upheld when there is
a fundamental breach of the contract so as to defeat the purpose for which it was
made.[4] In general this is a rule of construction so that where the clause is
inescapably applicable to the breach which has occurred it will be upheld.
However even that is subject to exception if the contract should be repudiated at
the earliest opportunity, rather than afformed by continuance. A still more
recent case suggests that this may be extended to apply to circumstances in which
there is no sensible alternative to carrying on.[5] If so, a substantial hole seems
to have been punched into the limitation of the doctrine to a rôle as a rule of
construction. The whole topic is rather obscure and in a state of constant
development from which it may be hoped that clearer rules may eventually emerge.
Under the various statutes the Unfair Contract Terms Act 1977 substitutes a new
form of section three of the Misrepresentation Act 1967 which in effect casts upon
any party seeking to rely upon an exclusion clause the burden of showing that such
term is fair and reasonable according to the guidelines set out in Schedule two.[6]
These same guidelines are also applied to the construction of notices purporting
to restrict liability in tort even in the absence of a contract.[7] The guidelines
themselves first appeared in the Supply of Goods (Implied Terms) Act 1973. This
Act will apply to software contracts to the extent that any goods are supplied to
the client pursuant to the contract. The Act invalidates any attempt to oust the
new implied warranty as to title.[8] In non-consumer sales, as many software
contracts will be, it further invalidates clauses seeking to exclude the new
warranties of merchantibility[9] and fitness for purpose[10] where it would not be
fair or reasonable to permit reliance upon it.[11] The factors to be considered in

---

1 10th Report of the Law Reform Committee "Innocent Misrepresentation" Cmnd. 1782
   (1962) para. 23
2 White v. J. Warwick & Co. Ltd. [1953] 1 W.L.R. 1285
3 Hollier v. Rambler Motors Ltd. [1972] 2 Q.B. 87
4 Suisse Atlantique Société D'Armament Maritime S.A. v. N.V. Rotterdamsche Kolen
   Centrale [1967] 1 A.C. 36], 393
5 Harbutt's 'Plasticine' v. Wayne Tank and Pump Co. [1970] 1 Q.B. 447
6 Sect. 8
7 Sect. 2(2)
8 New sect. 55(3) of the Sale of Goods Act 1893 enforcing new sect. 12
9 New sect. 14(2)
10 New sect. 14(3)
11 New sect. 55(4)

any such assessment are so relevant to software contracts as to justify
reproduction in full,[1]

> "In determining...whether or not reliance on any such term would
> be fair or reasonable regard shall be had to all the circumstances
> of the case and in particular to the following matters -
> (a) the strength of the bargaining position of the seller and
> buyer relative to each other, taking into account, among other
> things, the availability of suitable alternative products and
> sources of supply;
> (b) whether the buyer received an inducement to agree to the term
> or in accepting it had an opportunity of buying the goods or
> suitable alternatives without it from any source of supply;
> (c) whether the buyer knew or ought reasonably to have known
> of the existence and extent of the term (having regard among other
> things, to any custom of the trade and any previous course of
> dealings between the parties);
> (d) where the term exempts from all or any of the provisions of
> sections 13, 14 or 15 of this Act if some condition is not complied
> with, whether it was reasonable at the time of the contract to
> expect that compliance with that condition would be practicable;
> (e) whether the goods were manufactured, processed or adapted to
> the special order of the buyer."

Almost every one of these considerations will be found exemplified in the
arguments in software contract cases both in the United States and in the United
Kingdom.

It makes much more sense to assimilate contractual and tortious causes of
action in the United States since the remedies are much more comparable in the two
cases than they are in the United Kingdom. The most important consideration is
usually the measure of damages, though different limitation periods,[2] the relative
effectiveness of exemption clauses,[3] and the application of principles of
comparative or contributory negligence may also on occasion have critical effect.
Although there is recent authority to the contrary, the better British view now
appears to be similar to the American in holding that where a duty is owed
independently of contract its breach in a contractual context leaves the plaintiff
with a choice of remedy.[4] He can sue in tort or contract as he pleases. If the
plaintiff sues in tort the modern British view is that damages are recoverable in
deceit for fraudulent misrepresentation, or in negligence for negligent
misrepresentation. In both cases they are assessed on the basis of restoring the
plaintiff to the position he would have occupied had the misrepresentation not
been made.[5] This is usually regarded as being the difference between the price
paid and the value received, bearing in mind that the value may prove to be
negative. If the plaintiff sues in contract the British view is that he is
entitled to the value represented to him less the value he actually received. It
is thus apparent that if the plaintiff has made a bad bargain in the sense that
the value of the software as described by the salesman is greater than the price
he has paid for it, the plaintiff will gain more by suing in tort; whereas if he
has made a good bargain in the sense that the value of the software as described
by the salesman is greater than the price he has paid for it, then he will recover

---

1 New sect. 55(5)
2 As in Midland Bank Trust Co. v. Hett, Stubbs & Kemp [1978] 3 All E.R. 571
3 As in Clements Auto Co. v. Service Bureau Corporation 444 F. 2d 169 (CA8, 1971)
4 Esso Petroleum Co. v. Mardon [1975] Q.B. 819
5 Doyle v. Olby Ltd. [1969] 2 Q.B. 158 explicitly endorsing the views expressed
   in Mayne & McGregor on Damages (12th ed.)

more by suing in contract. It is assumed for these purposes that the application
of different rules as to remoteness of damage, and as to the relevance of special
and consequential damage can be ignored. In most of the jurisdictions in the
United States the situation is different because there the measure of damages for
fraud includes those for loss of the bargain. This no doubt helps to explain the
traditional laxity with regard to the form of the action.

Problems of proof and of damages constitute some of the most bitterly
contested issues in these cases. Where there is an allegation of fraud it is
sometimes intimated that an especially high standard of proof is required. Thus
in Strand v. Librascope Ltd.[1] it was asserted that proof should be "clear and
satisfactory".[2] In the United Kingdom however, despite occasional assertions to
the contrary,[2] the general view is that there is only one standard of proof in
civil cases, but that in relation to some issues, of which fraud is one, it takes
more evidence to satisfy that standard.

In general it is for the plaintiff to prove negligence in any case where
negligence is alleged.[3] In the United Kingdom this represents an advance upon the
older state of the law where by suing in trespass it was possible to cast upon the
defendant the onus of proving inevitable accident as a defence, or in effect of
disproving negligence.[4] The United States has however preserved a still earlier
form of British law in allowing, in some cases at least, a species of strict
liability by the use of an implied warranty. The precise extension of these
doctrines in both British and American law is however uncertain.

The party who bears the burden of proving negligence can sometimes rely upon
the doctrine of res ipsa loquitur. According to that doctrine the defendant must
assume the burden of explaining any accident which has occurred contrary to normal
expectations in a situation in which he might have been expected to be in control.
It has yet to be seen whether a court would be willing to hold that defective
software is contrary to normal expectation. Since so late as 1955 the courts were
not prepared to hold that the mid-air explosion of a jet engine was contrary to
normal expectation,[5] it is submitted that they would not. It is likely that in
time to come component failure will be regarded as an appropriate situation for
the application of the doctrine, but on present evidence it is likely to be a
very long time indeed before software, even proprietary software, achieves a
similar status. In fact so far from its being easy to establish negligence by
application of the doctrine, it may, for the reasons which prevent the doctrine
from applying, be very difficult to establish it by any means at all. So long as
people remain convinced that computers have an existence and personality of their
own different from that of their operators and programmers so will it be common to
ascribe error to the intrinsic fallibility, unpredictability and malevolence of
the machines themselves rather than to the negligence of the human beings
concerned. Such irrationality provides further justification for the American
partiality for strict liability.

Issues of causation and remoteness of damage are notoriously difficult in
relation to the tort of negligence. They are intimately related to the definition
of the duty which is owed and to the standard of care required. A typical case

---

1   197 F. Supp. 504 (D. Conn., 1975)
2   For example, per Denning L.J. in Bater v. Bater [1951] P. 35 at 37
3   Letang v. Cooper [1965] 1 Q.B. 232, Mulligan v. Atlantic Coast Line Railway Co.
    242 U.S. 620 (1917)
4   Stanley v. Powell [1891] 1 Q.B. 86, cf. Fowler v. Lanning [1959] 1 Q.B. 426
5   Williams v. U.S. 218 F. 2d 473 (CA5, 1955)

might involve a negligently designed and operated system, assuming this to be admitted, which prevents work from being done with the result that third parties suffer loss. There are really two different questions. First, did the designer of the system owe a duty of care to the third parties who have suffered loss to prevent its occurrence to them; and secondly, if he did owe a duty to take care to prevent some loss, is the loss which actually occurred sufficiently connected with it for the third party to be able to recover damages in respect of it ? The American treatment of some of these questions in the computer setting is illustrated by the case of Independent School District No. 454, Fairmont, Minnesota v. Statistical Tabulating Corporation.[1] The school district wished to value its schools for the purposes of insurance and hired a firm of surveyors to do the work. The surveyors took all of the relevant measurements, but sent the data to the defendant for computation of the values. The computation was inaccurate with the result that the plaintiff under-insured, and suffered loss when one of its schools was destroyed by fire. It was held that the defendant did owe a duty of care to the plaintiff. The conclusion was justified on the basis that they knew that their computation would be relied upon by the school board for insurance purposes, that liability was restricted to a relatively small group, that it was undesirable that an innocent reliant party should bear the loss, and that[2] "recovery by a foreseeable user will promote cautionary techniques among computer operators." The justificatory reasoning reveals the causes for concern about extending liability in such cases. It would clearly be a great hardship for software smiths to be under potential liability to an unlimited class for an unlimited time to an unlimited extent in respect of an unlimited range of applications. These dangers are particularly severe in this context as computers are highly flexible devices, and at the same time capable of such incredible rates of work. The volume and range of work capable of being done on a computer far exceeds that applicable to any other device. Nor in the case of any other device is it so difficult to guard against error. The transposition of two characters in an operating system could easily go undetected through all reasonable trials, and still in operation suddenly lead to disastrous errors causing enormous expense to users. In such circumstances the conditions laid down in the School Board case could easily be satisfied without the result seeming obviously just. The problem is exacerbated by the difficulty of establishing in relation to any new system how to set an appropriate insurance premium. Nor is it obvious that in such circumstances insurance is a fair way of distributing the loss.

It should be noted that in the School Board case the court was quite prepared to allow recovery despite a number of acts which intervened between the inaccurate computation and the loss to the board. Those events were separated by the board's acting on the valuation and the school's being burned down, among others. They may however have been disregarded since this very sequence of events was exactly that envisaged throughout the transaction. If the damage had been of a very different kind from that envisaged, say the suicide of the schools' manager in consequence of his learning that his schools were under-insured, it is much less certain that a common law court would permit recovery. A French court has however awarded the equivalent of over a quarter of a million dollars to a wholesaler who became bankrupt after his suppliers lost confidence in him when a computer error at his bank caused his cheques to be returned.[3] In England the test for

---

1 359 F. Supp. 1095 (N.D. Ill., 1973)
2 At 1098
3 Computerworld 16 August 1976 p. 16

remoteness of damage in tort appears to be whether the loss could reasonably have been foreseen[1] whereas in the United States it is more often cast in terms of proximity.[2] Neither test is very precise, and the law can be regarded as neither certain in formulation nor predictable in application. Certainly its application to software remains to be worked out in detail by the courts. It is submitted that the need to avert the dangers noted in connection with the range of possible plaintiffs applies equally strongly to the extent and variety of loss.

Further unpredictability is injected into the situation in the United States by the availability of punitive damages which have been awarded in a number of automobile repossession cases involving computers.[3] They might well also be awarded in cases involving fraudulent misrepresentation though despite having been occasionally claimed,[4] they seem not so far to have been awarded. In the United Kingdom the principle that damages are compensatory and not punitive has become even more strongly entrenched.[5] It seems that even if deceit is an appropriate cause of action for such an award the circumstances are hardly ever going to be such as to justify one.[6]

The conduct of the plaintiff is relevant to the assessment of damages in two different, though related, ways. In the United Kingdom,[7] in a few American states,[8] and under a number of special statutes,[9] a doctrine of comparative[10] negligence applies where the plaintiff's recovery is reduced in proportion to his blame for the occurrence. The concept is applied in a very impressionistic way so that the blame of the plaintiff might relate to such conduct as failure to test the software properly before relying upon it, failure to run a manual system in parallel for a time, failure to provide adequate back-up facilities, careless operation, or any of a multitude of other matters which the trier of fact might regard as sufficient justification for denying recovery in full. As yet these factors have received little attention from the courts. In those states where there is , so far, no doctrine of comparative negligence, the courts adopt the principle that the plaintiff is barred from any recovery at all if his fault has contributed to the loss, or if he has voluntarily assumed the risk. However there is evidence of some reluctance to accept defences which result in such stringent consequences for the plaintiff. Since in these jurisdictions comparative negligence is not available to reduce recovery, any inclination to do so must find some other outlet. It is quite often possible for the trier of fact to express his views about comparative blame in his application of a vague test like proximity of loss in an effort to achieve a fair apportionment. Another possibility is provided by the rule that the plaintiff must mitigate his loss as

---

1 Overseas Tankship (U.K.) v. Morts Dock and Engineering Co. (Wagon Mound No. 1)
   [1961] A.C. 388
2 Prosser 'Torts' (4th ed.) sect. 42
3 Price v. Ford Motor Credit Co. 5 C.L.S.R. 956 (Mo.; 1975), Ford Motor Credit Co.
   v. Swarens 447 S.W. 2d 533 (Ky., 1969), Ford Motor Credit Co. v. Hitchcock
   158 S.E. 2d 468 (Ga., 1967)
4 As in Catamore v. International Business Machines Inc. 548 F. 2d 1065 (CA1,
   1976)
5 Since Rookes v. Barnard [1964] A.C. 1129 per Lord Devlin
6 Mafo v. Adams [1970] 1 Q.B. 548
7 Law Reform (Contributory Negligence) Act 1945
8 For example, Alabama, Georgia, Hawai, Maine, Massachusetts, Minnesota,
   Mississippi, Nebraska, New Hampshire, South Dakota, Vermont and Wisconsin.
9 For example, Federal Employers Liability Act 45 U.S.C. sects. 51-59
10 In the United Kingdom it is still called 'contributory' negligence, but the
   American term is clearer.

far as possible. The plaintiff must take all reasonable steps to reduce his
losses once he has appreciated that they will occur. He can not just sit back and
wait for his damages to accrue. The plaintiff is usually in the best position to
assess his own loss, and to take steps to seek an appropriate remedy. He is,
however, entitled to the benefit of any doubt as to the application of the rule
that his actions must be reasonable, since otherwise the defendant who is <u>ex
hypothesi</u> the author of the plaintiff's dilemma would himself be put in an unduly
favourable position. Thus a balance must be struck, and one which it is often
extremely difficult to strike fairly. No where is this more true than in the case
of software, especially where the customer has been supplied with a malfunctioning
system. This very situation has occurred in a number of decided cases. The
question is often one of how long the plaintiff is entitled to wait for his
original supplier to put the system right before making a clean break and going on
to an alternative supplier. A complete change of supplier is always, and for many
different reasons, an unpalatable choice, and most customers will hang on as long
as possible with their original supplier in the hope that they can avoid the
disruption and expense of a change. They are always likely to be encouraged to
persist by their original supplier who will be reluctant to lose a customer. In
Clements[1] the system struggled along for four years at a cost of over a quarter of
a million dollars in rental payments throughout which time the system was
substantially useless, but was always represented as being on the brink of
complete efficiency. The contracts in that case ran for a year at a time and were
terminable on thirty days notice by either party. Thus the issue hinged upon
precisely when the plaintiff ought to have realised that the system would never
work, bearing in mind the disparity in expertise between the parties. The
appellate court took a view substantially more favourable to the defendant on this
point than the trial court had done, and reduced the damages accordingly. It
should be noted that in a number of cases where the claim is based both in tort
and in contract, similar factors may determine whether for the purpose of the
contract the buyer has affirmed the contract so as to lose the right to rescind.[2]
Here also the court is usually reluctant to find affirmation in a situation in
which a more experienced seller is pressing the buyer to retain the goods while
efforts are being made to put the system into a satisfactory state.

1 Clements Auto Co. v. Service Bureau Corporation 444 F. 2d 169 (CA8, 1971)
2 See Carl Beasley Ford v. Burroughs Corp. 361 F. Supp. 325 (E.D. Pa., 1973)
   and Public Utilities Commission of Waterloo v. Burroughs Business Machines Ltd.
   1 O.R. (2d) 257 (1974)

Performance Evaluation of Numerical Software, Fosdick (ed.)
© IFIP, North-Holland Publishing Company, 1979

DISCUSSION OF SESSION ON RELIABILITY AND WARRANTY
OF NUMERICAL SOFTWARE*

Wayne R. Cowell
Applied Mathematics Division
Argonne National Laboratory
Argonne, Illinois

Existing numerical software has originated, for the most part, in publicly funded
research and development projects at universities and government laboratories. It
has been shared informally within the research community, distributed by various
government-funded centers and, increasingly, is being incorporated into proprietary libraries and disseminated through business enterprises. Numerical software
is thus emerging as a product, which is represented as having certain characteristics by vendors, for the purpose of promoting its sale or licensing to users, for
whom the performance of the software may be critical. The attendant questions
about protection of intellectual property, warranty of performance, and liability
for misrepresentation become of concern to software suppliers and users. The
search for constructive approaches involves the law and computer science, both
highly technical subjects. When they become interdependent, one is struck by the
ways in which their respective complexities reinforce one another. It appears to
this discussant, upon reading the papers in this section, that the application of
the law to issues arising from software technology is at a very early stage of
development. The framework is there but progress toward the establishment of just
relationships between sellers and buyers will hinge on further experience in and
out of court, clarified by research in computer science.

Niblett discusses various means of protecting property rights in numerical software namely, patents, trade secrets (confidential information), copyrights, and
trademarks. These are applicable or not under various circumstances but it would
seem that experience in the courts is still too limited to permit easy prescrip-
tions. Thus, for example, the Patents Act of 1977 (U.K.) obviates the possibility
of patenting computer programs as such but does not settle the question of
patentability of a device which realizes a program in a novel way, such as a
microelectronic chip. Battiste notes, in his paper, that chips with elementary
function capability are nearly at hand, presaging a mathematical software to
mathematical firmware movement. Thus, it seems likely that the question of patenting programs on chips may soon attract considerable attention.

Niblett points out that the exposure of software required by thorough testing may
be incompatible with the protection of that software as a trade secret. He argues
that the most effective way, at least in the U.K., of asserting intellectual
property rights in numerical software is by means of copyright laws. Responding
to a question from the floor, Niblett stated that civil cases involving copyright
of programs have not yet been considered by the courts in the U.K., but there have
been a number of out-of-court settlements on the basis that copyright subsists in
a computer program and has been infringed.

The impression on this discussant is that the traditional means of protecting
intellectual property are not well adapted to the protection of computer programs.
Programmers do not yet enjoy the same degree of protection of their work as do in-
ventors and artists. Like the inventor's work, a computer program is sterile
unless it is used but its use (execution) requires that it be replicated, infringing, in some sense, the right to copy. Therefore protecting the program as if it
were a literary work (as Niblett says is possible) is not fully satisfactory.
Consider how much source code, intended for submission to a compiler (and thus

*Work performed under the auspices of the U.S. Department of Energy, with partial
support from the U.S. National Science Foundation.

copied), is published in books and journals for which the right to copy is
reserved by the publisher. At the same time, patent law is not applicable to
algorithms, and thus it appears that only firmware will have reasonable protection, in the traditional sense, for some time to come.

All three authors in this section treat the question of guaranteeing the performance of software products. Battiste's emphasis is on the computer science-
related difficulties of providing reasonable warranties. Tapper and Niblett
examine the legal ramifications of liability for misrepresentation of software.
The effect is sobering, especially if we accept that the body of law reflects so-
cietal expectations of software performance. Where these expectations are
unrealistic, it means that the public view of computing is unrealistic, probably
because computing experts have oversimplified the presentation of computing to the
public. (They might not otherwise have been heard at all.) These papers lead us
to believe that the use of the law will become more reasonable and just when the
public has a better understanding of computing realities.

Tapper illustrates this point in his discussion of the legal proof of negligence.
It is for the plaintiff to prove negligence when negligence is alleged. He can
sometimes rely on a doctrine which asserts that the defendent has the burden of
explaining any accident which has occurred contrary to normal expectations in a
situation in which he (the defendent) might have been expected to be in control.
Is defective software contrary to normal expectations? The question has not been
tested in court. Tapper goes on, "So long as people remain convinced that computers have an existence and personality of their own different from that of their
operators and programmers so will it be common to ascribe error to the intrinsic
fallibility, unpredictability, and malevolence of the machines themselves rather
than to the negligence of the human beings concerned."

As is apparent from other sessions at this conference, there is beginning to
emerge from research efforts a sense of what acceptable software performance
means. Unfortunately, much of this understanding is highly technical, quite
dependent on the area of computation, and difficult to communicate in lay terms.
The challenge is to translate this technical understanding into guidelines for
software performance that will enable reasonable people to have reasonable
expectations. We can hope that, eventually, the application of the law will
incorporate this increase in reason.

Tapper's paper considers liability for computer software misrepresentation both in
the law of contract and in that of tort. He cites a number of cases in which some
question of liability for misrepresentation arose and he discusses their implications for contracting and for seeking remedies for alleged wrongs. There are
guidelines here that will serve as indicators for users and vendors but it is
clear that legal counsel should be sought before signing, selling, or accusing.
Indeed, the Misrepresentation Act 1967 (U.K.) has been described as leaving the
situation "almost incredibly complex" as it provides "a gamut of five different
causes of action from which an action for damages might result, some sounding in
tort and others in contract according to the circumstances of the misrepresentation." One may encounter the need to distinguish between "recklessness" and
"negligence." Again, one might have entered into a contract with a representer
after a misrepresentation was made. Attempting to escape liability, the representer might claim that he believed "on reasonable grounds" that the statement was
true. Computer scientists, upon reading the paper, may ask what constitutes reasonableness in the representation of computer software. They will be impressed
(though, on reflection, not surprised) by the extent and intricacy of the legal
machinery that deals with liability. They will also recognize terms (for example,
"precise specification of the performance of a program") that must derive meaning
from their science if the machinery is to be successfully applied in their field.

The extent of liability is a serious concern. Tapper points out, for example,
that a transposition of two characters in an operating system could easily go

undetected through all reasonable trials [that word "reasonable" again!], and still in operation lead to disastrous errors causing enormous expense to users. This discussant was comforted when Tapper cited a case in which the court was very careful to focus the extent of liability narrowly, precisely because of the potential for almost unlimited liability through the use of faulty software. Too broad an extent of liability would place a heavy burden on programmers.

There appears, to this discussant, to be a danger that vendors will be hamstrung by capricious liability suits (or the threat of them) until their best efforts to represent complex software realities are distinguished from fraudulent, careless or stupid misrepresentation. Battiste, in his paper, expresses this danger and argues strongly for the development of a technical basis permitting the description of adequate software performance in terms suitable for reasonable warranties.

Battiste characterizes various areas of mathematical computation from special function approximation to multivariate analysis in terms of suitability for reasonable warranty. He singles out special function approximation as the area in which software performance is best understood by virtue of an advanced state of testing methodology. Therefore, he asserts, special function software may be the first for which reasonable warranties are offered. However, Battiste warns that this does not imply that warranty is possible for programs constructed from special function kernels. In this regard, he quotes J. T. Schwartz on the subject of integrating large collections of simple programs into coherently functioning wholes. Battiste also reminds us that the perfection of contracts is not the only purpose served by an ability to describe the performance of numerical software; the software that is cast in firmware will be the best only insofar as we are able to measure and describe what is best.

These papers should serve to raise the consciousness of software experts about the social implications of their work. The need to improve the quality of software and to express clearly what quality means take on new urgency when seen in terms of satisfying the demands of the law as well as in scientific terms. These papers should also encourage vendors and users to insist on well-drawn contracts that clearly define their responsibilities to each other, within the limits of technical understanding.

SESSION 5 : PERFORMANCE EVALUATION IN
ORDINARY DIFFERENTIAL EQUATIONS

Performance Evaluation of Numerical Software, Fosdick (ed.)
© *IFIP, North-Holland Publishing Company, 1979*

PERFORMANCE EVALUATION OF O.D.E. SOFTWARE

THROUGH MODELLING

Hans J. Stetter

Technical   University, Vienna

INTRODUCTION

Performance evaluation is a fundamental activity in engineering: In the design and the production of complex devices their expected as well as their actual performance must be evaluated. Normally, there are a number of design objectives which may be informally defined only and which may be partially conflicting. Performance evaluation concerns the actual achievement of these objectives in the prototype or in the manufactured product. A quantitative evaluation of the achievement is highly desirable since it permits a comparison between competing designs.

Software is a typical engineering product: It is meant to deliver specified services in an acceptable fashion in a variety of different situations under given restrictions of resources and cost. Therefore software designers face the same need for a reliable evaluation of the performance of their products as engineers, and they have to use basically the same approaches:

Testing: Check the actual behavior of the whole product or of specified components in sample situations.

Modelling: Design a mathematical model of the whole product or of specific components; analyze the behavior of the model analytically or through simulation.

The advantages and the limitations of these two fundamental approaches are well-known:

	Advantage	Limitation
Tests	Show actual behavior of complete product	Cover only a limited set of situations
Models	Give general insight; permit a quantitative performance evaluation for individual components relative to different design objectives	May neglect essential aspects

In this contribution we want to indicate some of the possibilities of the modelling approach in the performance evaluation of o.d.e. software.

A SIMPLE EXAMPLE: INTERPOLATION IN AN ADAMS PC-CODE

In initial value problem codes for o.d.e. the stepsize control strategy can only become effective if gridpoints are not selected for extraneous reasons, e.g. because of the need for output. Hence some codes disregard output requests in the stepsize selection process; instead they generate output values by interpolation. This approach is particularly natural in Adams PC-Codes where the basic integrator is essentially an interpolation procedure; see, e.g., [1], chapter 5 .

A while ago, I wanted to evaluate the performance of two alternative interpolation modules in a comparative fashion. Results of tests depended strongly on special circumstances and a statistical evaluation would have required an intricate normalization of the individual results. On the other hand, the situation was sufficiently simple to permit an appropriate model. Details of the following analysis appear elsewhere ([2]) in a slightly different context.

The main design objective of the interpolation module of a code is the generation of an approximate solution between grid values which behaves "like a solution of the differential equation"; see fig.1.

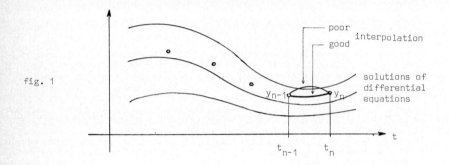

fig. 1

A quantitative performance criterion for this design objective may be defined thus:

In each grid interval $[t_{n-1}, t_n]$, the ideal interpolation function z is uniquely determined by

$$(2.1) \quad z(t_{n-1}) = y_{n-1}, \quad z(t_n) = y_n, \quad \max_{t \in [t_{n-1}, t_n]} \|z'(t) - f(t, z(t))\| = \min$$

(and $z'(t) - f(t, z(t)) = $ const if $\|..\|$ is the max-norm); cf. also [3], equ. (3). Therefore, the deviation

$$(2.2) \qquad q_n := \max_{t \in [t_{n-1}, t_n]} \|\tilde{z}(t) - z(t)\|$$

is a measure for the quality of the interpolation function $\tilde{z}$ generated by the code in $[t_{n-1}, t_n]$.

For a given problem and a given run with a given code, the q_n-sequence is a suitable description of the performance of a given interpolation routine attached to that code. Our task is to evaluate $\{q_n\}$ over "all problems and all runs".

Our model arises from simplifications in the analytic representation of q_n for a variable order, variable step Adams PC-code of the type described in [1]: If we consider only lowest-order-in-h_n terms and consider all derivatives of f evaluated along the true solution we have, in $[t_{n-1}, t_n]$,

$$\tilde{z}(t) - z(t) = M_n[f]v(t) + N_n[f]v^*(t)$$

where, for a given interpolation procedure, the scalar functions v and v^* depend on the local stepsize and order situation but not on the problem which enters only into the vectors M_n and N_n. By treating each component individually, by normalizing the deviation relative to the local error in this step and by normalizing the step to [0,1] we obtain a (componentwise) representation

$$(2.3) \qquad \frac{\tilde{z}(s) - z(s)}{L_n} = (1-N_n)V(s) + N_n V^*(s), \qquad s \in [0,1],$$

where the local behavior of the problem now enters only through the scalar quantities N_n. The interpolation procedure determines the functions V and V^* which also depend to some extent on the local run situation. Details of the derivation of (2.3) may be found in [2] where a slightly different notation has been used.

In our model the evaluation task is thus reduced to the following: For each of the two interpolation codes and for typical stepsize and order situations, plot

$$q(N) := \max_{s \in [0,1]} |(1-N)V(s) + NV^*(s)|, \qquad -\infty < N < +\infty,$$

and compare.

A typical result is shown in fig.2. A similar result is obtained for all run situations, thus the conclusion is immediate: Within our model and with respect to the design objective stated above, code A performs better than code B. Since the two interpolation codes require the same computational effort, code A should be chosen for implementation.

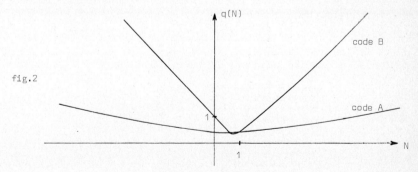

fig.2

The validity of this conclusion in actual computations may be judged by considering the simplifications which were introduced: The essential one is the neglection of higher-order-in-h terms, it should upturn our result only in rare cases and only locally. In my opinion, the sketched approach gives the assertion more reliability then could have ever been reached by testing.

Note that our analysis has established that the problem dependence of the performance may be essentially represented by the value of one <u>real parameter</u> (per component) locally. This is how a comparative evaluation "over all problems" became feasible.

Of course, this also proves that we have had a particularly simple situation which is furthermore simplified by the virtual independence between the interpolation routine and the underlying integration code. The following two sections will show that it is possible to use models for a comparative evaluation of specific aspects of o.d.e. codes also in less favorable situations.

PERFORMANCE EVALUATION OF STEPSIZE CONTROL PROCEDURES

In initial value problem codes for o.d.e., the stepsize control procedure constitutes a crucial module of the code. Nevertheless, its design is often strongly heuristic. It is almost impossible to devise tests which permit an isolated performance evaluation of the stepsize control procedure and to perform a comparative evaluation of different designs on this basis.

Some of the design objectives for the stepsize control are the following:

<u>Efficiency</u>: In smooth situations, an "optimal" stepsize sequence should be generated (whatever that is).

<u>Robustness</u>: Unusual situations should be recognized and coped with in a sensible way.

<u>Economy</u>: Rejections of steps should occur rarely; on the other hand, the prescribed tolerance should not be discounted too much.

In the following, we will restrict ourselves to fixed order (and fixed method) codes; it is not clear at the present time how the approach may be reasonably extended to variable order situations.

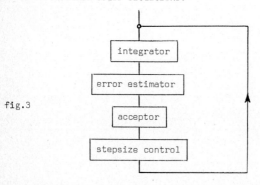

fig.3

In our model, we assume (cf. fig.3) that the stepsize control procedure is activated at the end of each "step", after an <u>error estimation procedure</u> has somehow generated an error estimate and after an <u>acceptance procedure</u> has decided whether the results of the step are to be accepted or not. Information from these two procedures is thus available for the computation of the stepsize to be used in the next step. (This "next step" is a new step or a recomputation of the previous one depending on the result of the acceptor.) This is a realistic assumption on the flow of control for most current o.d.e. codes.

The error estimation procedure delivers a quantity $L \in \mathbb{R}^S$ (s is the dimension of our problem), the subscript n refers to the step $t_{n-1} \to t_n$ which has just been executed. Our model assumption is

$$(3.1) \qquad\qquad L_n = c(\text{grid}) \; \varphi(t_n) \; h_n^p$$

where $\bar{\varphi}: [0,T] \to \mathbb{R}^S$ depends on the problem while the scalar quantity c may depend on the history of the computation, $h_n := t_n - t_{n-1}$, and p is a fixed integer. Actually, an expression like (3.1) correctly describes the lowest-order-in-h terms of the output of almost all currect error estimators in one-step or multi-step codes. φ normally consists of a linear combination of high order derivatives of f taken along the true solution of the problem. Note that the validity of the error estimate is irrelevant for the following.

Usually, only $\| L_n \|$ is used subsequently by the acceptor and the stepsize control. Here $\| .. \|$ denotes some $\mathbb{R}^S$-norm which may also depend on the local values of the approximate solution as in the case of relative error weighting. (3.1) implies the assumption

$$(3.2) \qquad\qquad \| L_n \| = c(\text{grid}) \, \bar{\varphi}(t_n) \, h_n^p$$

where now $\bar{\varphi}: [0,T] \to \mathbb{R}$ is a fixed function for a given code and a given problem.

Thus the set of problems enters into our model only through this scalar function $\bar{\varphi}$: The properties of $\bar{\varphi}$ are the only aspects of the system of o.d.e. which influence the step sequence which is generated. It is also easy to show that for each (non-pathological) choice of $\bar{\varphi}$ and a given code there exist problems with that $\bar{\varphi}$; hence "all problems" is equivalent to "all $\bar{\varphi}$" in our model.

The acceptance procedure is assumed to decide on the basis of

$$(3.3) \qquad a(h_n, \text{grid}, \text{state}) \, \| L_n \| \quad \begin{cases} \leq \text{TOL} \;\to\; \text{accept} \\[4pt] > \text{TOL} \;\to\; \text{reject} \end{cases} ;$$

the result of the decision may alter the "state" of the computation. This adequately represents virtually all acceptance procedures; the factor a may be used to accomodate the error-per-step or error-per-unit-step issue and also changes of strategy in abnormal situations.

Our object of investigation proper, the stepsize control procedure, will be assumed to evaluate

$$(3.4) \qquad\qquad h_{new} := H(h_{old}; \| L_n \|, \text{TOL}, \text{grid}, \text{state})$$

where, except in emergency situations

$$H(h_{old}; \ldots) = h_{old} \, H(\ldots).$$

Again, this does not seem to exclude any important implementations; we may further admit that the "state" of the computation may be altered as a side effect.

The coefficients c and a are chosen in correspondence with a given code and remain fixed. The function H is determined so that it represents the design of a given stepsize control procedure. Different designs may be evaluated comparatively by using different functions H.

fig.4

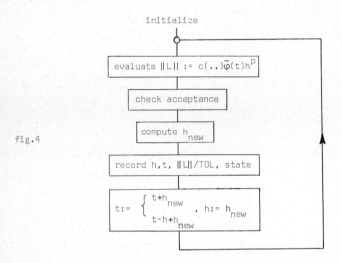

Now we may choose a function $\bar{\varphi}$ and simulate a run of the code for a specified value of TOL and an initial stepsize h_1. In simple cases, the simulation may be carried out analytically; normally we will have to use a computer to determine the sequence of steps and states which arise. A recording of the values of t_n, h_n, $\|L_n\|/TOL$, and state will constitute the trace of a simulation run which may then be checked for its agreement with various design objectives.

In this fashion we are able to evaluate the performance of different stepsize control strategies expressed by different functions H. In particular, we may check how a certain control reacts under typical situations modelled by

> $\bar{\varphi}$ constant
> increasing ⎫ ⎧ linearly
> decreasing ⎭ ⎩ exponentially
> periodic
> discontinuous

Also the case

which frequently occurs in scalar problems (change of sign in the local error estimate) will lead to different reactions from different control strategies.

Interesting aspects of the performance will be:

- Is $\|L_n\|$/TOL approximtely stationary? At which value?
- How is the reaction to typical changes in $\bar{\varphi}$?
- How frequently are steps rejected and in which situations?

Some of these aspects may be evaluated quantitatively or at least comparatively for different control designs.

The following result has been deduced analytically and confirmed by simulation runs: Let

$$(3.5) \qquad H(h;\|L\|,TOL) = \left(r \frac{TOL}{\|L\|} \right)^{1/p} h, \qquad r<1 \text{ a constant,}$$

(except for very large or small values of $\|L\|$/TOL) and let c=const in (3.2); then it follows, for a smooth function $\bar{\varphi}$, that h is controlled such that

$$(3.6) \qquad \|L_n\|/TOL = (1 + h_n \frac{\bar{\varphi}'(t_n)}{\bar{\varphi}(t_n)} + O(h_n^2)) \; r$$

if rejections are neglected. If the action of the acceptor, with a=1 in (3.3), is taken into account, the trace (3.6) of $\|L_n\|$/TOL remains essentially unchanged except that peaks are cut off.

(3.6) implies approximate proportionality of $\|L_n\|$ to TOL. Note that, in order appear as a factor in (3.6), r must be under the root in (3.5) (otherwise the factor is r^p in (3.6)). More details and further results will be presented in a separate publication.

Note that, in our model, the performance of the stepsize control is completely independent of the integrator and almost indpendent of the error estimator and the acceptor. Thus we are able to trace the effects of the stepsize control design to their source.

On the other hand, this model cannot be used for the performance evaluation of a code w.r.t. design objectives which demand an interaction of the stepsize control with the integrator and the error estimator as is the case, e.g., in the achievement of some global optimality.

EVALUATION OF INTEGRATOR/ERROR ESTIMATOR EFFICIENCY

The design of the integrator for o.d.e. has always been based on mathematical models; in fact, the numerical analysis of o.d.e. is almost exclusively concerned with the analysis of integrator models. However, these models often neglect the interaction with other modules of an o.d.e. code, particularly the error estimator and the stepsize control. Results obtained for constant and "sufficiently small" steps are often not adequate to describe the behavior of the integrator at large values of TOL.

Also, from the point of view of code design, the integrator and the error estimator appear strongly coupled: In a tolerance controlled integration, high accuracy reduces the effort only if recognized and low accuracy is intolerable if not recognized. Also the numerical computation in the integrator and the error estimator are strongly interrelated ("embedded RK-methods", PC-methods; see, e.g. [4],[1]).

On the other hand, one would like to judge the potential performance of an integrator/error estimator pair independently of the acceptor and the stepsize control which "drive" the computation. In testing, it is difficult to attribute an insufficiency to a particular module of the code.

A project to judge the efficiency of integrator/error estimator pairs via modelling has recently been published by scientists of Toronto University ([3]); I will shortly discuss the features of this "Toronto Model" because it nicely displays the advantages and the limitations of modelling in performance evaluation of o.d.e. software. For details, the publication [3] should be consulted.

The intention to consider higher-order-in-h terms throughout enforced a serious restriction of the scope of the model: The class of problems was restricted to

$$(4.1) \qquad\qquad y' = A\,y\,, \qquad A \text{ a constant matrix.}$$

To permit the formation of extrema "over all problems", the problem space was made compact by assuming $\|A\| \leqslant 1$, $\|y_0\| \leqslant 1$ (the 1's only amount to a scaling).

As formulated in [3], the model applies only to one-step methods. For given values of A, y_{n-1}, and h_n, the integrator and the error estimator deliver vectors y_n and L_n resp. The quality of this output is assessed through artificial acceptance and stepsize control procedures:

The acceptor is of type (3.3), with a coefficient function a depending on h only: For each h, the value a(h) is the minimal value which guarantees rejection if the true local error (see below) satisfies $\|u_n\| > TOL$, uniformly over "all problems".

The "true local error" $u_n \in \mathbb{R}^s$ is defined as the constant vector which solves (cf.(2.1))

$$(4.2) \qquad z(t_{n-1}) = y_{n-1}, \quad z(t_n) = y_n, \quad z'(t) = Az(t) + u_n.$$

Note that $\|u_n\| = \min_{z} \max_{t \in [t_{n-1}, t_n]} \|z' - Az(t)\|$ over all functions z which reach the solution value y_n generated by the integrator from y_{n-1}.

Since the influence of the problem (i.e. of A) has been eliminated by considering worst cases, the performance of the integrator/error estimator pair over "all problems" is fully reflected by the function H(TOL) which represents, for this pair, the largest admissible step at a given tolerance. The expression

$$(4.3) \qquad\qquad cost(TOL) = \frac{\text{no. of f-evaluations per step}}{H(TOL)}$$

is therefore a measure of the efficiency of the integrator/error estimator pair.

The development of the Toronto Model marks a great achievement: It has made it possible to study, in a realistic context, the eventual deterioration of the efficiency of high order integrators at less and less restrictive tolerances and to assess the relative efficiencies of high and low order integrators at such tolerances. A number of interesting results have been presented in [3].

On the other hand, the restriction of the problem class to (4.1) makes the scope of the model very narrow. Although one would intuitively expect that the results carry over to more general problem classes at least qualitatively, it is extremely difficult to give a rational basis for this expectation:

(4.1) can be taken as a local approximation of a general problem only for small h; but we are interested in the effects at large h. Even worse, for codes with "local extrapolation" there exists no fixed relation at all between L_n and u_n except for very restricted problem classes like (4.1); this eliminates the possibility of tuning the acceptor to the integrator/error estimator pair which is the central idea of the Toronto Model.

In order to obtain some insight into the sensitivity of the results of the model w.r.t. changes in the problem class, Mrs. Weinmüller of the author's group at Vienna has extended the approach to the problem class

(4.4)
$$y' = (A_0 + tA_1)y,$$

A_0 a fixed matrix,

A_1 arbitrary, with $\|A_1\| \leq 1$.

It has been possible to define a properly tuned acceptor and an associated stepsize control function H(TOL) for (4.4), at least for commuting matrices A_0 and A_1; thus a cost function may be defined via (4.3) as previously. First evaluations for $A_0 = I$ seem to indicate that, qualitatively at least, no dramatic changes occur in the transition from (4.1) to (4.4).

VALIDITY OF PERFORMANCE EVALUATION BY MODELLING

The crux in modelling is the <u>validity question</u>: Are the results valid in "real life"?

Restrictions of the validity may often be recognized by a scrutiny of the model assumptions: E.g., if only lowest-order-in-h terms have been considered in the model the results may (but need not) be invalid in a situation of rapid change as in the transient phase in stiff problems.

Another approach to a sensitivity analysis could consist in a careful modification of certain model assumptions and a study of the influence on the results. The consideration of the problem class (4.4) in connection with the Toronto Model is an example of this approach which may not always be feasible.

The validity question posed above hinges also on the definition of "real life problems". Very little is known about the features of the problems which are most frequently solved numerically with the aid of o.d.e. software.

(This holds as well for other areas of numerical computation. Studies like the one described in [5] are quite rare and meaningful results seem not to have been compiled anywhere so far!)

In the modelling approach to performance evaluation, the problem set is often represented by a set of parameters or functions which may have a rather indirect relation to the actual problems. Here it would be important to know which values of the parameters or types of functions occur most frequently in "real life".

So far, a tentative answer can only be obtained by compiling traces of these parameters or functions over sets of test problems. But this brings us back to the unsatisfactory situation which prevails whereever test sets are used. While a test set should be called representative if the distribution of a certain model parameter or function in the set were closely similar to the distribution in real life problems, we deduce the "real life distribution" from a test set.

Nevertheless, the insight gained through modelling into the aspects of a problem which determine the performance of a certain module of a code could partially relieve this <u>test set dilemma</u>: In experiments concerning this module, we would certainly try to use a test set in which the determining aspect would vary appropriately.

In any case, modelling and testing must not be seen as competitive or even conflicting approaches to performance evaluation but as complementary ones: The judicial use of both approaches will help us to understand more fully the functioning of numerical software and to design better codes.

REFERENCES

[1] L.F.Shampine, M.K. Gordon: Computer solution of ordinary differential equations, the initial value problem. W.H. Freeman, San Francisco, 1975.

[2] H.J. Stetter: Interpolation and Error Estimation in Adams PC-codes; to appear in SINUM, vol.16(1979) No.3.

[3] K.R. Jackson, W.H. Enright, T.E. Hull: A theoretical criterion for comparing Runge-Kutta formulas; SINUM, vol.15(1978) 618-641.

[4] D.G. Bettis: Efficient embedded Runge-Kutta methods; in: Numerical Treatment of Differential Equations, Lecture Notes in Mathematics, vol.631(1978) 9-18.

[5] H.J. Stetter, Ch.W. Öberhuber: Proposal for population studies in numerical quadrature; J. for Comp. and Appl. Math., vol.1 (1975) 213-215.

Performance Evaluation of Numerical Software, Fosdick (ed.)
© IFIP, North-Holland Publishing Company, 1979

COMPARISON OF ALGORITHMS FOR SYSTEMS OF ORDINARY
DIFFERENTIAL EQUATIONS ORIGINATING FROM PARABOLIC
INITIAL-BOUNDARY VALUE PROBLEMS IN TWO DIMENSIONS

P.J. van der Houwen & J.G. Verwer
Mathematisch Centrum, Amsterdam

1. TIME INTEGRATORS

When an initial-boundary value problem for a partial differential equation is
discretized with respect to its space variables (method of lines), an initial value
problem results for a system of ordinary differential equations. Let us write this
system in the form

$$(1.1) \qquad \frac{d\vec{y}}{dt} = \vec{f}(t,\vec{y}),$$

where $\vec{y}$ is a vector of which the components approximate the solution of the partial
differential equation at the grid points used to discretize the space variables
(note that the boundary conditions are lumped into the right hand side function $\vec{f}$).
In cases where (1.1) originates from a *parabolic* differential equation, the eigenvalue spectrum of the Jacobian matrix $\partial\vec{f}/\partial\vec{y}$ is usually located in a long strip
along the negative axis. We shall call such equations *parabolic* systems of ordinary
differential equations. The coupling in parabolic systems is very specific. When
(1.1) corresponds to a partial differential equation in two space dimensions, we
can often arrange the vector $\vec{f}$ in a two-dimensional array in which each component
of $\vec{f}$ is only coupled to its neighbours. We will be concerned with the case where
either the 4 *direct* neighbours are coupled *(five-point coupling)* or with the case
where *all* 8 neighbours are involved *(nine-point coupling)*.

Three classes of integration methods may be distinguished: (i) *explicit*,
(ii) *partially implicit* and (iii) *fully implicit* methods. At the Mathematical
Centre in Amsterdam a number of comparative experiments have been carried out for
methods from the first and second class [7, 8].

1.1 Explicit Stabilized Runge-Kutta methods

The advantage of explicit methods is the easy application to a wide class of
problems. In our case of parabolic ordinary differential equations, however, their
finite stability interval forces the algorithm to take small integration steps, in
contrast with partially or fully implicit methods which may be unconditionally
stable. The application of implicit methods, however, may be very time-consuming

per integration step (in particular the fully implicit ones) because of the large systems involved in the case of two or higher dimensional problems. Since the order of magnitude of the length of the eigenvalue spectrum of the parabolic system is the same for one-, two- and higher dimensional problems, explicit methods become more attractive as the number of space variables increases. The reason why we included explicit methods in the algorithms to be tested was not because we expected the methods to be competitive with the implicit methods, but rather to get insight into what price has to be paid for the easy applicability and the robustness of such methods.

The method we selected consists of a family of 11 second order, three-step Runge-Kutta formulas of the form

$$\vec{y}_{n+1}^{(0)} = \vec{y}_n,$$

$$\vec{y}_{n+1}^{(j)} = (1-b_j)\vec{y}_n + b_j\vec{y}_{n-1} + c_j\tau f(t_{n-1},\vec{y}_{n-1}) +$$

(1.2)
$$\lambda_j\tau f(t_n + \mu_{j-1}\tau, \vec{y}_{n+1}^{(j-1)}), \quad \tau = t_{n+1} - t_n, \quad j = 1(1)m,$$

$$\vec{y}_{n+1} = d\vec{y}_{n+1}^{(m)} + (1-d)\vec{y}_{n-2}, \quad d = 1/(c_m + \lambda_m),$$

where $\mu_0 = 0$, $\mu_j = -b_j + c_j + \lambda_j$ and $m = 2,3,\ldots,12$. The coefficients λ_j, b_j and c_j are determined by the condition that (1.2) is second order accurate and has a *maximal real stability boundary* $\beta(m)$ (cf. [9]). The stability boundaries are rather large, i.e.

$$\beta(m) \cong 2.29m^2,$$

hence the "stability interval per function evaluation" or "effective" stability interval has length 2.29m which varies between 4.58 and 27.48. In comparison with the standard one-step fourth order Runge-Kutta method, which has a real effective stability limit of magnitude .70, the stabilized methods (1.2) are a considerable improvement. The error constants corresponding to the third order Taylor terms are fixed at almost constant values .44 and 1.27 (cf. [9]), hence we need not fear a "Du Fort-Frankel inconsistency". Finally, it should be noted that (1.2) requires a relatively small number of storage arrays when implemented on a computer (only 6 arrays).

In the application of (1.2) the appropriate value of m was determined according to the formula

$$m = \text{entier}(1 + \sqrt{\frac{\tau\sigma}{2.29}}) \cong \text{entier}(1 + .66\sqrt{\tau\sigma}),$$

where σ denotes the spectral radius of $\partial\vec{f}/\partial\vec{y}$, an estimate of which was obtained during the integration process by means of the power method.

A FORTRAN and ALGOL 68 implementation of (1.2), including automatic error control and estimation of σ, are presented in [10] and [5], respectively.

1.2 One-step splitting methods

The alternating direction implicit (ADI) and hopscotch methods are familiar examples of partially implicit integration methods. In such *splitting methods* the implicitness is distributed over several stages within a single integration step. In the presentation of integration methods via the method of lines, splitting methods can be given a general and unified formulation. In the construction of algorithms for large classes of parabolic systems a general and unified description is very convenient. Let us first consider the case where $\vec{f}$ does not depend on t and let $\vec{F}(\vec{v},\vec{w})$ be a function such that $\vec{F}(\vec{y},\vec{y}) = \vec{f}(\vec{y})$. Then we may define the *second order* consistent *splitting formula* [8].

$$\text{(1.3)} \qquad \begin{aligned} \vec{y}_{n+1}^{(1)} &= \vec{y}_n + \tfrac{1}{2}\tau\vec{F}(\vec{y}_{n+1}^{(1)},\vec{y}_n) \\ \vec{y}_{n+1} &= \vec{y}_{n+1}^{(1)} + \tfrac{1}{2}\tau\vec{F}(\vec{y}_{n+1}^{(1)},\vec{y}_{n+1}) \end{aligned} \qquad .$$

This scheme is *unconditionally stable* if the matrices $\partial\vec{F}/\partial\vec{v} + (\partial\vec{F}/\partial\vec{v})^T$ and $\partial\vec{F}/\partial\vec{w} + (\partial\vec{F}/\partial\vec{w})^T$ are non-positive definite. Furthermore, when these matrices are simply structured, that is when $\vec{y}_{n+1}^{(1)}$ and $\vec{y}_{n+1}$ are easily solved in (1.3) by Newton iteration, then (1.3) is a computationally attractive scheme. The function $\vec{F}$ is called the *splitting function*. Once a function $\vec{F}$ is specified we shall speak of a *splitting method*. For an example, consider in a domain Ω the diffusion equation

$$\frac{\partial u}{\partial t} = \frac{\partial^2 u}{\partial x_1^2} + \frac{\partial^2 u}{\partial x_2^2} \; ,$$

with homogeneous Dirichlet boundary conditions on the boundary $\partial\Omega$, then the method of lines will lead to the system of ordinary differential equations

$$\frac{d\vec{y}}{dt} = A_1\vec{y} + A_2\vec{y},$$

where A_1 and A_2 are tridiagonal matrices corresponding to the three-point finite difference replacements of $\partial^2/\partial x_1^2$ and $\partial^2/\partial x_2^2$, respectively. In this case an obvious splitting function is given by

$$\vec{F}(\vec{v},\vec{w}) = A_1\vec{v} + A_2\vec{w}.$$

Substitution into the splitting formula (1.3) yields the classical ADI method of
Peaceman-Rachford [3]. We call (1.3) a splitting formula of *Peaceman-Rachford type*.

It is possible to construct many splitting formulas (for a survey see [8]).
Here, we report the performance of formula (1.3) combined with the *line hopscotch splitting* of Gourlay and McGuire [2] and a *non-linear ADI splitting*. In order to
represent these splitting functions in a compact way we rearrange the components of
$\vec{f}$ in a two-dimensional array and divide them into four sets according to the figure
below

$$
\begin{array}{ccccc}
\bullet & \circ & \bullet & \circ & \bullet \\
+ & \times & + & \times & + \\
\bullet & \circ & \bullet & \circ & \bullet
\end{array}
$$

Furthermore, we denote by $\vec{f}_\circ$ the array obtained by replacing all components of $\vec{f}$ by
zero if they do not belong to the set of type o (and similar definitions of $\vec{f}_\bullet$, $\vec{f}_+$
and $\vec{f}_\times$). The line hopscotch splitting is then given by [8]

$$(1.4) \qquad \vec{F}(\vec{v},\vec{w}) = \vec{f}_\circ(\vec{v}) + \vec{f}_\bullet(\vec{v}) + \vec{f}_+(\vec{w}) + \vec{f}_\times(\vec{w}),$$

leading to *tridiagonally implicit* equations for $\vec{y}_{n+1}^{(1)}$ and $\vec{y}_{n+1}$ in (1.3) for all *nine-point coupled* functions $\vec{f}$. The non-linear ADI splitting is given by [8]

$$(1.5) \qquad \vec{F}(\vec{v},\vec{w}) = \vec{f}_\circ(\tfrac{1}{2}\vec{v}_\circ + \vec{v}_\bullet + \vec{w}_\times + \tfrac{1}{2}\vec{w}_\circ) + \vec{f}_\times(\tfrac{1}{2}\vec{v}_\times + \vec{v}_+ + \vec{w}_\circ + \tfrac{1}{2}\vec{w}_\times) +$$

$$\vec{f}_\bullet(\tfrac{1}{2}\vec{v}_\bullet + \vec{v}_\circ + \vec{w}_+ + \tfrac{1}{2}\vec{w}_\bullet) + \vec{f}_+(\tfrac{1}{2}\vec{v}_+ + \vec{v}_\times + \vec{w}_\bullet + \tfrac{1}{2}\vec{w}_+)$$

which leads to tridiagonally implicit equations for all *five-point coupled* $\vec{f}$. The
splitting function (1.5) may be considered as an extension of the original linear
ADI splitting of Peaceman and Rachford to *general non-linear five-point coupled
functions* $\vec{f}$.

In the application of (1.3), the implicit equations were solved by Newton iteration; only one or two iterations were performed depending on the system to be linear or non-linear, respectively. The Jacobian matrices were updated (if necessary)
at the beginning of each integration step. Further details about the implementation
are given in [7].

2.3 Multistep splitting method

In an attempt to improve the order of accuracy of splitting methods, we applied
the idea of splitting to higher order multistep formulas. For instance, the four-
step backward differentiation formula

$$\vec{y}_{n+1} = \frac{1}{25} [18\vec{y}_n - 36\vec{y}_{n-1} + 16\vec{y}_{n-2} - 3\vec{y}_{n-3}] + \frac{12}{25} \tau \vec{f}(\vec{y}_{n+1})$$

is fourth order accurate and, because it has excellent stability properties for parabolic differential equations, it is a good starting point for the construction of higher order and stable splitting methods. Consider the scheme [6]

$$\vec{y}_{n+1}^{(0)} = \vec{y}_n,$$

(1.6)
$$\vec{y}_{n+1}^{(j)} = \frac{1}{25} [18\vec{y}_n - 36\vec{y}_{n-1} + 16\vec{y}_{n-2} + 3\vec{y}_{n-3}] + \frac{12}{25} \tau \vec{F}(\vec{y}_{n+1}^{(\ell)}, \vec{y}_{n+1}^{(2j-1-\ell)}),$$

$$j = 1(1)m,$$

$$\vec{y}_{n+1} = \vec{y}_{n+1}^{(m)},$$

where $\ell = j$ for j odd and $\ell = j-1$ for j even. This scheme may be interpreted as the iterative solution of the fully implicit backward differentiation formula, in which the implicitness is split up and disctributed over the successive iterations. We have tested the second order (m=2)-formula and the fourth order (m=4)-formula. For model problems where $\partial F/\partial v$ and $\partial F/\partial w$ have a common eigensystem with negative eigenvalues, these formulas can be shown to be unconditionally stable [6]. For $\vec{F}$ we selected the hopscotch splitting function (1.4). The implicit equations in (1.6) were solved by applying just one Newton iteration in each stage. The Jacobian matrices were updated (if necessary) at the beginning of each integration step (for further details we refer to [7]).

For a second attempt to improve the accuracy of splitting methods we refer to [11] where the effect of iterated defect correction (cf. [12]) applied to the LOD method is investigated.

2. PERFORMANCE EVALUATION

Until now our first concern in the performance evaluation of the algorithms discussed in the preceding section was: (i) the sensitivity of the splitting methods to the t-dependency in the right hand side $\vec{f}$ introduced by time dependent boundary conditions; (ii) the sensitivity to non-linearly coupled space derivatives; and (iii) a comparison of the behaviour of the stabilized Runge-Kutta methods and the splitting methods by (computational effort-accuracy)-plots. In the near future we intend to test strategies for the choice of the step length and for updating the Jacobian matrices or the spectral radius σ.

In the tables given below we denote by A the *accuracy* in terms of the number of correct digits, i.e. $A = -^{10}\log | maximum\ of\ the\ absolute\ error\ in\ the\ end\ point |$ and by E the number of *evaluations* of the right hand side function per integration step.

When we draw conclusions from the (A,E)-values produced by the various methods, we have to bear in mind that the computational effort involved to evaluate σ, to compute Jacobian matrices, to perform LU-decompositions, etc., is not taken into account. This part of the integration process, however, is strongly related to error and stepsize control, and is completely left out of consideration in this paper.

2.1. The effect of time dependent boundary conditions

Consider the equation [8]

$$(2.1) \qquad \frac{\partial u}{\partial t} = \Delta u + [\frac{\partial \tilde{u}}{\partial t} - \Delta \tilde{u}], \qquad 0 \le t \le 1, \qquad (x_1, x_2) \in \Omega,$$

where

$$(2.2) \qquad \tilde{u} = \sin(2\pi t)\sin(x_1 x_2 (1-x_1)(1-x_2)).$$

For Ω we chose the following squares: $\Omega_j = \{(x_1, x_2) \mid \omega_j \le x_1, x_2 \le 1 + \omega_j\}$, where $\omega_j = 0, \frac{1}{10}$ and 1, respectively. Initial and Dirichlet boundary conditions follow from choosing the function $1 + \tilde{u}$ as the exact solution. Thus, on Ω_1 the initial-boundary value problem has constant boundary conditions, on Ω_2 they are "almost" constant and on Ω_3 they are distinctly time dependent.

The problem is converted into a system of 81 ordinary differential equations by replacing Ω with a grid Γ of square meshes with sides 1/10, by semi-discretizing (2.1) on this grid using standard symmetrical differences, and by substituting the boundary conditions where boundary values of u occur in the difference formulas.

It is well-known [1] that splitting methods are sensitive to time variations in the boundary conditions. We illustrate this by applying the Peaceman-Rachford type method {(1.3), (1.5)}. However, this method should be first adapted to non-autonomous equations (1.1) in order to apply it to (2.1). At first sight this might be done by simply adding in the autonomous splitting function (1.5) the variable t, i.e.

$$(1.5') \qquad \vec{F}(t,\vec{v},\vec{w}) = \vec{f}_o(t, \tfrac{1}{2}\vec{v}_o + \vec{v}_\bullet + \vec{w}_x + \tfrac{1}{2}\vec{w}_o) + \vec{f}_x(t, \tfrac{1}{2}\vec{v}_x + \vec{v}_\bullet + \vec{w}_o + \tfrac{1}{2}\vec{w}_x) +$$

$$\vec{f}_\bullet(t, \tfrac{1}{2}\vec{v}_\bullet + \vec{v}_o + \vec{w}_+ + \tfrac{1}{2}\vec{w}_\bullet) + \vec{f}_+(t, \tfrac{1}{2}\vec{v}_+ + \vec{v}_x + \vec{w}_\bullet + \tfrac{1}{2}\vec{w}_+),$$

and by replacing the splitting formula (1.3) by

$$\vec{y}_{n+1}^{(1)} = \vec{y}_n + \tfrac{1}{2}\tau\vec{F}(t_n + \tfrac{1}{2}\tau, \vec{y}_{n+1}^{(1)}, \vec{y}_n),$$

(1.3')

$$\vec{y}_{n+1} = \vec{y}_{n+1}^{(1)} + \tfrac{1}{2}\tau\vec{F}(t_n + \tfrac{1}{2}\tau, \vec{y}_{n+1}^{(1)}, \vec{y}_n).$$

We shall call {(1.3'), (1.5')} the *symmetric* Peaceman-Rachford method (SPR method).
The results listed in table 2.1 clearly show the drop in accuracy when the

Table 2.1.

(A,E)-values by SPR for (2.1)-(2.2)

Ω	$\tau = \dfrac{1}{10}$	$\tau = \dfrac{1}{20}$	$\tau = \dfrac{1}{40}$
Ω_1	(2.31,2)	(2.92,2)	(3.53,2)
Ω_2	(2.26,2)	(2.75,2)	(3.34,2)
Ω_3	(0.79,2)	(1.23,2)	(1.79,2)

Table 2.2.

(A,E)-values by SPR for (2.1)-(2.3)

Solution u	$\tau = \dfrac{1}{10}$	$\tau = \dfrac{1}{20}$	$\tau = \dfrac{1}{40}$
$\tilde{u}$	(3.13,2)	(3.64,2)	(3.98,2)
$\tilde{\tilde{u}}$	(1.52,2)	(2.24,2)	(2.88,2)

boundary conditions become time dependent. Even on Ω_2 where the boundary conditions
are "almost" constant the effect is already noticeable. However, one may object
that the loss of accuracy is due to increased variations with time in the inhomogeneous term of the partial differential equation itself. Therefore, in the second
example [8], we fix both the inhomogeneous term and the domain Ω, that is we choose
in (2.1)

(2.3.a) $\tilde{u} = \exp[-tx_1 x_2 (1-x_1)(1-x_2)], \quad \Omega = \Omega_1,$

and define the initial and boundary conditions such that first the exact solution
is given by $u = \tilde{u}$ and secondly by $u = \tilde{\tilde{u}}$, where

(2.3b) $\tilde{\tilde{u}} = \tilde{u} + e^{-2t}\sin(x_1 + \tfrac{1}{2})\sin(x_2 + \tfrac{1}{2}).$

Thus, the boundary values are constant and time dependent, respectively.

The results listed in table 2.2 make it unavoidable to conclude that the non-
autonomous version {(1.3'), (1.5')} does not work. On second thought this is not
surprising because information about the boundary conditions is used nowhere, that
is the inhomogeneous terms in the partial differential equation and in the boundary
conditions are both treated in exactly the same manner.

In [8] this problem is discussed and the following adaptation to the boundary
conditions is proposed. Let $\vec{y}$ correspond to the values of the solution of the partial differential equation at the *internal* grid points and let $\vec{b}$ correspond to the
values at the *boundary* grid points. Furthermore, let the method of lines yield the
system of ordinary differential equations

(1.1a) $\dfrac{d\vec{y}}{dt} = \vec{f}(t,\vec{y},\vec{b})$,

and let $\vec{b}$ be expressed in the values at the internal grid points according to

(1.1b) $\vec{b} = \vec{g}(t,\vec{y})$.

Then we may define the non-autonomous ADI splitting function

(1.5'') $\vec{F}(t,t_v,t_w,\vec{v},\vec{w}) = \vec{f}_0(t,\tfrac{1}{2}\vec{v}_0 + \vec{v}_\bullet + \vec{w}_x + \tfrac{1}{2}\vec{w}_0,\vec{g}_\bullet(t_v,\vec{v}) + \vec{g}_x(t_w,\vec{w})) +$

$\qquad\qquad\qquad\qquad \vec{f}_x(t,\tfrac{1}{2}\vec{v}_x + \vec{v}_+ + \vec{w}_0 + \tfrac{1}{2}\vec{w}_x,\vec{g}_+(t_v,\vec{v}) + \vec{g}_0(t_w,\vec{w})) +$

$\qquad\qquad\qquad\qquad \vec{f}_\bullet(t,\tfrac{1}{2}\vec{v}_\bullet + \vec{v}_0 + \vec{w}_+ + \tfrac{1}{2}\vec{w}_\bullet,\vec{g}_0(t_v,\vec{v}) + \vec{g}_+(t_w,\vec{w})) +$

$\qquad\qquad\qquad\qquad \vec{f}_+(t,\tfrac{1}{2}\vec{v}_+ + \vec{v}_x + \vec{w}_\bullet + \tfrac{1}{2}\vec{w}_+,\vec{g}_x(t_v,\vec{v}) + \vec{g}_\bullet(t_w,\vec{w}))$.

The difference with (1.5') is the splitting up of the time-variable. The first variable t in $\vec{F}$ refers to the time dependency introduced by the partial differential equation itself, whereas the variables t_v and t_w refer to the time dependent boundary conditions. The splitting formula (1.3') transforms into

(1.3'')

$$\vec{y}_{n+1}^{(1)} = \vec{y}_n + \tfrac{1}{2}\tau\vec{F}(t_n + \tfrac{1}{2}\tau, t_n + \tfrac{1}{2}\tau, t_n, \vec{y}_{n+1}^{(1)}, \vec{y}_n),$$

$$\vec{y}_{n+1} = \vec{y}_{n+1}^{(1)} + \tfrac{1}{2}\tau\vec{F}(t_n + \tfrac{1}{2}\tau, t_n + \tfrac{1}{2}\tau, t_n + \tau, \vec{y}_{n+1}^{(1)}, \vec{y}_{n+1}).$$

Note that the second and third time variable of $\vec{F}$ correspond to the levels of consistency of the fourth and fifth argument of $\vec{F}$, respectively. We shall call {(1.3''), (1.5'')} the *asymmetric* Peaceman-Rachford method (APR method). The computational effort of the SPR and APR method is the same in most cases. (The adaptation to the boundary conditions presented by the APR method can presumably be improved further by applying certain ideas of Fairweather and Mitchell [1] to the splitting function (1.5'').) In the tables 2.1' and 2.2' the accuracies obtained by the APR method are given for the problems specified above. A comparison with the accuracies delivered by the SPR method clearly illustrates that the APR method improves the SPR method

Table 2.1'

A-values by APR for (2.1)-(2.2)

Ω	$\tau = \dfrac{1}{10}$	$\tau = \dfrac{1}{20}$	$\tau = \dfrac{1}{40}$
Ω_1	2.31	2.92	3.53
Ω_2	2.29	2.90	3.51
Ω_3	0.72	1.30	1.89

Table 2.2'

A-values APR for (2.1)-(2.2)

Solution u	$\tau = \dfrac{1}{10}$	$\tau = \dfrac{1}{20}$	$\tau = \dfrac{1}{40}$
$u = \tilde{u}$	3.13	3.64	3.98
$u = \tilde{\tilde{u}}$	3.13	3.62	3.94

when the time dependency of the boundary conditions is significant with respect to the time dependency of the right hand side of the partial differential equation. To test this conclusion two further experiments were carried out [8]. In (2.1) we chose respectively

$$(2.4) \qquad \tilde{u} = e^{-t}[(1-8x_1^2 + 8x_1^4)(1-8x_2^2 + 8x_2^4) - 1], \qquad \Omega = \Omega_1,$$

$$(2.5) \qquad \tilde{u} = e^{-2t}\sin(x_1 + \tfrac{1}{2})\sin(x_2 + \tfrac{1}{2}), \qquad \Omega = \Omega_1,$$

and we took $u = 1 + \tilde{u}$ as the exact solution. It is easily verified that (2.4) leads to larger time variations in the inhomogeneous term of the partial differential equation than in the inhomogeneous term of the boundary conditions, whereas (2.5) leads to a zero-inhomogeneous term. Thus, we expect no improvement in the first case and a considerable improvement in the second case. The results in the tables 2.3 and 2.4 support this expectation.

Table 2.3

A-values for (2.1) - (2.4)

method	$\tau = \dfrac{1}{10}$	$\tau = \dfrac{1}{20}$	$\tau = \dfrac{1}{40}$
SPR	1.53	2.14	2.44
APR	1.50	2.11	2.50

Table 2.4

A-values for (2.1) - (2.5)

method	$\tau = \dfrac{1}{4}$	$\tau = \dfrac{1}{8}$	$\tau = \dfrac{1}{16}$
SPR	0.69	1.27	2.03
APR	4.17	5.15	5.11

2.2 The effect of non-linearly coupled space derivatives

Consider the two-parameter family of equations [8]

$$(2.6) \qquad \frac{\partial u}{\partial t} = \frac{1}{1+t} \Delta u + \omega \left[\frac{\partial u}{\partial x_1} \frac{\partial u}{\partial x_2} \right]^\nu + \frac{(x_2^2 - x_1^2)(1+t)^{2\nu-2} - \omega(-4x_1x_2)^\nu}{(1+t)^{2\nu}},$$

$$0 \le t \le 1, \ \Omega = \Omega_1$$

with exact solution $u = 1 + (x_1^2 - x_2^2)/(1+t)$ from which initial and Dirichlet boundary conditions follow. The parameters ω and ν control the presence of non-linearly coupled space derivatives in the equation. We observe that to this type of equation (with $\omega \neq 0$, $\nu \neq 0,1$) the ADI methods in the form usually presented in the literature cannot be applied. The generalized ADI method, however, which is obtained by the method of lines approach and by defining the splitting function (1.5") formally applies to any partial differential equation which can be discretized to a five point coupled system of ordinary differential equations. Therefore, we included (2.6) in our tests to illustrate the performance of the generalized non-linear ADI method. As before, (2.6) is converted into a system of 81 ordinary differential equations by discretization on a grid with square meshes of width 1/10.

For several values of ω and ν we have compared the performance of the APR method and the line hopscotch method (LH method) of Gourlay [2] which in our notation is defined by the splitting formula (1.3") and the splitting function

(1.4") $\vec{F}(t, t_v, t_w, \vec{v}, \vec{w}) = \vec{f}_{o\bullet}(t, \vec{v}, \vec{g}(t_v, \vec{v})) + \vec{f}_{+x}(t_w, \vec{w}, \vec{g}(t_w, \vec{w}))$.

This non-autonomous version of the line hopscotch splitting (1.4) shows a similar splitting up of the time variable as the ADI splitting function (1.5"). It should be noted that the line hopscotch splitting method {(1.3"), (1.4")} admits the so-called *fast form* when implemented on a computer [2]. This means that per step the LH

Table 2.6

(A,E)-values for (2.6)

(ω,ν)	$\tau = \frac{1}{10}$	$\tau = \frac{1}{20}$	$\tau = \frac{1}{40}$	method	$\tau = \frac{1}{10}$	$\tau = \frac{1}{20}$	$\tau = \frac{1}{40}$	(ω,ν)
	(1.94,4)	(2.61,4)	(3.23,4)	SPR	(1.94,4)	(2.61,4)	(3.23,4)	
(0,1)	(3.19,4)	(3.80,4)	(4.40,4)	APR	(3.19,4)	(3.80,4)	(4.40,4)	(0,2)
	(3.19,2)	(3.68,2)	(4.28,2)	LH	(3.19,2)	(3.68,2)	(4.28,2)	
	(1.90,4)	(2.58,4)	(3.20,4)	SPR	(1.93,4)	(2.60,4)	(3.22,4)	
(1,1)	(3.00,4)	(3.61,4)	(4.21,4)	APR	(3.14,4)	(3.78,4)	(4.38,4)	(1,2)
	(3.17,2)	(3.67,2)	(4.26,2)	LH	(3.11,2)	(3.65,2)	(4.25,2)	
	(1.77,4)	(2.46,4)	(3.08,4)	SPR	(1.91,4)	(2.57,4)	(3.19,4)	
(5,1)	(2.52,4)	(3.13,4)	(3.74,4)	APR	(2.14,4)	(2.84,4)	(3.45,4)	(5,2)
	(2.55,2)	(3.71,2)	(4.30,2)	LH	(1.53,2)	(2.39,2)	(4.21,2)	

method requires half the number of right hand side evaluations as required by the APR (and SPR) method (cf. table 2.6). When we compare the accuracies produced by the APR and LH method which are listed in table 2.6, we see that in the linear case ($\omega = 0$) the APR method is slightly more accurate but that an increase of the non-linearities in (2.6) by increasing ω and ν, does affect the APR method more than the LH method for small integration steps and vice versa for larger integration steps. On the whole, our conclusion is that the non-linear ADI splitting (1.5") is a satisfactory non-linearization of the classical linear ADI splitting of Peaceman and Rachford when compared with the behaviour of the LH method for increasing non-linearities. However, when the computational effort per integration step is also taken into account, the LH method, due to its fast form implementation, clearly is the superior one. Moreover, the LH method is applicable to nine-point coupled $\vec{f}$-functions, whereas the APR method requires five-point coupled right hand side functions (which excludes mixed derivative terms and curvi-linear grid lines).

Finally, we observe that the results of the SPR method, which we included in table 2.6 as a further illustration of the preceding subsection, show the effect of a careful treatment of boundary conditions in splitting methods.

2.3 Accuracy-computational effort plots

From the preceding section we may conclude that the line hopscotch method is presumably the most efficient method in the class of single step splitting methods. Therefore, in the comparative tests we have selected this method, i.e. {(1.3"), (1.4")}, as the representative of this class. Furthermore, we have chosen the stabilized Runge-Kutta (SRK) method (1.2) and the four-step method (1.6) with $\vec{F}$ defined according to (1.4") (the FLH(m) method).

In all examples (which were taken from [2,7]), the integration interval was $0 \le t \le 1$, the domain Ω was the L-shaped region Ω_4 of figure 2.1, and the grid consisted of square meshes of width 1/21 which yields a system of 292 ordinary differential equations. Below, we present the partial differential equation, the exact solution (from which the Dirichlet boundary conditions and initial condition are derived), and the (A,C)-plots where C denotes the total number of right hand side evaluations. The SRK, LH, FLH(2) and FLH(4) curves are indicated by 1, 2, 3 and 4.

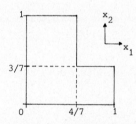

Fig. 2.1 L-shaped region Ω_4

(2.7) $\dfrac{\partial u}{\partial t} = u^2(\Delta u - \dfrac{3}{2}\dfrac{\partial^2 u}{\partial x_1 \partial x_2}) + au^2 + b,$

$a = 2t^2(x_2 - x_1 - \sin 2\pi t),$

$b = t^2[(x_1^2 + x_2)(\dfrac{2\sin 2\pi t}{t} +$

$2\pi \cos 2\pi t + 2\dfrac{x_1 x_2^2}{t}],$

$u = 1 + t^2[(x_1^2 + x_2)\sin 2\pi t + x_1 x_2^2].$

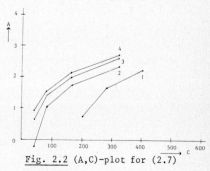

Fig. 2.2 (A,C)-plot for (2.7)

(2.8) $\dfrac{\partial u}{\partial t} = \dfrac{1}{1+t}(2\dfrac{\partial^2 u}{\partial x_1^2} + \dfrac{\partial^2 u}{\partial x_2^2}) + 5\dfrac{\partial u}{\partial x_1}\dfrac{\partial u}{\partial x_2} +$

$2(1-u)\dfrac{\partial^2 u}{\partial x_1 \partial x_2} + a,$

$a = \dfrac{(16x_1 x_2 - 1)(x_1^2 + 2x_1^2 x_2^2 - x_2^2) - 4(1+2x_2^2)^2}{(1+t)^2}$

$+ \dfrac{2(2x_1^2 - 1) + 20x_1 x_2 (1+2x_2^2)(2x_1^2 - 1)}{(1+t)^2},$

$u = 1 + \dfrac{x_1^2 + 2x_1^2 x_2^2 - x_2^2}{1+t}$

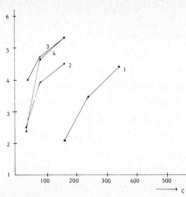

Fig. 2.3 (A,C)-plot for (2.8)

(2.9) $\dfrac{\partial u}{\partial t} = d\left[a\dfrac{\partial^2 u}{\partial x_1^2} + 2b\dfrac{\partial^2 u}{\partial x_1 \partial x_2} + c\dfrac{\partial^2 u}{\partial x_2^2}\right],$

$a = \tfrac{1}{2}x_1^2 + x_2^2,$

$b = -(x_1^2 + x_2^2),$

$c = x_1^2 + \tfrac{1}{2}x_2^2,$

$d = [1 + x_1 x_2 (x_1 + x_2)e^{-t}]^{-1},$

$u = x_1 x_2 (x_1 + x_2)e^{-t}$

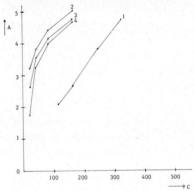
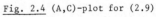

Fig. 2.4 (A,C)-plot for (2.9)

(2.10) $\dfrac{\partial u}{\partial t} = \sqrt{u}\,\dfrac{\partial^2 u}{\partial x_1^2} - \dfrac{1}{1+t}\dfrac{\partial^2 u}{\partial x_1 \partial x_2} +$

$\sqrt{u}\,\dfrac{\partial^2 u}{\partial x_2^2} + \dfrac{2u}{1+t} - 2u\sqrt{u},$

$u = \dfrac{e^{-x_1 - x_2}}{\sqrt{1+t}}$

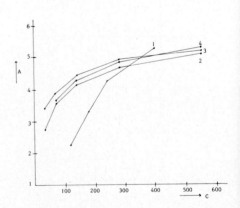
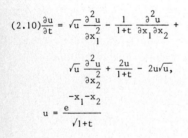

Fig. 2.5 (A,C)-plot for (2.10)

Further experiments, including Von Neumann type boundary conditions, are reported in [7]. From these and the preceding experiments we may conclude that the three line hopscotch methods are competitive and several times cheaper than the stabilized Runge-Kutta methods. The higher overhead costs of the LH methods (up dating of the Jacobian matrix, LU-decompositions, etc.) will reduce this advantage, but the SRK methods still require a high price for their robustness and easy applicability. In three-dimensional problems, however, the SRK methods might be competitive with implicit methods.

The four-step method FLH(4) does not show its fourth order consistency in the accuracy range we tested, that is its effective order is less than 4. Some additional experiments showed that the fourth order behaviour is demonstrated by the FLH(m) methods for *larger* m-values. Such methods, however, require too many function evaluations per integration step. In this connection we mention a second attempt to derive high order splitting methods presented in [11], where *iterated defect correction* [12] is applied to the LOD method. Although the iterated method is more efficient than the single LOD method, it is more or less comparable with the LH method, and again the reason is that the effective order in the relevant accuracy range is lower than the "asymptotic" order of consistency.

Our final conclusion is that the LH method in its fast form is one of the most attractive methods for integrating parabolic equations in two-dimensions.

ACKNOWLEDGEMENTS

The authors are indebted to Mr. B.P. Sommeijer for carrying out all numerical experiments and for many constructive remarks during the preparation of this paper.

REFERENCES

[1] Fairweather, G. & A.R. Mitchell: A new computational procedure for ADI methods, SIAM J. Numer. Anal. 4, 163-170, (1967).

[2] Gourlay, A.R. & G. McGuire: General hopscotch methods for partial differential equations, J. Inst. Math. Applics. 7, 216-227, (1971).

[3] Peaceman, D.W. & H.H. Rachford Jnr.: The numerical solution of parabolic and elliptic equations, J. Soc. Ind. Appl. Math. 3, 28-41, (1955).

[4] Sommeijer, B.P.: An ALGOL 68 implementation of two splitting methods for semi-discretized parabolic differential equations, Note NN 15/77, Mathematisch Centrum, Amsterdam, (1977).

[5] _____: Parabolic PDE, an ALGOL 68 implementation for the time integration of semi-discretized parabolic differential equations, Note NN 16/78, Mathematisch Centrum, Amsterdam, (1978).

[6]   Van Der Houwen, P.J.: Multistep splitting methods of high order for initial
      value problems, extended version of Report NW 49/77, Mathematisch Centrum,
      Amsterdam, (prepublication) (1978).

[7]   _____, B.P. Sommeijer & J.G. Verwer: Comparing time integrators for parabolic
      equations in two space dimensions with a mixed derivative, J. of Comp. and
      Appl. Math. 5, No. 2, (1979).

[8]   _____ & J.G. Verwer: One-step splitting methods for semi-discrete parabolic
      equations, Report NW 55/78, Mathematisch Centrum, Amsterdam, (submitted for
      publication in Computing) (1978).

[9]   Verwer, J.G.: A class of stabilized three-step Runge-Kutta methods for the
      numerical integration of parabolic equations, J. of Comp. and Appl. Math. 3,
      155-166, (1977).

[10] _____: An implementation of a class of stabilized, explicit methods for the
      time integration of parabolic equations, Report NW 38/77, Mathematisch Centrum,
      Amsterdam, (prepublication) (1977).

[11] _____: The application of iterated defect correction to the LOD method for
      parabolic equations, Report NW 58/78, Mathematisch Centrum, Amsterdam, (in a
      condensed form this report will appear in BIT) (1978).

[12] Frank, R. & C.W. Ueberhuber: Iterated defect correction for the efficient solu-
      tion of stiff systems of ordinary differential equations, BIT 17, 146-159,
      (1977).

Performance Evaluation of Numerical Software, Fosdick (ed.)
© IFIP, North-Holland Publishing Company, 1979

USING A TESTING PACKAGE FOR THE AUTOMATIC ASSESSMENT
OF NUMERICAL METHODS FOR ODE'S*

W. H. Enright
Dept. of Computer Science
University of Toronto

The need and requirements for software tools to
aid in the assessment of numerical methods for
ODE's is discussed. The need for including
measures of reliability and efficiency is emphasized, and a collection of test programs that
have been developed and extensively used is
described. Examples are given of the application
of these test programs to assess and compare
methods for both stiff and nonstiff problems.

INTRODUCTION

During the past several years, a group of us at Toronto have been
involved in the development, assessment and comparison of numerical
methods for ODE's. These investigations have convinced us of the
need for a collection of testing programs to aid in the evaluation
of numerical methods. The design and use of such a package is the
subject of this paper. The collection of programs actually consists of two very similar packages: one for the assessment of nonstiff methods and another for the assessment of stiff methods.
Before we discuss the test programs in more detail, we must describe the type of applications we envisage for it and distinguish
two points of view that will be taken when interpreting the results.

With the number of new methods being developed and considered for
publication, it is a formidable task for an author or a referee to
evaluate a new method. In order to publish an algorithm in a
technical journal, the author is expected to provide evidence of
its performance characteristics compared to that of other methods.
This evidence all too often takes the form of results of the new
method relative to out-of-date methods on a few carefully selected
problems. Refereeing such submissions is difficult, and a referee
is often reluctant to ask for more evidence since this often would
require that an author obtain and implement a more appropriate
reference method and run several new test problems. On the other
hand, if a catalogue of standard benchmark results were available
for existing software, and if these results could be duplicated and
results for new methods readily obtained, then we could require
much stronger evidence of performance for new methods. Another use
of the testing package would be that of a potential user with a
particular type of problem. He may wish to determine which method

---

* This work was supported in part by the National Research Council
of Canada and in part by the University of Texas at Austin while
the author held a visiting appointment in the Department of Aero-
space Engineering and Engineering Mechanics.

is most suitable for him by benchmarking the performance of available methods on a set of his own model problems.

When assessing the performance of a numerical method, two points of
view may be taken. One may be primarily interested in the evaluation or assessment of the performance of an individual method, or
alternatively, he may be interested in comparing the relative
performances of different methods. In the former case, it is
essential that the performance criteria and performance monitors
introduced have no effect on the method being assessed. That is,
we must ensure that the method is not modified in any way, as such
modifications might degrade the performance of the method. In
order to satisfy this condition, the testing package must be flexible and able to assess a particular method on how well it (the
method) is accomplishing what it was designed to accomplish.

On the other hand, when one is comparing methods, the various
measures of reliability and efficiency can be compared only if the
methods being compared are accomplishing the same task. One way to
ensure that this is the case is to modify the methods so that they
are all attempting to accomplish the same task. This was the
approach adopted in Hull, et al (1972), Enright, et al (1975) and
Enright, et al (1976). This approach allows one to compare the
relative performances of different techniques or approaches. It
must be acknowledged, however, that modified methods are compared,
and these modifications might have a greater effect on some methods
than others.

An alternate approach, which also permits a meaningful comparison
of reliable methods, is used in our current testing package. Rather than modify methods before testing, we modify or "normalize"
the cost or efficiency statistics after the assessment of a method
to account for the fact that different methods will be accomplishing similar, but different, tasks. Before we can be more precise
about what a reliable method is and how we normalize the efficiency
results, we will describe the minimum requirements that must be met
before a method can be assessed using our package. These requirements involve the identification of a basic error control option
that must be available as well as an understanding of how the
accuracy parameter, TOL, is to be interpreted. We assume only that
control of absolute error is one of the options available and that
the method is attempting to ensure that the global error is proportional to the prescribed tolerance.

We must emphasize that it is essential that one assess the reliability of a method as well as its cost or efficiency. In fact, we
believe that it is only appropriate to discuss the efficiency of a
reliable method. The technique that we use to normalize the cost
or efficiency measures is, strictly speaking, only appropriate for
reliable methods. Our measures of reliability are based on assessing how well a method is able to achieve proportionality between
global error and the prescribed tolerance, and we normalize the
efficiency results to account for the fact that methods will
satisfy the proportionality requirements differently.

In the next section, we will briefly describe how the testing
package is used and what options are available. In the final two
sections, we will describe typical applications of the nonstiff and
stiff testing packages and give some guidance in how to interpret
the results.

OVERVIEW OF THE TESTING PACKAGE

In this section, we will describe and motivate the measures of reliability and efficiency that are produced by our testing package.
Our collection of test programs actually consists of two very
similar packages of programs: one for assessing methods for nonstiff equations and another for assessing methods for stiff equations. Although these packages share a common organization and
structure, they are based on a different set of test problems, and
the package for stiff methods determines some additional efficiency
measures. The complete collection of test programs is documented
and described in more detail in Enright and Hull (1979).

The assumption is made that the method being assessed is attempting
to satisfy:

$$\text{global error} \cong C \cdot TOL^E \quad , \tag{1}$$

where it is understood that the appropriate constants C and E
will be method- as well as problem-dependent, but that the exponent
E should always be close to 1. Most users assume a relationship
of this type, and most software that has been introduced in the
past few years (either explicitly or implicitly) satisfies this
assumption. It is clear that one will achieve this proportionality
only over an appropriate range of error tolerances since, at re-
laxed values of TOL, the error estimate will become unreliable,
while at stringent values of TOL, round-off errors will dominate
and destroy this proportionality. The measure of global error used
in (1) will depend on the type of error control specified, and we
will assume that one is able to prescribe absolute error control.
The testing package will allow the option of using either the endpoint error or the maximum observed error (over all steps of the
integration).

A user will invoke the testing package by selecting a set of problems and a set of values of tolerance to be used for each problem.
The user must also specify the level of detail desired for the
standard assessment statistics as well as whether he would lke the
results normalized (to permit comparisons). For each of the selected problems, the testing package will solve the problem at the
specified error tolerances, TOL_i (i = 1,2...NTOL), monitoring and
printing the appropriate standard assessment statistics as well as
the corresponding global $error_i$ (i = 1,2...NTOL). The validity of
the relationship (1) will then be determined by computing the
values of C and E which minimize

$$RES = \sum_{i=1}^{NTOL} [\ln(\text{global error}_i) - \ln(C) - E \cdot \ln(TOL_i)]^2 \quad , \tag{2}$$

a method will be considered to have been reliable on a problem
if the corresponding value of E is near one and the value of
RES/NTOL is small.

Once the values of C and E have been computed, it is possible
to define the expected accuracy at a tolerance, TOL, to be $C \cdot TOL^E$.
(This is all the accuracy a user can expect in the ideal situation
that he knows the appropriate values of C and E .) One can then
"normalize" the cost vs. accuracy profile by tabulating expected
accuracy vs. cost rather than TOL vs. cost or global error vs. cost.

The normalized tables that are produced by the testing package are approximations to the expected accuracy vs. cost curves. These normalized tables are produced using piecewise linear interpolation of the actual recorded values of the cost (to solve the problem at TOL_i) vs. the corresponding expected accuracy, $C \cdot TOL_i^E$, (for i = 1,2...NTOL). This piecewise interpolant is evaluated at a predetermined set of expected accuracy values to permit comparisons of tables of costs for the same expected accuracies.

The lowest level of detail possible for the standard assessment statistics is the most inexpensive to apply and consists of the following reported results:

> TIME   - the total time (in seconds) required to
>   solve the problem,
>
> OVHD   - the total time (in seconds) excluding the
>   time used in evaluating the derivative,
>
> FCN CALLS   - the total number of derivative evaluations
>   required to solve the problem,
>
> NO OF STEPS   - the total number of steps required to solve
>   the problem
>
> GLB ERR AT E PT   - the global error at the endpoint, measured
>   in the max-norm and expressed in units of
>   TOL.

For the stiff package, two additional statistics are output at this level:

> JAC CALLS   - the number of Jacobian evaluations required
>   to solve the problem,
>
> EQ MAT FACT   - a count of the number of $O(n^3)$ operations
>   (when n is the size of the system) ex-
>   pressed as a multiple of the number of L-U
>   decompositions.

After solving the problem at the specified error tolerances, the corresponding values of C, E and RES for the endpoint error are also output (C-END, E-END and RES-END). The second level of detail includes, in addition to the statistics of level 1, the maximum observed global error, MAX GLB ERR . To determine this, the testing package must compute the "true" solution at the steps determined by the method being assessed. Since this is done using a reliable and accurate reference method, the cost of applying the testing package with this level of detail is larger than that with the lowest level of detail. When the second level of statistics is requested, the testing package will also output the values of C, E and RES corresponding to the maximum observed error.

The final level of detail possible includes, in addition to the statistics of the second level, an assessment of the local reliability of the method being tested. If the method being assessed is attempting to control the global error by monitoring and controlling the local error, then the testing package can compute the "true" local error on each step and compare it to the bound that was assumed satisfied. Of course, this bound will be method-

dependent and can change from step to step. Since this third
level of detail requires that the testing package perform two
parallel integrations (for the "true" local and "true" global
solutions), it is the most expensive option to apply.

If level one statistics have been specified, then normalized
efficiency statistics are determined, if requested, using the
values of C-END and E-END. If level two or level three has been
specified, then the user can request that normalized efficiency
statistics be determined using either endpoint error or maximum
observed error. We will see that this latter option is particularly useful when assessing methods for stiff equations.

A sample output from the nonstiff package, typical of the table
produced for each problem is presented in Table 1. Notice that we
have requested level three detail as well as normalized efficiency
results corresponding to the endpoint error. An example of the
reported results from the stiff package with level three detail
requested, as well as normalized results corresponding to the
maximum observed error, is presented in Table 2. In addition to
results such as these for each problem, the testing package also
outputs summary statistics, after each problem class, and overall
summaries after all problems have been completed. These summaries
are not normalized, though, and would be inappropriate for comparisons.

In the next section we will describe some applications of the
nonstiff testing package and give some advice on interpreting the
results. In the final section, we will do the same for the stiff
package.

APPLICATIONS OF THE NONSTIFF PACKAGE

The nonstiff package contains a set of 30 standard differential
equations divided into six problem classes. These problems include
the five classes of problems described in Hull, et al (1972) as
well as an additional class of five differential equations with
discontinuities. The user may select any subset of the standard
problems or he may choose to add his own class of problems to the
package. The way in which problems are specified and new problems
defined is described in detail in Enright, et al (1979). To
illustrate typical applications of the package, we have selected a
subset of six problems and have run tests on most of the available
nonstiff numerical methods. To illustrate the type of behavior to
be expected and the care that must be exercised when interpreting
the results, we will discuss the relative performances of five of
the better methods assessed. A complete assessment of all available methods on the 30 problems has been catalogued and the results
stored on tape. It is our plan to distribute our testing packages
to others so that our results can be duplicated and new results
obtained.

The first five problems we have chosen for this assessment are A3,
B1, C2, D4 and E2 of Hull, et al (1972). The five methods we have
chosen to use are the Runge-Kutta methods RKF45 of Shampine and
Watts (1978) and DVERK of Hull, et al (1976) (also in the IMSL
software library, IMSL (1977)); an extrapolation code DODES,
developed by Schryer (1975) and included in the Bell Labs PORT
library; and the variable-order Adams methods DE/STEP of Shampine
and Gordon (1975), and VOAS of Sedgwick (1974).

Table 1

Sample output from nonstiff package

CLASS B1   DESTEP

TOL	TIME	OVHD	FCN CALLS	NO OF STEPS	GLB ERR AT E PT	MAXIMUM GLB ERR	MAXIMUM LOC ERR	FRACTION LOC DECV	FRACTION BAD DECV
$10^{**}-2$	0.052	0.050	187	88	41.06	252.286	6.49	0.102	0.011
$10^{**}-4$	0.117	0.115	375	183	2.92	29.407	5.81	0.060	0.005
$10^{**}-6$	0.184	0.180	577	284	14.50	29.020	3.08	0.028	0.0
$10^{**}-8$	0.283	0.277	849	420	52.03	226.776	2.43	0.017	0.0
$10^{**}-10$	0.436	0.427	1228	604	2.01	36.691	1.77	0.013	0.0

SMOOTHNESS	C-END		E-END		RES-END		C-MAX		E-MAX	RES-MAX
	29.0380		1.0685		7.8727		122.1496		1.0394	4.6405

NORMALIZED EFFICIENCY - ENDPOINT ERROR

B1 EXP ACC	TIME	OVHD	FCN CALLS	NO OF STEPS
$10^{**}\ 0$	0.031	0.030	127	58
$10^{**}-1$	0.062	0.060	215	102
$10^{**}-2$	0.092	0.090	303	146
$10^{**}-3$	0.123	0.120	392	191
$10^{**}-4$	0.154	0.151	487	239
$10^{**}-5$	0.187	0.182	583	287
$10^{**}-6$	0.233	0.228	710	350
$10^{**}-7$	0.279	0.273	838	414
$10^{**}-8$	0.348	0.341	1011	498
$10^{**}-9$	0.420	0.412	1188	584

Table 2

Sample output from stiff package

CLASS B4   SECDER

TOL	TIME	OVHD	FCN CALLS	JAC CALLS	MAT FACT	NO OF STEPS	GLB ERR AT E PT	MAXIMUM GLB ERR	MAXIMUM LOC ERR	FRACTION DECEIVED
$10^{**}-2$	0.141	0.110	67	36	20	33	0.00	0.22	0.141	0.0
$10^{**}-4$	0.255	0.220	127	64	21	63	0.01	0.64	0.374	0.0
$10^{**}-6$	0.393	0.353	211	106	23	105	0.02	0.46	0.243	0.0
$10^{**}-8$	0.626	0.575	327	164	28	163	0.01	0.49	0.195	0.0

SMOOTHNESS	C-END		E-END		RES-END		C-MAX		E-MAX	RES-MAX
	0.0010		0.8356		0.5464		0.2474		0.9539	0.4275

NORMALIZED EFFICIENCY - MAXIMUM GLB ERROR

B4 EXP ACC	TIME	OVHD	FCN CALLS	JAC CALLS	MAT FACT	NO OF STEPS
$10^{**}-2$	0.111	0.080	51	28	20	25
$10^{**}-3$	0.170	0.138	82	42	20	40
$10^{**}-4$	0.229	0.196	113	56	20	55
$10^{**}-5$	0.296	0.260	152	76	21	75
$10^{**}-6$	0.369	0.330	196	98	22	97
$10^{**}-7$	0.475	0.431	251	126	24	125
$10^{**}-8$	0.597	0.547	311	156	26	155

Each method was run with level three standard statistics at tolerances of 10^{-2}, 10^{-4}, 10^{-6}, 10^{-8}, and 10^{-10}, and normalized efficiency statistics were requested. Hence, the format of the results reported for each method on each problem is that of Table 1. A summary of the reliability statistics is presented in Table 3. The summaries of Table 3 verify that the methods assessed are reliable and able to obtain proportionality between achieved accuracy and requested tolerance over a wide range of error tolerances.

Table 3

Summary of reliability statistics on the
five problems A3, Bl, C2, D4 and E2

	RKF45	DVERK	DODES	DE/STEP	VOAS
Average E-END	.94	1.04	1.12	1.07	1.04
Average RES-END	4.70	10.30	4.90	4.18	1.70
Average E-MAX	.92	.96	1.13	1.01	.98
Average RES-MAX	1.20	2.10	5.50	3.00	2.40

Notice that the average values of E-MAX and E-END for DODES are a little larger than might be expected and that the value of average RES-END for DVERK is relatively large. In all cases, the local reliability statistics revealed that the methods were able to keep the magnitude of the local error within a small multiple of the appropriate bounds.

The normalized efficiency statistics can be used to compare the relative cost vs. accuracy profiles of each method for each measure of cost. For nonstiff methods, the most appropriate measures of cost are total time and the number of derivative evaluations. It is well known that the low overhead and fixed order of the Runge-Kutta techniques makes them most appropriate if derivative evaluations are inexpensive and the order of the Runge-Kutta method is chosen to match the desired accuracy. On the other hand, the variable order Adams methods are designed to keep function evaluations to a minimum and match the order to the requested tolerance. As a result, these methods will be most suitable when derivative evaluations are expensive to evaluate or a wide range of tolerances is to be used. It is also clear that, while the individual normalized tables of each method on each problem can be compared, superimposing plots of the corresponding cost vs. accuracy profiles allows a quicker overview of the relative performances (especially if several methods are to be evaluated). This approach has the disadvantage that only one measure of cost can be compared in each plot. We have chosen to use such plots of FCN CALLS vs. expected accuracy to present the relative performance on three of the problems listed above. In interpreting the results, then, it is important to realize that the overhead of the methods is neglected. Figure 1 shows the normalized cost curves for problem Bl, a standard nonstiff problem (see the appendix for the definition of this

problem). This example is included to indicate the typical
behavior observed on standard nonstiff problems.

Two difficulties which can arise in the application of nonstiff
methods, and with which good software should cope, are mild stiffness and discontinuities in the derivative. Figures 2 and 3,
showing the relative performances on problems C2 and F2, demonstrate the reaction of the five methods to these difficulties
(see the appendix for the definition of these problems). It is
interesting to note that on the mildly stiff problem C2, all codes
were able to retain proportionality of global error to prescribed
tolerance. It is also clear that after the initial transient
region, over which the transient components were accurately repro-
duced and the significant errors introduced, the stepsize for the
Runge-Kutta and Adams methods was determined by the stability
regions associated with the particular formulas. As a result, the
cost was almost independent of the tolerance, and this is reflected
in the cost curves. The stability properties of the extrapolation
techniques are more difficult to analyze, but it is clear that on
mildly stiff problems, it is not competitive with the other methods.

Problem F2 has several discontinuities in the derivative over the
range of integration. We have run the above-named methods on this
problem with tolerances of 10^{-2}, 10^{-4}, 10^{-6} and 10^{-8}; with level 1
standard assessment statistics; and with a request for normalized
results. The reliability measures determined for this problem are
given in Table 4. The method VOAS was unable to solve the problem
at any tolerance since its error control is based on keeping an
estimate of the error-per-unit-step less than the tolerance.

Table 4

Results on Problem F4

	RKF45	DVERK	DODES	DE/STEP
E-END	.70	.77	.81	.68
RES-END	9.50	2.30	.20	1.70
Total time (in seconds) for expected accuracy = 10^{-5}.	.49	.96	3.50	.90

The other methods are based on a local error strategy that attempts
to ensure that the error-per-unit step of the "extrapolated" solution is proportional to the prescribed tolerance by keeping an
estimate of the error per step of the unextrapolated solution less
than the tolerance. (This idea is discussed more fully in Shampine
(1977).) As might be expected, these methods are able to cope with
discontinuities, although the respective values of E-END will be
less than one. Also, since all formulas reduce to first order when
applied to this problem, the lower-order methods will out-perform
the high-order methods as they have a lower per-step cost. To
illustrate the relative overhead costs of these methods when dealing with discontinuities, Table 4 also contains the normalized
timing results for the four methods for an expected accuracy of
10^{-5}.

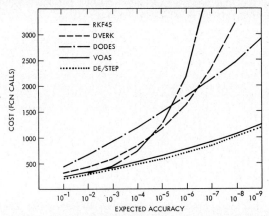

Figure 1.   Normalized results on problem Bl

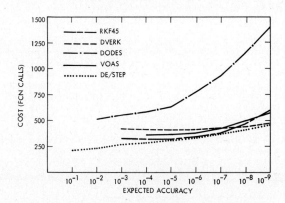

Figure 2.   Normalized results on problem C2

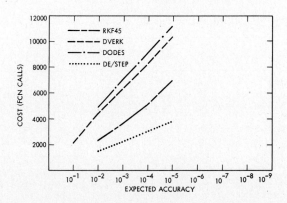

Figure 3.   Normalized results on problem F2

APPLICATION OF THE STIFF PACKAGE

Although the overall design and appearance to the user of the stiff
test package is almost identical to that of the nonstiff package,
there are a few differences that must be understood when using the
package. First of all, the size of the system, n, is very important for stiff systems since, to a large extent, this value determines what the significant cost will be when solving a stiff
problem. That is, of the efficiency statistics that are produced,
it will generally be the case that, although the respective counts
should be relatively independent of the size of the system, the
relative importance of the different statistics will not. It is
clear that the $O(n^3)$ operations will dominate the total execution
time for large n, and hence the most suitable measure of cost
would be the EQ MAT FACT. On the other hand, for small-to-moderate
values of n, the FCN CALLS and JAC CALLS would be the most suitable
measures of cost.

The stiff test package contains a set of 30 standard stiff ODE's
consisting of the 25 problems of Enright, et al (1975) and five
additional chemical kinetics problems of Enright, et al (1976). As
with the nonstiff package, the user may select any subset of the
standard problems, or he may add his own class of problems.

We have assessed the performance of most available methods for
stiff equations on the standard problems, and the results are
available on tape. To illustrate the use of the package, we have
run five methods with level-three statistics and a request for
normalized results at tolerances 10^{-2}, 10^{-4}, 10^{-6} and 10^{-8}. The
subset of problems selected were A2, B4, C4, D3 and E3 of Enright,
et al (1975) and CHM 9 of Enright, et al (1976). The methods
assessed were two implementations of a multistep method based on
the backwards differentiation formulas, GEAR (developed by Hindmarsh (1974)) and EPISODE (developed by Byrne and Hindmarsh
(1975)); a blended multistep method, SKEEL (developed by Skeel and
Kong (1977)); a composite multistep method, STINT (developed by
Tendler, et al (1978)); and a second derivative multistep method,
SECDER (developed by Addison (1979)). A summary of the respective
reliability results appears in Table 5.

Table 5

Summary of reliability results on six stiff problems

	GEAR	EPISODE	SKEEL	STINT	SECDER
Average E-END	.80	.81	1.03	.78	.78
Average RES-END	4.20	3.00	8.60	4.80	3.70
Average E-MAX	.90	.84	.92	.87	.94
Average RES-MAX	.20	.20	.50	.70	.60

In all cases, the local reliability statistics indicated that the
methods were effective in keeping the local error within the
appropriate bound. From Table 5, one can observe that the value of

E is usually much closer to one, and the corresponding value of RES is smaller for the maximum observed error than for the endpoint error. This has generally been observed to be the case for all stiff methods tested. Of course, it is also true that the value of C-MAX is generally quite a bit larger than that of C-END. The reason for this behavior can be understood when one recognizes that the maximum error usually occurs in the initial transient region where accuracy of the transients is important, while the endpoint error is made up of two components: the numerically damped initial errors in the transient components and the errors of the "smooth" or steady-state components. Although the initial transient errors and the errors of the smooth components should both be proportional to TOL, the amount of dampening will depend on the average stepsize and will increase as the stepsize decreases. It is also clear that if a large proportion of the steps has been constrained for other than accuracy reasons (for example, the method may have a swing factor or the stepsize could be restricted to hit the endpoint), then the proportionality will also be lost. We have chosen to normalize the results according to the maximum observed global error.

Because of the number of different measures of cost that are determined by the stiff package and because their relative importance depends on the size and complexity of the system, it is clear that the complete normalized tables would have to be compared for a thorough comparison. Table 2 illustrates the format of the normalized tables that are produced for each problem.

To clarify the presentation and to provide an indication of the relative performances of the methods for small-to-moderate-sized systems, we have plotted a cost-vs.-accuracy profile based on normalized statistics with the measure of cost being the sum of FCN CALLS plus JAC CALLS. Figures 4, 5 and 6 illustrate the behavior of the methods on problems B4, D3 and E3, respectively (see the appendix for a definition of these problems).

Problem B4 is a linear problem with eigenvalues -0.1, -0.5, -1, -4, and $-10 \pm 25i$. It is well known that methods based on backwards differentiation formulas will experience difficulty when there are eigenvalues close to the imaginary axis, and Figure 3 illustrates the type of behavior that can be expected. It must be emphasized that both EPISODE and GEAR retain reliability as the eigenvalues approach the imaginary axis; they just become less efficient, and the cost becomes unpredictable.

Problem D3 is a nonlinear problem with a Jacobian matrix whose eigenvalues remain real during the integration, while problem E3 is also a nonlinear problem; but the corresponding Jacobian matrix has some nonreal eigenvalues. The results demonstrate that the good variable order methods have very similar cost-vs.-accuracy profiles over the range of tolerances we have chosen.

We feel that these testing packages will prove to be extremely valuable in the development of new methods as well as in the assessment of new and existing methods. Hopefully, the distribution of testing aids, such as the one discussed in this paper, will make it possible to quantify more effectively, and with very little effort, the advantages and disadvantages of new methods as they are introduced.

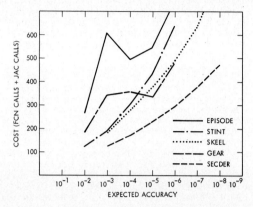

Figure 4.   Normalized results on stiff problem B4

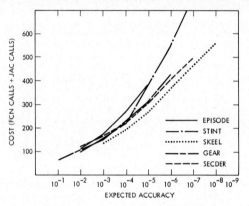

Figure 5.   Normalized results on stiff problem D3

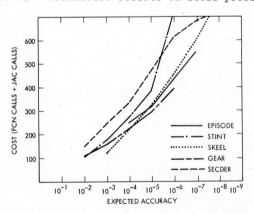

Figure 6.   Normalized results on stiff problem E3

REFERENCES

|1| C. Addison: User's Guide for SECDER - a subroutine for solving
stiff systems of ODE's, Dept. of Comp. Science Tech. Rep.
(1979) (in preparation).

|2| G. Bjurel, G. Dahlquist, B. Lindberg, S. Linde, and L. Oden:
Survey of stiff ordinary differential equations, Rep. NA
70.11, Dept. of Information Processing Royal Inst. of Tech.,
Stockholm (1970).

|3| G. D. Byrne and A. C. Hindmarsh: EPISODE: An experimental
package for the integration of systems of ordinary differential equations, Rep. UCID-30112, Lawrence Livermore Lab.,
Livermore, California (1975).

|4| W. H. Enright and T. E. Hull: Comparing numerical methods
for the solution of stiff systems of ODE's arising in chemistry, in "Numerical Methods for Differential Systems,"
(L. Lapidus and N. Schiesser, eds.), Academic Press, New
York (1976a), pp. 45-65.

|5| W. H. Enright and T. E. Hull: Test results on initial value
methods for nonstiff ordinary differential equations, SIAM J.
Numer. Anal. 13, No. 6 (1976b), pp. 944-961.

|6| W. H. Enright and T. E. Hull: A collection of programs for
assessing initial value methods, Dept. of Comp. Science Tech.
Rep., University of Toronto, Toronto (1979) (in preparation).

|7| W. H. Enright , T. E. Hull and B. Lindberg: Comparing
numerical methods for stiff systems of ODE's, BIT 15 (1975),
pp. 10-48.

|8| C. W. Gear: The automatic integration of stiff ordinary
differential equations, Proceedings IFIP Congress 1968, North-
Holland Publishing Company, Amsterdam (1969), pp. 187-193.

|9| A. C. Hindmarsh: GEAR: Ordinary differential equation system
solver, Rep. UCID-30001, Lawrence Livermore Lab., revision 3,
Livermore, California (1974).

|10| T. E. Hull, W. H. Enright, B. M. Fellen, and A. E. Sedgwick:
Comparing numerical methods for ordinary differential equations, SIAM J. Numer. Anal. 9 (1972), pp. 603-637.

|11| T. E. Hull, W. H. Enright and K. R. Jackson: User's guide for
DVERK - a subroutine for solving nonstiff ODE's, Dept. of
Comp. Science, Tech. Rep. No. 100, University of Toronto,
Toronto (1976).

|12| IMSL: IMSL Library 3 reference manual, edition six, IMSL,
Houston, Texas (1977).

|13| N. L. Schryer: A user's guide to DODES, a double precision
ordinary differential equation solver, Bell Labs Comp. Science
Tech. Rep. No. 33 (1975).

|14| L. F. Shampine: Local error control in codes for ordinary
differential equations, Appl. Math. and Comp. 3 (1977), pp.
189-210.

|15| L. F. Shampine and M. K. Gordon: "Computer solution of ordinary differential equations: the initial value problem," W. H. Freeman and Company, San Francisco (1975).

|16| L. F. Shampine and H. A. Watts: The art of writing a Runge-Kutta code, Part II, Appl. Math. and Comp., to appear (1978).

|17| R. Skeel and A. Kong: Blended multistep methods, ACM TOMS 3, No. 4 (1977), pp. 326-345.

|18| J. M. Tendler, T. A. Bickart and Z. Picel: A stiffly stable integration process using cyclic composite methods, ACM TOMS, to appear (1978).

APPENDIX

A) <u>Nonstiff Problems</u>

B1: $y_1' = 2(y_1 - y_1 y_2)$ $y_1(0) = 1$

$y_2' = -(y_2 - y_1 y_2)$ $y_2(0) = 3$

range = [0,20],

(the growth of two conflicting populations).

C2:
$$
\begin{bmatrix} y_1' \\ y_2' \\ \\ \cdot \\ \cdot \\ \\ y_{10}' \end{bmatrix}
=
\begin{bmatrix}
-1 & & & & & 0 \\
1 & -2 & & & & \\
& 2 & -3 & & & \\
& & & \cdot & & \\
& & & & \cdot & \\
& & & & -9 & \\
0 & & & & 9 & 0
\end{bmatrix}
\begin{bmatrix} y_1 \\ y_2 \\ \cdot \\ \cdot \\ \cdot \\ \\ y_{10} \end{bmatrix}
, \; y(0) =
\begin{bmatrix} 1 \\ 0 \\ \cdot \\ \cdot \\ \cdot \\ \\ 0 \end{bmatrix}
$$

range = [0,20],

(a radioactive decay chain).

F2:
$$
y' =
\begin{cases}
\dfrac{E-y}{C \cdot R} - \dfrac{y}{C \cdot r_0} & \text{for } t \; \varepsilon \; [t_i, \; t_i + D] \\
\\
\dfrac{E-y}{C \cdot R} & \text{for } t \; \notin \; [t_i, \; t_i + D]
\end{cases}
$$

$y(0) = E$,

where $E = 110$, $D = 1$, $R = 2$, $C = 1$, $r_0 = 1$, $t_i = 0,2,4,\ldots,18$;

range = [0,20]

(a gas discharge tube: Shampine and Gordon (1975, p. 277)).

B) Stiff Problems

B2: $y_1' = -10y_1 + 25y_2$ $y_1(0) = 1$
     $y_2' = -25y_1 - 10y_2$ $y_2(0) = 1$
     $y_3' = -4y_3$ $y_3(0) = 1$
     $y_4' = -y_4$ $y_4(0) = 1$
     $y_5' = -.5y_5$ $y_5(0) = 1$
     $y_6' = -.1y_6$ $y_6(0) = 1$

                    range = $[0,20]$.

D3: $y_1' = y_3 - 100y_1y_2$ $y_1(0) = 1$
     $y_2' = y_3 + 2y_4 - 100y_1y_2 - 2 \times 10^4 y_2^2$ $y_2(0) = 1$
     $y_3' = -y_3 + 100y_1y_2$ $y_3(0) = 0$
     $y_4' = -y_4 + 10^4 y_2^2$ $y_4(0) = 0$

                    range = $[0,20]$

     (chemical problem: Bjurel, et al (1970)).

E3: $y_1' = -(55 + y_3)y_1 + 65y_2$ $y_1(0) = 1$
     $y_2' = .0785(y_1 - y_2)$ $y_2(0) = 1$
     $y_3' = .1y_1$ $y_3(0) = 0$

                    range - $[0,500]$

          (physics: Gear (1969)).

Performance Evaluation of Numerical Software, Fosdick (ed.)
© *IFIP, North-Holland Publishing Company, 1979*

DISCUSSION OF SESSION ON PERFORMANCE
EVALUATION IN ORDINARY DIFFERENTIAL EQUATIONS*

L. F. Shampine
Applied Mathematics Research Department
Sandia Laboratories
Albuquerque, New Mexico, U.S.A.

INTRODUCTION

The three presentations given in the session on performance evaluation in ordinary differential equations show the wide range of activity in this area. A few comments will be made about each in the order they appear. We hope these comments will help place the talks in perspective. There was a lively discussion period from which we have selected a few of the more pertinent questions and their responses so as to amplify the presentations.

CONTRIBUTION OF H. J. STETTER

The author can speak from his experience in saying that the construction of an item of software for the solution of ODEs involves a great deal of art. The insight that the kind of modeling proposed by Stetter can provide is very much needed for this task.

The presentation makes some of the issues seem deceptively simple. As an example, let us look at the question of interpolation in an Adams PC-code. The design objective stated and the formal criterion for this objective seem so natural that many would be surprised to learn there are alternatives. The fact is that <u>none</u> of the Adams and backward differentiation formula (BDF) codes were written with this objective or criterion in mind. As explained in the texts of Gear |1| and Shampine and Gordon |2|, both the Adams and BDF methods are based on interpolation. In the first case the underlying polynomial interpolates to the first derivative of the solution at mesh points and to the current solution value. In the second case the interpolation is to the solution values at mesh points and to the current first derivative value. Evaluation of these polynomials off the mesh is used for change of step size and the like. From this point of view, the natural definition of a solution between mesh points is the underlying polynomial; this choice is made in all the current codes. However, the definition does not yield a C^1 interpolant. For Adams methods the approximate derivative is continuous, but the solution is not. The solution has a jump (in general) at mesh points of the size of the error tolerance. For BDF methods the jump is in the derivative. Error in the solution must occur somewhere, and this approach piles it up at a mesh point. The approach suggested by Stetter has some advantages. Our aim here is just to explain why such obvious requirements on the approximate solution were not imposed on the present generation of codes.

One of the most important difficulties of performance evaluation is to select representative test cases. If, for example, one were to compare codes only on linear problems, he would find Merson's method behaved completely differently from a comparison including nonlinear problems. Other examples of methods are known which have special behavior for large classes of problems. Thus the worrisome possibility exists of an unusual correlation between a method being

---

* Supported in part by the U. S. Army Research Office.

tested and the test set. Modeling affords the advance of treating classes
instead of individual problems. The price seems to be a limitation to very
restricted classes.

CONTRIBUTION OF P. J. VAN DER HOUWEN AND J. G. VERWER

In recent years there has been increasing attention given to solving partial
differential equations (PDEs) by approximation with ODEs. The more common
approach (and the one studied in this talk) is to discretize the space variables
and to consider the solution in time of the resulting ODEs. Of course this
solution proceeds via subsequent discretization of the time variable. The
resulting fully discrete problem and its analysis is not different from the
methods considered directly by those interested in the solution of PDEs. The
approach and viewpoint are, however, quite different and have contributions to
make to the area.

The solution of ODEs has focused on very general problems and is customarily done
with variation of the time step and estimation of the local error. When applied
to parabolic problems of one space variable, general purpose ODE codes have been
rather successful. They provide a convenience, reliability, and generality not
common in the direct solution of PDEs. For problems of two and more space dimensions the same advantages are possible, but it is necessary to specialize the
solver so as to cope with the size of the problem. Van der Houwen and Verwer
devote much of their attention to splittings of the problem so as to reduce the
effect of dimensionality. This kind of splitting is that customary in the
solution of PDEs, but they are attempting it in the greater generality customary
in the solution of ODEs. They show that the effect of the handling of the
boundary can be quite significant and propose ways of improving the situation.

It is already very difficult to compare the effectiveness of solution of ODE
codes and the situation becomes much worse when they are coupled with the solution of PDEs. There are many algorithmic possibilities introduced in the discretization of the spacial variables, the splitting of the equations, and the
handling of the boundary conditions. The paper notes, for example, the very
important effect of using or not the "fast form" of the line hopscotch method.
There are in addition important computing issues of storage and the costs of
various kinds of factorizations to be considered. For these reasons one cannot
expect definitive comparisons; one can at best illuminate important aspects of
the task as van der Houwen and Verwer have done.

CONTRIBUTION OF W. H. ENRIGHT

The kind of tool provided by Enright is necessary for the production of mathematical software. It's convenient to have it available rather than for each
serious code writer to put together his own version. The tool is also useful
for evaluation, but this is a more delicate task. In the author's experience
with codes for non-stiff problems, the most serious difficulty was unquestion-
ably that of non-comparability of codes. The mere presence of an evaluation
tool urges some standardization in the task attempted by the codes. The
situation is still more serious in the case of stiff problems. Then the evaluation of a method is made difficult by the important role played by the numerical
treatment of problems of linear algebra. The evaluation of a code is also made
difficult because machine independent measures of cost, like the number of calls
to the derivative routine, often do not reflect actual computing costs. Despite
the caution required by these observations, people must still choose which code
they are to use, and the information provided by a tool such as Enright's is of
obvious importance.

An assumption fundamental to Enright's approach is that the code produces a
global error which behaves like $C \cdot TOL^E$. There is no doubt that this is an

important design objective in the construction of the present generation of codes, but it is not the primary objective. The primary objective is at each step to advance the solution a step of length h by producing a solution approximation which an associated error estimate says is as accurate as the user specified. For efficiency one tries to use an h about as large as possible. This attempt is what produces proportionality. In no code known to this author is a successful step rejected in order to use a larger value and so obtain better proportionality. We emphasize that the codes do try to get proportionality, and they usually do pretty well, which accounts for the success of Enright's approach, but that a lack of proportionality has been regarded as acceptable in achieving efficiently the primary objective.

In the discussion John Rice of Purdue asked, "How do you account for complete program failures in the summaries of program performance that you prepare." Enright replied, "This is a difficulty. Currently a message is output (identifying the type of failure) at the point of failure. Overall summaries are only over all successful integrations. When interpreting the summaries one must therefore check that no failures have occurred. Since the summaries are inappropriate for comparison, in any event, we don't feel this is a serious difficulty." Alan Curtis of Harwell asked, "While understanding the difficulty about controlling anything other than absolute error in comparative testing of programs, I think it gives misleading results on some of the test problems actually used for stiff methods. There seems to be no good reason to insist on absolute error when testing for certification purposes." Enright answered, "If other error controls are used (and we have certainly considered them) it is not clear what we should define as the measure of global error."

References

|1| C. W. Gear, Numerical Initial Value Problems in Ordinary Differential Equations (Prentice-Hall, Englewood Cliffs, NJ, 1971).

|2| L. F. Shampine and M. K. Gordon, Computer Solution of Ordinary Differential Equations: the Initial Value Problem (W. H. Freeman, San Francisco, 1975).

SESSION 6 : PERFORMANCE EVALUATION IN
OPTIMIZATION AND NONLINEAR EQUATIONS

Performance Evaluation of Numerical Software, Fosdick (ed.)
© IFIP, North-Holland Publishing Company, 1979

PERFORMANCE EVALUATION FOR OPTIMIZATION SOFTWARE

Philip E. Gill and Walter Murray

Division of Numerical Analysis and Computer Science
National Physical Laboratory
Teddington, Middlesex, England

1 Introduction

In many ways, the evaluation of the performance of optimization software is analogous to the evaluation of new products by a consumer association. In a consumer test on washing machines the tester is concerned with two aspects. He is concerned with how well the machine performs its stated function and whether or not it is safe for the user to use. Frequently consumer organizations refuse to consider testing some products because they fail to comply with national or international standards of safety. In testing software we are unable to draw on such standards, although frequently we would like to brand a routine as "unsafe". No one is going to be electrocuted while using optimization software (which is probably the reason why no standards exist) but someone else might be if incorrect results were relied upon as being correct. One of the consequences of using a "bad" routine is that it creates within the user community an atmosphere of mistrust which is rarely dispelled by subsequent improvements.

Recently at the National Physical Laboratory we have solved three practical problems where the consequences of incorrect results would be grave. The first problem was to ascertain whether a particular location within the vicinity of a nuclear reactor was safe for a human operator. Although monitoring equipment would register dangerous levels of radiation, the information would be too late to avert the ill effects. In the second problem, also concerned with a nuclear reactor, it was necessary to determine the maximum and minimum temperatures within a gas-water heat exchanger in order to determine a safe cross-over point from one type of construction material to another. If the computed solution were incorrect, catastrophic corrosion could occur because of the rapid flow of the gas and steam/water mixture. Apart from the considerable expense involved in such an eventuality there is a danger to human life. The third problem concerned the determination of minute concentrations of substances such as insulin, hormones etc in human blood samples by radio-immunoassay techniques. The danger here is obvious. This particular program may be used in hundreds of hospitals and there is the prospect that it will be programmed into a micro-processor and used by operators who may not realize that an optimization routine is involved. These examples may

sound dramatic but they are far from exceptional and serve as a reminder of the
responsibilities that rest on the shoulders of those who seek to distribute
software.

2   Background

We shall be concerned with optimization problems of the form

$$(P1) \quad \text{minimize } F(x), \quad x \epsilon E^n$$

$$\text{subject to } c_i(x) = 0 \quad i = 1, 2, \ldots, m'$$

$$c_i(x) > 0 \quad i = m'+1, \ldots, m,$$

where F and $\{c_i\}$ are given real-valued functions. The function F is normally termed the "objective function" and the set $\{c_i\}$ is the set of "constraint
functions". The efficient solution of P1 by a single algorithm which disre-
gards the specific form of $F(x)$ and $\{c_i(x)\}$ is not currently possible. Consequently optimization algorithms are designed to solve particular categories of
problems, where each category is defined by properties of the objective and constraint functions, as illustrated below.

Properties of $F(x)$	Properties of $\{c(x)\}$
Linear	No constraints
Sum of squares of linear functions	Upper and lower bounds
Quadratic	Linear
Sum of squares of nonlinear functions	Sparse linear
Nonlinear	Nonlinear

(We could extend this categorization further by considering nonlinear functions
with specific continuity properties etc). The choice of algorithm depends not
only on the type of problem, but also on the available information about the dimension of the problem, the problem functions, the cost of evaluating them, and so
on.

Despite the wide variety of optimization algorithms most of them are iterative and
compute a sequence $\{x^{(k)}\}$ such that

$$x^{(k+1)} = x^{(k)} + \alpha^{(k)} p^{(k)},$$

where $x^{(k)}$ denotes the kth approximation to the solution of (P1), $p^{(k)}$ is the
kth search direction and $\alpha^{(k)}$ is a step length chosen to satisfy some criterion
such $F(x^{(k+1)}) < F(x^{(k)})$.

Throughout this paper we shall use $g(x)$ to denote the gradient vector of first
derivatives of $F(x)$, $G(x)$ to denote the Hessian matrix of second derivatives of
$F(x)$ and $A(x)$ to denote the Jacobian matrix of first derivatives of the constraint vector $c(x)$.

For most algorithms, convergence to a solution x^* of (P1) is guaranteed only in the limit (if it is guaranteed at all).   Notable exceptions to this rule are the algorithms for linear and quadratic programming, but even in these cases, the upper limit on the number of iterations which could be taken is prohibitively large except for small problems.

3   Objectives in algorithmic performance

Before the performance of a particular algorithm can be evaluated it is necessary to define those aspects of performance with which we are concerned.   The objectives are very similar to those for other areas of numerical analysis, but owing to the nature of optimization algorithms, the detail and emphasis are different.

Efficiency

In optimization algorithms we are concerned with two forms of efficiency, the computational overhead per iteration (which may involve complicated numerical processes such as finding the singular-value decomposition of a matrix) and the cost associated with the calls of the user-supplied functions.   For most nonlinear problems of small or moderate size the number of entries to the user-supplied routines dominates the computational effort.   This should not be taken to imply that the overheads which are mainly concerned with numerical linear algebra are not worth considering.   The question "What percentage of the computer time is actually involved in solving linear algebra problems?" once met with the reply: "A lot less than would be the case if they were not solved in the most efficient manner.".   As the dimension of the problem increases so usually does the significance of the overhead costs, although it is a rare problem for which the time spent within user-supplied routines is of negligible importance.

Assessing the likely number of entries to user-supplied routines is a difficult task about which more will be said later.   It may be thought that assessing overhead costs is straightforward since this is just a matter of counting the number of operations.   Unfortunately this is not the case since the overheads are not independent of the number of entries to the user-supplied routines.   They are related, for example, through the number of iterations.   Moreover, assessing the number of operations per iteration is not straightforward, since this depends on the number of constraints currently active and whether or not a constraint is added to, and/or deleted from, the active set within the iteration.

Success rate

Since no optimization routine is guaranteed to work on all problems, the success rate of a routine is usually the single most important consideration of users.

Confidence in the solution obtained

Given that the algorithm reports a successful termination, the degree of confidence that can be placed in the solution obtained varies from algorithm to algorithm.

Consistency

Given two algorithms in which the mean number of function calls in the solution of a large set of problems is equal, users will prefer the algorithm with the smaller variance about this mean.

Stability

Optimization algorithms frequently include parameters which enable a user to "tune" the algorithm to a particular problem.   It is important that the perform-ance of an algorithm is not erratic when these parameters are altered.   Moreover, it should be possible to _predict_ the behaviour of an algorithm as these parameters are varied.

An example of a parameter which has a predictable effect upon the performance of an algorithm is the tolerance which determines how accurately the step length $\alpha^{(k)}$ approximates the minimum of $F(x)$ along $p^{(k)}$.   In general, as the accu-racy of the linear search is slackened fewer function evaluations will be required to find x^*.   However, the _decrease_ in the number of function evaluations is accompanied by an _increase_ in the number of iterations.   Consequently we try to choose $\alpha^{(k)}$ with a degree of accuracy that will optimize between overhead costs, which decrease, and entries to user-supplied routines which become more numerous with the reduction in number of iterations.   Likewise in problems for which finite-differences of the gradient vector are required, the determination of the search direction becomes more expensive as the dimension of the problem increases, and, in these circumstances a tighter linear search is justified.

Qualitative output

Algorithms vary in the information they provide concerning the quality of their solution.   For example, algorithms utilizing the Hessian matrix of the objective function can be expected to converge to a _local minimum_ of $F(x)$ and provide a good estimate of the condition number of the Hessian matrix at the solution. However, conjugate-gradient algorithms using only first derivatives can be expec-ted to converge only to a _stationary point_ of $F(x)$.

The different objectives in performance are not necessarily in conflict.   For example, superlinearly-convergent algorithms are usually more efficient than linearly-convergent algorithms and they usually supply more qualitative informa-tion about the solution.   Similarly, the objectives of a high success rate and an efficient algorithm are often compatible.   An obvious exception is the omission of a check to verify that a stationary point is a minimum; this will clearly reduce the amount of computational effort on those problems for which the point is a minimum but will cause a true minimum not to be found otherwise.

4   Test problems

One common method of measuring an algorithm's performance is by applying it to a set of test problems.   The provision of suitable test problems for optimization is extremely difficult.   For example, suppose we judge that algorithms ought to be able to successfully move away from the neighbourhood of a saddle-point.   With some ingenuity we could design a test problem which is such that all algorithms of current interest would generate a point in their iterative sequence which is in the neighbourhood of a saddle-point.   We could not, however, guarantee that all algorithms (including those not yet invented) would generate a point lying in the region of the saddle-point.

Large sets of test problems exist for nearly all categories of optimization activity, but with a few exceptions the test problems fall short of being ideal. (However, it is far better for an algorithm to be tested on an unsatisfactory test set than not tested at all!).

Test problems are available to measure how well routines satisfy all the perform-ance objectives listed above.   In the measurement of efficiency, however, test problems are very frequently abused.   The most common fault is to draw conclu-sions from test sets which are patently too small, typically three to five prob-lems.   In our view at least ten problems should be used for testing, as many as fifty being required under some circumstances.   One useful criterion which can be applied when drawing conclusions about algorithm testing is the following.

Suppose that a set of test problems is used to compare the efficiency of two algorithms and that we conclude algorithm A is superior to algorithm B.   If a subset of these test problems exists for which an alternative conclusion could be drawn then we should still be reasonably confident of the original conclusion if the number of problems in this subset were less than 25% of the whole.

Perhaps the greatest deficiency of current test problems is that they are almost all analytic functions of a simple algebraic type.   Moreover there is a prepond-erance of test problems in which the objective function is a sum of squares (that is, $F(x) = \sum_{i=1}^{m} \left\{ f_i(x) \right\}^2$ with the optimal value $F(x^*) = 0$.   One consequence of this is that $F(x)$ and its derivatives can be evaluated more accurately than is common for real problems and this has a tendency to disguise deficiencies in the purely numerical aspects of the optimization routines being tested.   For example, unrepresentative accuracy is often obtained in the solution of such test problems. To illustrate our point we use the famous test problem due to Rosenbrock (1960);

$$\text{minimize} \left\{ F(x) = 100(x_1 - x_2^2)^2 + (1-x_1)^2 \right\}$$

The solution is $x^* = (1,1)^T$, at which $F(x^*)$ is zero.   Ironically this problem was introduced because it was considered difficult for minimization;   yet

Rosenbrock's function has favourable properties enjoyed by very few real problems.
When solving this problem it is not uncommon to obtain the exact value of $F(x)$ at
the solution, or to obtain $F(x^{(k)}) = O(10^{-2t})$ on a machine with t decimal
digits relative precision. This is a direct consequence of $F(x)$ being a sum of
squares with $F(\overset{*}{x})$ zero. Merely having $F(\overset{*}{x})$ equal to zero enables a routine
to distinguish between two estimates, say, 10^{-18} and 10^{-16} when on the same
machine it is impossible to distinguish between $1 + 10^{-18}$ and $1 + 10^{-16}$. Moreover the error in computing $F(x)$ when $F(x)$ is close to zero is much smaller
than one might normally expect. If an upper bound on the error in computing
$f_i(x)$ is ε then to first order, an upper bound on the error in computing $F(x)$
is $2 \sum_{i=1}^{m} |f_i(x)| \varepsilon$. Test problems such as Rosenbrock's function can be improved
by the simple expedient of adding unity to $F(x)$.

Real-life problems

In our view there is no substitute for testing routines on real-life problems.
Unfortunately, although practical problems are immensely useful for in-house
testing, there are a number of obstacles to making such problems generally available. For example, there may not be a succinct mathematical definition of the
problem or the user-supplied routines may not have been written in a portable
code.

The main reason that real-life problems are not useful for general testing is that
they invariably require considerable CPU time to evaluate the problem functions.
(The extra in-house computational effort can be justified because the results are
of practical value).

Testing by the user community

Optimization problems in industry or elsewhere are rarely of the one-off type.
Usually a whole sequence of problems is solved, arising from the adjustment of
various parameters in the model. If a user is already using optimization software from a particular source he is usually willing to try either modifications to
this software or replacement routines obtained from the same source. It is
important, therefore, to foster such liaison between user and source since this,
in the long term, provides the most promising method of assessing efficiency and
other aspects of performance.

5   The use of an efficiency index

There is a growing trend towards measuring efficiency by translating the number of
arithmetic operations, number of entries to various user-supplied routines, etc.
into a single performance index. In our view the application of this technique
to optimization is misleading and therefore inappropriate.

A common error is to combine the number of entries to the routines to evaluate $F(x)$, $g(x)$ and (if appropriate) $G(x)$, with unrealistic weights. The value of this single number is then made the basis of comparison between algorithms. The argument usually advanced is that $g(x)$ is a vector of n nonlinear functions and consequently it takes n times as much effort to evaluate $g(x)$ as $F(x)$. A similar argument is used to demonstrate that the effort to evaluate $G(x)$ is $n(n + 1)/2$ times that to evaluate $F(x)$. Although it is conceivable that this analysis is true for some problems, it is rarely (if ever) true for either real-life problems or test problems. In practice there is a considerable overlap in the computation of $F(x)$, $g(x)$ and $G(x)$ and frequently $F(x)$ and $g(x)$ can be computed for almost the same amount of effort as $F(x)$ alone. It is implicitly assumed in the design of some algorithms that the effort to evaluate $g(x)$ is comparable to that for $F(x)$ and, one would never use a gradient linear search along $p^{(k)}$ if this were not the case. This is perfectly reasonable on the part of the designer since a user may check the validity of the assumption before solving a problem. Algorithms for which $F(x)$ and its higher derivatives require approximately the same amount of computer effort will always be penalized when assessed using a single performance index which weights them otherwise.

Another poor index of the efficiency of algorithms is the CPU time taken. The results of comparisons between algorithms based upon CPU time have rarely been of value owing to differences in compiler, hardware, operating conditions and, in the case of nonlinear problems, an untypical preponderance of overhead costs over the effort required to evaluate the test functions. The user needs to be informed of the relative strengths of a routine with respect to various aspects of efficiency so that he may judge how these aspects should be weighted in his own situation.

6   Termination criteria and points of assessment

All optimization software requires a criterion for terminating the computation of the sequence $\{x^{(k)}\}$. In modern software, this criterion is often complicated, involving such quantities as the projected gradient and Lagrange-multiplier estimates. There is no universal agreement on what is the best termination criterion for a given situation. Indeed, it is rarely discussed and in the theoretical definition of algorithms, unrealistic termination criteria are often used. Variations in termination criteria may be due not only to differences of opinion but also to fundamental differences in what is computed by various algorithms and this presents significant difficulty for the comparative assessment of optimization software. Ideally, if we wish to measure comparative efficiency of routines we should set a uniform termination criterion for all the routines and then compute the cost of the minimization, in terms of the number of function evaluations for instance. However, setting the same tolerance in two different

algorithms with the same termination criteria can still result in a wide variation
in the accuracy of the solutions obtained.   The question remains, therefore, as
to the point at which we should assess the efficiency of optimization software.
A commonly-used criterion for unconstrained optimization software when the deriva-
tives of $F(x)$ are available is the first point for which $\|g(x^{(k)})\| < \varepsilon$, where
$g(x)$ is the gradient vector of $F(x)$, $\|.\|$ denotes the Euclidean norm of a
vector and ε is some prescribed tolerance.   There are a number of objections to
this criterion: it is not universally applicable because the gradient is not
always available; the quantity $\|g(x^{(k)})\|$ is not normally monotonically decreas-
ing, and it may, moreover, satisfy the criterion at points well removed from x^*;
and finally there is difficulty in deciding a constant value of ε for all the
problems of interest.

Another suggestion in the literature is the use of

$$\|x^{(k)} - x^*\| < \varepsilon (1 + \|x^*\|) .$$

Again, this is not a consistent test.   Moreover there is no possibility of making
a sensible uniform choice of ε.

The assessment criterion used at the National Physical Laboratory is to take the
first point $x^{(k)}$ for which

$$F(x^{(k)}) - F(x^*) < \varepsilon (1 + |F(x^*)|) . \tag{1}$$

This criterion can be applied to all software for unconstrained optimization and
software for constrained problems which retain feasibility.   In both these cases
$F(x^{(k)}) < F(x^{(k-1)})$ and consequently there is no difficulty with consistency or
premature assessment.   Moreover, for different categories of algorithms it is
relatively easy to make a uniform choice of ε.   Some authors have argued against
the use of (1) because it includes $F(x^*)$ which is unknown on real problems.   We
believe that such authors are confusing an assessment criterion, where the use of
$F(x^*)$ is legitimate, with a termination criterion where it is not.

If the criterion (1) is to give a realistic assessment of the performance of an
algorithm, the choice of ε must give a point $x^{(k)}$ which is close to a final
estimate of x^* obtained using a realistic termination criterion.   For example,
an iterate $x^{(k)}$ would not be an adequate point of assessment if it were such
that

$$F(x^{(k)}) - F(x^*) < 10^{-1}(1 + |F(x^*)|)$$

(ie $F(x^{(k)})$ is correct to within ten per cent) since $F(x^{(k)})$ could not be confirmed as being close to $F(x^*)$ by termination criteria which do not involve
$F(x^*)$.   In practice we must consider only small values of ε such as $\varepsilon < 10^{-4}$.

An exception to this rule occurs when testing routines for minimizing general non-
differentiable functions.   In this case it is not possible to confirm whether any

point is in a close neighbourhood of $\overset{*}{x}$.   Since routines for this category of
problem converge exceedingly slowly, a value of ε equal to 10^{-3} or 10^{-2} is
appropriate.

The relative performance of algorithms with superlinear convergence is almost
invariant with the choice of ε and a stringent termination criterion is recommended.   Consequently, on machines which evaluate $F(x)$ to more than 11 decimal
digits accuracy, a reasonable choice for ε is 10^{-10}.   For algorithms with a
linear rate of convergence the performance can vary widely with the choice of ε.
It is not unusual for the number of function evaluations to be three times greater
for $\varepsilon = 10^{-10}$ than for $\varepsilon = 10^{-5}$.   Only a moderately stringent termination
criterion is recommended for such routines and 10^{-5} is an appropriate choice of ε.

If two routines with different rate of convergence are being compared, the larger
value of ε (associated with the routine with the slower rate of convergence)
should be applied.

The choice of an assessment criterion for a constrained nonfeasible-point algorithm is more difficult.   The problem is that any sensible criterion does not
behave consistently – even in the neighbourhood of the solution.   It is also
difficult to choose a criterion that does not favour one type of algorithm.   The
fairest resolution of these conflicts has yet to be decided.   However, it is
clear that a single assessment criterion is inadequate.   A possible criterion is
to choose the first point $x^{(k)}$ for which

$$|F(x^{(k)}) - F(\overset{*}{x})| < \varepsilon_1 (1 + |F(\overset{*}{x})|)$$

and

$$\|c(x^{(k)})\| < \varepsilon_2,$$

where $c(x)$ is the vector of constraints active at the solution and $x^{(k)}$ satisfies the remainder of the constraints.   In Figure 1 we illustrate that this criterion is biased against certain types of algorithms.   The iterative path depicted
by Algorithm A is typical of the class of "projected-" or "reduced-"gradient
methods.   Essentially such methods approach the solution along the constraints
surface (although in doing so, nonfeasible points are sometimes computed).

Algorithm B is typical of the class of "penalty function" or augmented-Lagrangian" methods.   It is clear that the assessment criterion we have proposed favours
Algorithm A since

$$|F(\overset{*}{x}) - F(x_A)| < |F(\overset{*}{x}) - F(x_B)|$$

and

$$\|c(x_A)\| << \|c(x_B)\|.$$

It could be argued that x_A is indeed a better estimate of $\overset{*}{x}$ than x_B but it is evident from Figure 1b that

$$\|x_B - \overset{*}{x}\|_2 \ll \|x_A - \overset{*}{x}\|_2 .$$

Moreover, x_B may satisfy the optimal condition that $g(x_B)$ is a positive linear combination of the columns of $A(x_B)$ (the Jacobian matrix of $c(x)$) which is true of $\overset{*}{x}$, where x_A does not.

The choice of norm for the criterion involving $c(x^{(k)})$ is complicated by the

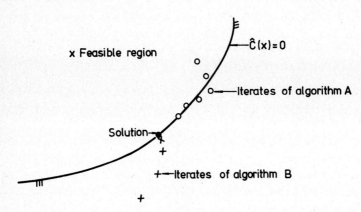

Figure 1a

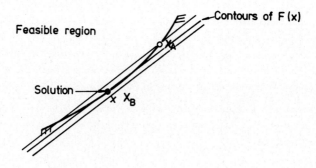

Figure 1b

fact that within some algorithms the constraints are dynamically scaled. For instance, $c_i(x)$ may be premultiplied by a varying positive weight to ensure that the length of the gradient of the scaled constraint is unity. This scaling may prevent the original constraint being satisfied to some prescribed tolerance in the assessment criterion. Therefore it would seem necessary to choose a norm of the form $\|y\| = (y^T D y)$, where D is a diagonal matrix whose diagonal elements represent the scaling of the constraint vector. One further complication in the nonlinear-constraint case is the problem of determining at which stage within an iteration we should apply the assessment criterion. We have this problem to a lesser degree in the unconstrained case: the point $x^{(k)}$ may not be the first point that satisfies the assessment criterion since other trial points are computed in the linear search subproblem. In practice we choose not to test at these other points because termination always takes place after a complete linear search and we prefer points of assessment which are also candidates for termination. Fortunately the difference in performance which would occur if we altered this strategy is negligible. However, in the nonlinear-constraint case, the subproblem in which trial points are computed may be extensive. Moreover, auxiliary quantities such as Lagrange-multiplier estimates, which would be required for a sensible termination criterion are computed only on the completion of the subproblem. For this reason, algorithms which pose substantial subproblems have the disadvantage that either they will be assessed at unrealistic points or they will appear not to be competitive with other algorithms.

7   Testing methodology

Testing in the proper mode

A mistake commonly made in the testing of a new idea concerning a specific aspect of an algorithm, is to compare it to alternatives in the wrong algorithmic setting. For example, suppose a new quasi-Newton updating formula is proposed and a comparison with existing alternative formulae is required. It is important when making such comparisons to use the formulae within a properly constituted routine. What frequently happens is that an improvised routine is patched together for comparative tests of the two formulae. This type of testing may simply show up the susceptibility of an updating formula to deficiencies elsewhere in the routine. For example, if the approximation to the inverse Hessian is recurred, one updating formula may be significantly more susceptible to rounding errors made during the updating process. In a properly constituted routine in which the triangular factors of the Hessian are recurred, this adverse behaviour may disappear. Similarly an unsophisticated linear search may apparently improve the relative performance of an updating formula that has a tendency to produce a better _initial_ step for the linear search.

Evaluating variants of a routine

One of the most common reasons for evaluating performance is to ascertain whether
or not some minor or major variation in a routine will improve performance.
Usually only one or two performance criteria are affected in such circumstances
and consequently the task of determining the relative merits of alternative routines is simplified.   An example of such a situation occurred recently at the
National Physical Laboratory.   The linear search within a nonlinear least-squares
routine was altered to take specific advantage of the structure of the objective
function.

It is important to realize that any change to an optimization routine may result
in the altered routine's converging on a problem for which the original routine
failed and vice versa.   Even in the simplest situations one of the difficulties
with the performance evaluation for routines is the effect that chance has on the
result.   It could be argued that the effects of chance will be eliminated if
sufficient testing is carried out.   However, if the change is small and the
effects are minor, such testing may be prohibitively expensive or unwarranted.
This difficulty is particularly acute for routines for discontinuous functions
where the variation in performance due to random effect is large.   It is not uncommon to remove an error from an optimization routine and find that the overall
performance on a set of test problems has deteriorated!

One method of eliminating virtually all of these random effects is to examine the
"local" consequences of a change to a routine.   In the case of the nonlinear
least-squares linear search mentioned earlier, this involved comparing the number
of entries to the user-supplied functions for each linear search.   Clearly this
comparison can be applied only as long as $x^{(k)}$ and $p^{(k)}$ are identical for both
techniques.   Normally this occurs only on the first iteration unless an exact
linear search is made.   (To increase the number of occasions when a comparison
might be made, the alternative linear search could also be carried out at every
iteration but only one being used to update $x^{(k)}$).   Comparison of the linear
searches indicated that the improved algorithm invariably required fewer calls of
the user-supplied functions per iteration.   This local improvement nearly always
resulted in a global improvement, that is the total number of calls necessary to
find x^* was reduced.   Unfortunately this degree of consistency is rare and in
general a local improvement will not necessarily result in an overall improvement.
An example of this less satisfactory behaviour occurred during tests of a variant
of the polytope algorithm for non-differentiable functions.   (This algorithm is
also known as the simplex algorithm; the reader is referred to Nelder and Mead,
1965, for further details).   A change was made to the method of generating the
first vertex after the construction of a polytope.   The new method always resulted in a better vertex, but the global variation in performance was so large that an

effective comparison from global effects was impossible.

Comparative testing and the development of algorithms

Inevitably, results obtained on a set of problems suggest improvements to an algorithm and consequently the development of an optimization routine proceeds simultaneously with its testing.   In this event it is important that there should be a sufficient number of diverse problems to prevent peculiarities of the test functions from being unfairly exploited in algorithm design.   It is not uncommon for an author to adjust the free parameters or to choose a strategy in an algorithm so that it will give the best results on a few special examples.   (Naturally we distinguish between the success of an algorithm at coping with a particular problem only and its ability to cope with a general difficulty illustrated by the problem.)   If performance is always assessed on a large number of dissimilar test problems, the algorithm designer will be less able to appear successful by "tuning" his algorithm to cater for specialized problems only.

Testing by analogy

One feature of optimization algorithms which we can use to our advantage when assessing the performance of optimization software is that distinct algorithms often have large amounts of code in common.   This implies that much of the testing of one algorithm applies directly to another.   For example, a quasi-Newton method for minimizing a function for which $F(x)$ and $g(x)$ can be evaluated at any point is very similar to a quasi-Newton method which computes finite-difference approximations to $g(x)$.   One would expect a double-length version of the finite-difference routine to have the same numerical behaviour as the single-length derivative routine.   Similarly, differences between two alternative derivative routines may be assumed to carry over to the finite-difference case.

The same principles apply when an algorithm is generalized to solve a more complex problem category.   For example, an algorithm for linearly-constrained optimization solves a sequence of transformed unconstrained problems and conclusions concerning the performance of unconstrained algorithms can be extended to the linear-constraint case.   Clearly the more complex algorithm will have additional features which need to be tested, but the testing process will be considerably shortened.

The conclusions drawn from this "testing by analogy" may not be straightforward. For example, if a conjugate-direction algorithm is generalized to the linear-constraint case there are theoretical reasons to suppose that it will compare less favourably with a quasi-Newton method than it would in the unconstrained case.

8   Closing remarks

We are compelled to draw the conclusion that many aspects of the current performance evaluation of optimization software are far from satisfactory. Fortunately the difference between algorithms is usually sufficiently large for even crude or biased evaluation techniques to be adequate.

In the past, researchers have regarded the comparative efficiency of algorithms as being of prime importance (despite the fact that the term "efficiency" was often inadequately defined).   This largely reflected an interest in algorithms and methods rather than software.   Although the efficiency of a routine is still important there are other aspects and other facets of performance assessment which in the actual implementation are equally important.

Acknowledgement

We are grateful to our colleagues Miss Enid Long and Dr David Martin for their careful reading of the manuscript and a number of helpful suggestions concerning presentation.

References

Nelder, J.A. and Mead, R. (1965). "A simplex method for function minimization", Comput. J. 7, 308-313.
Rosenbrock, H.H. (1960). "An automatic method for finding the greatest or least value of a function", Comput. J. 3, 175-184.

Performance Evaluation of Numerical Software, Fosdick (ed.)
© IFIP, North-Holland Publishing Company, 1979

A NUMERICAL COMPARISON OF OPTIMIZATION PROGRAMS
USING RANDOMLY GENERATED TEST PROBLEMS

Klaus Schittkowski
Institut für Angewandte Mathematik und Statistik
Universität Würzburg
Würzburg, Germany (Fed. Rep.)

Abstract: This paper presents a comparative
study of optimization programs. The test
problems are randomly generated with predetermined solutions allowing to measure accuracy, efficiency, global convergence, and
reliability. These performance criteria are
evaluated for six qualified and widely distributed optimization programs. The numerical
results are obtained by 240 test runs.

1. Introduction

We intend to test and compare optimization software for solving
constrained nonlinear programming problems of the kind

$$\min \quad f(x)$$

$$x \in \mathbb{R}^n: \quad \begin{aligned} g_j(x) &= 0 \;, \quad j=1,\ldots,m_e \\ g_j(x) &\geq 0 \;, \quad j=m_e+1,\ldots,m \\ x_l &\leq x \leq x_u \end{aligned}$$

with continuously differentiable functions f and $g_1,\ldots,g_m$. Several
well-recommended and widely distributed NLP programs are submitted
for the comparative study. In contrast to earlier studies (Colville
[4], Stocker [14], Tabak [15], Miele, Tietze, Levy [8], Eason and
Fenton [5], Staha [13], Asaadi [2], and Sandgren [10]), the test
problems are randomly generated with predetermined solutions. This

allows to measure accuracy, efficiency, global convergence and
reliability of each optimization program under consideration.

The following section describes the construction of randomly
chosen test problems, section 3 the performance criteria, and
section 4 contains the numerical results obtained by six optimization codes and 240 test runs.

2. Randomly generated test problems

The construction of randomly chosen test problems is based on
signomials, generalized polynomial functions of the kind

$$s(x) := \sum_{j=1}^{k} c_j \prod_{i=1}^{n} x_i^{a_{ij}} \quad , \quad x > 0,$$

where the coefficients c_j and the exponents a_{ij} are real numbers.
Since each signomial is completely described by the data c_j and a_{ij},
it is possible to produce these data randomly using predetermined
bounds. Let $s_0, s_1, \ldots, s_m$ be $m+1$ randomly generated signomials.
These functions are to be used now to construct an optimization
problem of the desired form such that a randomly chosen x^* with
$x_l < x^* < x_u$ defines a local minimizer.

To ensure that x^* is feasible with exactly m_a active constraints,
define the restrictions by

$$g_j(x) := s_j(x) - s_j(x^*) \quad , \quad j=1,\ldots,m_e+m_a$$
$$g_j(x) := s_j(x) - s_j(x^*) + \mu_j \quad , \quad j=m_e+m_a+1,\ldots,m$$

with randomly chosen $\mu_j \in (0,m)$. The objective function is given by

$$f(x) := s_0(x) + \frac{1}{2} x^T H x + c^T x + \alpha$$

with an $n \times n$ matrix H, $c \in \mathbb{R}^n$, $\alpha \in \mathbb{R}$. The Lagrangian function of the
problem is defined by

$$L(x,u) := f(x) - \sum_{j=1}^{m} u_j g_j(x)$$

for $x \in \mathbb{R}^n$, $x_l < x < x_u$, $u = (u_1,\ldots,u_m)^T \in \mathbb{R}^m$. Optimal Lagrange
multipliers u_j^*, $j=1,\ldots,m_e+m_a$, are randomly chosen, the remaining
ones are set to zero. Using a randomly generated positive definite
matrix P, we set

$$H := -D_x^2 s_0(x^*) + \sum_{j=1}^{m} u_j^* D_x^2 g_j(x^*) + P$$

$$c := -D_x s_0(x^*) - Hx^* + \sum_{j=1}^{m} u_j^* \, D_x g_j(x^*)$$

$$\alpha := -s_0(x^*) - \frac{1}{2} x^{*T} Hx^* - c^T x^* \, .$$

It is easy to see that these definitions guarantee the validity of the Kuhn-Tucker conditions in x*, the second order condition that $D_x^2 L(x^*,u^*)$ is positive definite, and the condition $f(x^*) = 0$. This implies that x* defines an isolated local minimizer of the constructed problem, confer [11].

Using the test problem generator as described so far, 80 test problems are computed and combined in ten different classes. Each class contains 5 or 10 test problems, respectively; each example has to be solved from three randomly chosen starting points. The dimensions vary from 4 to 20, the number of constraints from 3 to 18, the number of active constraints from 3 to 12. There are classes with equality constraints only, with inequality constraints only, classes with dense and sparse problems, etc. A more detailed description of the data describing the test problems is given in [12].

3. The performance criteria

Since it is not possible to guarantee that the predetermined point x* defines a global minimizer of the test problem, we have to decide if a current computed solution approximates x* or not. Let $\bar{x}$ be the termination point of a test run and define

$$r(\bar{x}) := \sum_{j=1}^{m_e} |g_j(\bar{x})| - \sum_{j=m_e+1}^{m} \min(0, g_j(\bar{x}))$$

$$h(\bar{x}) := \|D_x L(\bar{x}, u^*)\|_2$$

$$e(\bar{x}) := \frac{1}{n} \sum_{i=1}^{n} \gamma_i$$

$$\text{with} \quad \gamma_i := \begin{cases} -\log_{10} |(\bar{x}_i - x_i^*)/\bar{x}_i| & , \ \bar{x}_i \neq x_i^* \ , \ \bar{x}_i \neq 0 \\ -\log_{10} |x_i^*| & , \ x_i^* \neq 0, \ \bar{x}_i = 0 \\ 12 & , \ \text{otherwise.} \end{cases}$$

Now we are able to distinguish between three categories of test runs:

1) $\overline{x}$ is called a successful solution, if

$$|f(\overline{x})| < \eta_1,\ r(\overline{x}) < \eta_2,\ h(\overline{x}) < \eta_3,\ e(\overline{x}) > \eta_4.$$

2) $\overline{x}$ is called a global solution, if

$$f(\overline{x}) \leq -\eta_1,\ r(\overline{x}) < \eta_2$$

or

$$f(\overline{x}) < 0,\ r(\overline{x}) < \eta_2,\ (h(\overline{x}) \geq \eta_3\ \text{or}\ e(\overline{x}) \leq \eta_4).$$

3) $\overline{x}$ is called a non-successful solution otherwise.

The tolerances $\eta_1,\ldots,\eta_4$ are predetermined. Furthermore, we denote by SS, GS, and NS the corresponding numbers of successful, global, and non-successful solutions computed by a battery of test runs.

The performance measures are explained now in more detail together with the corresponding abbreviations which are used in the subsequent tables.

a) <u>Accuracy</u>: To compare the achieved accuracy, we consider only successful runs.

FV : Geometric mean of the absolute values of f at a termination point $\overline{x}$, i.e. of numbers $|f(\overline{x})|$.

VC : Geometric mean of the sums of constraint violations at a termination point $\overline{x}$, i.e. of numbers $r(\overline{x})$.

KT : Geometric mean of the Euclidean norms of Kuhn-Tucker vectors with respect to a termination point $\overline{x}$, i.e. of numbers $h(\overline{x})$.

ED : Arithmetic mean of the numbers of exact digits at a termination point $\overline{x}$, i.e. of the numbers $e(\overline{x})$.

b) <u>Efficiency</u>: The efficiency is measured by execution time and number of function and gradient calls. In all cases, the arithmetic mean is used with respect to the successful runs.

ET : Mean value of the execution time in seconds.

NF : Mean value of the function calls of f.

NG : Mean value of the function calls of the constraints (Each restriction counted).

NDF : Mean value of the gradient calls of f.

NDG : Mean value of the gradient calls of the constraints (Each restriction counted).

c) Global convergence: To give more insight into the global convergence behavior, not only the percentage of global solutions is reported, but also the corresponding objective function values. This allows distinguishing between failures in approximating x* and solutions below zero.

PGS : Percentage of global solutions, i.e. PGS:= GS·100/(SS+GS).

GVC : Geometric mean of the sums of constraint violations.

GFV : Geometric mean of the objective function values of the global solutions.

d) Reliability: Besides the percentage of non-successful runs, the objective function values and the sums of constraint violations of the non-successful solutions are presented. So it is possible to distinguish between attempts to reach the solution x* and divergence of the program.

PNS : Percentage of non-successful solutions, i.e.
PNS:= NS·100/(SS+GS+NS).

FFV : Geometric mean of the positive function values of f with respect to the non-successful runs.

FVC : Geometric mean of the sums of constraint violations with respect to the non-successful runs.

F : Number of problems which could not be solved due to overflow, programming errors, etc.

4. Numerical results

This section presents the numerical results computed by the following six optimization programs:

VF02AD (Powell [9]): Quadratic approximation of the Lagrangian with linear inequality constraints.

OPRQP (Biggs [3]): Quadratic approximation of a penalty function with linear equality constraints.

GRGA (Abadie [1]): Generalized reduced gradient method.

VF01A (Fletcher [6]): Multiplier method.

FUNMIN (Kraft, DFVLR, Oberpfaffenhofen): Multiplier method.

FMIN (Kraft): Penalty method; FORTRAN-translation of the ALGOL 60
   procedure MINIFUN of Lootsma [7].

All programs are written in FORTRAN with about 500 - 4000 statements and are transformed into single precision subroutines, if necessary. In all cases, analytic first derivatives are provided. As
far as possible, all subroutines are executed with those tolerances
which are predetermined or recommended by the authors.

The numerical tests are performed on a Telefunken TR440 computer
at the Rechenzentrum of the Würzburg University with the same
FORTRAN compiler. The test problems are computed with double precision arithmetic, the calculations within the driving program are
carried out in single precision with more than 10 correct digits
(35 - 38 bit mantissa). The tolerances $\eta_1, \ldots, \eta_4$ defining a successful, global or non-successful solution are set to

$$\eta_1 := .001 \; , \quad \eta_2 := .001 \; , \quad \eta_3 := .1 \; , \quad \eta_4 := 2 \; .$$

Table 1 shows the computed results determining accuracy, efficiency, global convergence, and reliability of an optimization code
under consideration obtained by 240 test runs. In order to relate
the efficiency to the reached accuracy, we denote by

$$A := .25 \cdot (-\log_{10} FV - \log_{10} VC - \log_{10} KT + ED)$$

the accuracy of a program. The five items determining the efficiency
are divided by these values and subsequently scaled such that the
best code obtains the value 1. The corresponding results are presented in table 2, where the same abbreviations are used.

To give a more comprehensive impression of the performance of an
optimization program, it is possible to evaluate the average range
numbers of the three criteria efficiency related to accuracy (E),
global convergence (G), and reliability (R), confer table 3. Since
all items are equally weighted and the amounts of the differences
are not considered, the reader should exploit table 3 very cautious-
ly, especially the total mean values of all range numbers determining the "quality" of a program (Q).

Code	VF02AD	OPRQP	GRGA	VF01A	FUNMIN	FMIN
FV	.36E-8	.11E-7	.84E-5	.43E-8	.65E-9	.11E-4
VC	.40E-10	.51E-8	.30E-11	.13E-8	.44E-9	.87E-7
KT	.35E-5	.41E-6	.75E-3	.62E-6	.88E-5	.74E-3
ED	6.64	7.31	4.37	7.26	6.18	3.94
ET	31.5	22.6	37.7	42.2	98.8	118.8
NF	16	58	204	158	519	737
NG	179	599	2946	1595	5023	7027
NDF	16	40	67	158	112	158
NDG	179	418	378	603	1097	1300
PGS	25.6	19.2	28.6	10.5	17.9	16.0
GVC	.23E-8	.16E-7	.60E-10	.12E-7	.76E-5	.91E-5
GFV	-.11E1	-.46	-.10E2	-.44	-.19E2	-.58
PNS	6.2	23.2	12.1	23.9	32.9	35.0
FFV	.41E-1	.27E-7	.12E-2	.49E-6	.25E-2	.35E-7
FVC	.86E-4	.18E1	.22E-7	.49E1	.98E-1	.16E-1
F	5	1	3	13	5	0

Table 1: Accuracy, efficiency, global convergence, reliability.

Code	VF02AD	OPRQP	GRGA	VF01A	FUNMIN	FMIN
ET	1.2	1	1.8	1.6	3.7	7.1
NF	1	4.0	14.6	9.1	28.8	64.0
NG	1	4.1	19.4	8.7	26.3	57.5
NDF	1	2.6	4.8	9.1	6.3	13.9
NDG	1	2.6	2.5	3.2	5.8	10.8

Table 2: Efficiency related to accuracy.

Remark: The numerical tests presented so far are the first step of
a major project to compare optimization software. Up to date, 26
different programs are submitted for the comparative study and will
be tested in the same way. Besides the test problem generator, we
have gathered and coded about 120 test examples which were used in
the past to test optimization programs (for example the Colville

problems). The corresponding numerical tests are in preparation, they will be performed by W. Hock.

Acknowledgement: The author would like to thank Mr. W. Schliffer and Dr. G. Schuller from the Rechenzentrum of the University of Würzburg for their support making it possible to implement this extensive competing study.

Code	VF02AD	OPRQP	GRGA	VF01A	FUNMIN	FMIN
E	1.20	2.00	3.40	3.60	4.80	6.00
G	2.33	4.00	1.33	5.00	3.33	5.00
R	3.25	2.75	2.50	4.75	4.50	3.00
Q	2.17	2.75	2.58	4.33	4.33	4.75

Table 3: Average range numbers.

References:

[1]   J. Abadie, Methode du gradient reduit generalise: Le code GRGA, Note HI 1756/00, Electricité de France, Paris, 1975.

[2]   J. Asaadi, A computational comparison of some non-linear programs, Mathematical Programming, Vol.4 (1973), 144-154.

[3]   M.C. Biggs, Constrained minimisation using recursive quadratic programming: Some convergence properties, J.Inst.Maths.Applcs., Vol.21, No.1 (1978), 67-82.

[4]   A.R. Colville, A comparative study of nonlinear programming codes, Report 320-2949, IBM New York Scientific Center, 1968.

[5]   E.D.Eason, R.G. Fenton, Testing and evaluation of numerical methods for design optimization, UTME-TP 7204, University of Toronto, 1972.

[6]   R. Fletcher, An ideal penalty function for constrained optimization, in: Nonlinear Programming 2, O.L. Mangasarian, R.R. Meyer, S.M. Robinson eds., Academic Press, New York, 1975, 121-163.

[7]   F.A. Lootsma, The ALGOL 60 procedure MINIFUN for solving non-
        linear optimization problems, Dept. of Mathematics, University
        of Technology, Delft, Netherlands.

[8]   A. Miele, J.L. Tietze, A.V. Levy, Comparison of several
        gradient algorithms for mathematical programming problems,
        Aero-Astronautics Report No.94, Houston, Rice University, 1972.

[9]   M.J.D. Powell, A fast algorithm for nonlinearly constrained
        optimization calculations, in: Proc. of the 1977 Dundee Con-
        ference on Numerical Analysis, Lecture Notes in Mathematics,
        Springer-Verlag, 1978.

[10]   E. Sandgren, The utility of nonlinear programming algorithms,
        Thesis, Purdue University, West Lafayette, 1977.

[11]   K. Schittkowski, Randomly generated NLP test problems with
        predetermined solutions, Preprint No.35, Institut für Ange-
        wandte Mathematik und Statistik, Universität Würzburg, 1978.

[12]   K. Schittkowski, A numerical comparison of optimization soft-
        ware using randomly generated test problems - An intermediate
        balance, Preprint No.43, Institut für Angewandte Mathematik
        und Statistik, Universität Würzburg, 1978.

[13]   R.L. Staha, Constrained optimization via moving exterior
        truncations, Thesis, The University of Texas at Austin, 1973.

[14]   D.C. Stocker, A comparative study of nonlinear programming
        codes, M.S.thesis, The University of Texas, Austin, 1969.

[15]   D. Tabak, Comparative study of various minimization techniques
        used in mathematical programming, IEEE Transactions on Auto-
        matic Control, AC-14 (1969), 572.

Performance Evaluation of Numerical Software, Fosdick (ed.)
© IFIP, North-Holland Publishing Company, 1979

DEVELOPING MODULAR
SOFTWARE FOR UNCONSTRAINED
OPTIMIZATION

Robert B. Schnabel
Department of Computer Science
University of Colorado at Boulder
Boulder, Colorado U.S.A.

Abstract. In this paper we support the use of modular software
in constructing medium to large size numerical algorithms or
systems of algorithms. First we discuss the advantages of modular numerical software, which include ease of development, testing, use and modification. Then we suggest a way of progressing
from the initial high-level design of the modular system to the
computer code through a series of descriptions which also serve
as an important aid to understanding the system. As an example
we use our recently developed system of quasi-Newton algorithms
for unconstrained optimization. Our final pre-code description
stage causes us to be interested in programming languages which
allow user specification and efficient execution of basic data
operations.

1. INTRODUCTION

This paper discusses the modular development of a system of algorithms for unconstrained optimization written recently by the author as part of a forthcoming book
by Dennis and Schnabel [2]. The purpose of the paper, however, is not to talk
about this specific numerical application, but rather about the issue of modular
development of numerical software, its advantages, and some interesting considerations which arise when developing modular numerical software. The issues we discuss have been recognized and discussed by computer scientists and numerical analysts in the last ten years, but the application of these ideas to numerical software is still rather limited, and we hope to add a little impetus.

The activity we discuss, modular development of numerical software, precedes the
performance evaluation of numerical software which is the topic of these Proceedings. However, we will point out that two important advantages of modular development are that it greatly facilitates controlled testing, and that it leads naturally to good documentation. Thus modular development enhances performance evaluation although its advantages are not limited to this.

Our paper is divided into two sections. In the first (Section 2) we describe what
we mean by a modular system of algorithms, and discuss the important advantages of
modular development. As an example we use the system of quasi-Newton algorithms
developed by the author for the unconstrained minimization problem

$$\text{min } f: \ R^n \to R$$
$$x \in R^n \tag{1.1}$$

in the case when f is assumed twice continuously differentiable. However, it is
important to realize that most of our discussion applies equally well to most
other complex numerical problems, especially to other iterative algorithms such as
O. D. E. solvers. In Section 3 we discuss the communication (to implementors and

users) and computer implementation of a modular system, again using our system as an example. There are interesting questions concerning how to describe a modular system so that others can understand, code, use or modify it. We propose developing a description at two levels, a system description and an algorithmic description, before generating the actual code. Our algorithmic description leads us to be interested in facilities for defining primitive data operations, which are now available in a number of research computer languages but not yet in any production language.

2. DESCRIPTION AND ADVANTAGES OF THE MODULAR SYSTEM

In this section we describe the structure of our modular system for unconstrained optimization, and use it to illustrate the advantages of modular numerical software in general.

To discuss our modular system, one needs to know just a little bit about algorithms for problem (1.1). A user employing software to solve a particular unconstrained optimization problem will supply the function f, a starting point x_0, and various tolerances. Our system is concerned with the common case when f is also twice continuously differentiable, although the user may or may not be able to supply ∇f or $\nabla^2 f$. (For simplification in this paper only, we assume ∇f is available.) The type of algorithm favored for such a problem is a "quasi-Newton" algorithm, which we view as having the following form. (For background on quasi-Newton methods, see Dennis and Moré [1].)

> Algorithm 2.1 -- Quasi-Newton framework for unconstrained optimization
>
> I.   Initialization
>
> II. Iteration
>> given: $x_i \in R^n$, $\nabla f(x_i)$, $\nabla^2 f(x_i)$ or an approximation to it
>>
>> find: $x_{i+1} \in R^n$ a better approximation to the minimum
>> 1. Calculate "Newton-ish step" p_n,
>>   using an appropriate local model of f(x)
>> 2. Using p_n, calculate x_{i+1} ("Global strategy")
>> 3. Decide whether to Stop; if not
>> 4. Calculate or approximate $\nabla^2 f(x_{i+1})$

The structure of Algorithm 2.1 is all we wished to say about unconstrained optimization algorithms themselves -- just the fact that the algorithms have an initialization step followed by an iterative step with four distinct parts. But we should mention immediately that some researchers in unconstrained optimization take issue with our type of description, because they feel the pieces within the iteration cannot be completely separated. Their beliefs preclude a modular system. However, if one accepts the above (or any similar structure) a modular system follows easily. We give a diagram of our entire system of algorithms for the iteration step below.

Fig. 2.2 -- <u>Modular iteration for unconstrained optimization</u>

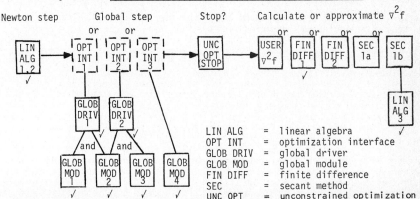

The four sections in Figure 2.2 correspond to the four steps in Algorithm 2.1. The solid boxes are <u>modules</u>. The Newton step and stopping criteria correspond to two and one rather simple modules, respectively. The other two parts are more interesting, and serve to illustrate some capabilities of modular numerical software. For the global step our system provides three alternative strategies, as there is a real difference of opinion in the field as to which is the best way to proceed. The first two are "model trust region" strategies which are seen to be related because they share a module. They each use a driver which does little more than call the two modules (1 and 2, or 3 and 2, respectively) alternately until a satisfactory next point is found. (The dashed "interface" box is explained at the end of this section.) The third strategy is a line search. What is important to note is that in a particular run of our system, a certain one of these three strategies would be selected and used exclusively. However, a software system developed from our description could contain all three with the user choosing one at run time, or it could contain just one of the three strategies if the implementor has a strong personal preference.

The five alternative modules for the "$\nabla^2 f$" step are used a bit differently. The choice of which to use depends on whether $\nabla^2 f$ is provided analytically, and if not, on how expensive calculation of f and ∇f is. Therefore a *system* of algorithms should contain all five; again, a particular run will choose just one.

From this description, the advantages of modular development become evident. We identify them in three categories:

a) <u>Development</u>: A modular structure allows naturally for top-down development, that is, starting at the highest level and successively adding levels of detail. This has become acknowledged by most computer scientists as the most effective manner to develop code. In Section 3 we will see that another advantage of our manner of top-down development is that it leads naturally to good documentation *at an early stage*.

b) <u>Testing and research</u>: A modular system is generally preferred for testing or verifying a large system, as it allows this to be done in manageable parts. It is also excellent for research, because it provides a controlled environment for testing alternative strategies. For example, in unconstrained optimization research one often wants to compare a new global strategy (Step 2, Alg. 2.1) or a new derivative approximation strategy (Step 4) to an existing one. If one changes only this portion of a system like Fig. 2.2, the differences in performance are clearly attributable to the new strategy. This is in sharp contrast to many test results reported

today, where the two strategies being compared are embedded in two different codes,
severely confusing the interpretation of the test results. (A common difference
between codes which is acknowledged to often hinder the drawing of conclusions from
test results is different stopping criteria.)

c) Software library: A modular system provides two additional advantages when
used as part of a software library. One is the convenient provision of alternative
capabilities or algorithms, as was discussed above. The second is that routines
for related problems may share many of the same modules. Indeed, this was a prime
motivation in our development of a modular system for unconstrained optimization.
We have concurrently developed a modular system for solving systems of nonlinear
equations,

$$\text{given } F: \ R^n \to R^n, \ \text{find } x \in R^n \ \text{such that } F(x) = \underline{0} \qquad (2.3)$$

which are also solved by a quasi-Newton algorithm with the same four-part iteration
as Alg. 2.1. It makes use of the ten modules indicated by checks ($\sqrt{}$) in Fig. 2.2,
including all the global modules which are the bulk of the system. Thus the development of the second system is *far* less work than the first. This also explains
the purpose of the three interfaces in Fig. 2.2: while the systems for unconstrained optimization and nonlinear equations use the same global modules, they call these
modules with parameters particular to their application. The interfaces simply
give the correspondences of the calling sequences of the main programs to the
parameter sequences of the modules.

3. COMMUNICATION AND COMPUTER IMPLEMENTATION OF THE MODULAR SYSTEM

At this point one may be convinced of the advantages of a modular system, but wary
of the difficulty involved in transforming it into code, and in supplying documentation which will make it reasonably easy for an outsider to use, and for a numerical analyst to understand or modify. In this section we discuss the resolution of
these problems in a medium to large size numerical system such as ours. We show
that we are led rather naturally to two intermediate descriptive steps between the
modular diagram of Fig. 2.2 and the ultimate code: a "system description" and then
an "algorithmic description." The latter leads to an interesting consideration in
programming language design.

Perhaps the best way to start is to ask what our final optimization system corresponding to Fig. 2.2 will consist of. The modules will probably each be subroutines or procedures, each with certain specified input, output and global variables.
Note that the main driver (which along with the initialization module was omitted
from Fig. 2.2) is itself such a module, with its input and output going from and to
the user respectively. The interfaces will essentially disappear, having specified
calling sequences in the code. The documentation which will be required is: the
modular diagram Fig. 2.2 and guidelines on how to choose between its options, the
input requirements and guidelines on specifying the input, and perhaps guidelines
on interpreting the output. It is important to note that this is *all* the documentation required.

The question remaining is how one gets the modules to fit together. To accomplish
this easily is one reason why we suggest a system description as the next stage
after the modular diagram of Fig. 2.2. Two related reasons are that this is sensible top-down development, and that it leads to a very understandable high-level description of the system.

By a system description we mean a level of description specifying the purpose and
arguments of each piece of Fig. 2.2, and also describing how the entire system is
to be used. Ours consists of three parts: high level module descriptions, interfaces and guidelines. Each module description at this level is simply a list of
input, output and global variables used by the module, and a very brief statement
of its function. An example is the top third (down to "Output") of Fig. 3.1. Each
interface is just a list matching the parameter sequence of a module with the

calling sequence it will be invoked with. The guidelines are precisely those for module selection, input and output mentioned above. Note that this amount of information completely specifies the linkage and use of our system. In addition, <u>the documentation has naturally been completed</u> at an early stage of system development.

What remains is to specify the algorithm of each module. One could of course just go and code each one. However, we believe a second intermediate stage is highly desirable, both for the initial development, and so that afterwards people can learn about the algorithms without having to go immediately to the code. Therefore we have produced a description of the algorithm of each module before the actual code. This level added to the system description forms our algorithmic description.

We have actually decided to describe the algorithm of each module in two stages. The first is a rather brief (typically 5-10 line) description in prose or outline form which relates the main points but not all the details of the algorithm. An example is the "Description" in the middle of Fig. 3.1 But the most interesting part is that then, rather than going directly to the code as would be expected, we have specified each algorithm in complete detail in "PASCALish" code, the first few lines of which for one module are shown below (under "Implementation"). We now discuss the reasons for this.

Fig. 3.1 -- <u>Description of global module 3</u>

Global variables: D_S - diagonal scaling matrix for x, (described in Guideline 2)

Input: $n \in R$, $x \in R^n$, $g \in R^n$, lower triangular $L \in R^{n \times n}$, $p_N \in R^n$ (where $p_N \triangleq -M^{-1} g$, $M \triangleq LL^T$), $\delta \in R$

Find: $x_+ \in R^n$ which approximates the solution to

$$\min f_Q(x_+) = f(x) + (x_+ - x)^T g + \frac{1}{2}(x_+ - x)^T M(x_+ - x)$$

$$\text{subject to } \| D_S(x_+ - x) \| \le \delta.$$

(by finding the x_+ which solves this problem among all points on the double dogleg curve)

Output: $x_+ \in R^n$, $\delta \in R$.

Description: If $\| D_S p_N \|_2 \le \delta$, $x_+ = x + p_N$ (p_N is the Newton step to the minimum of the quadratic model). Otherwise x_+ is chosen to be the (unique) point on the double dogleg curve such that $\| D_S(x_+ - x) \| = \delta$. The double dogleg curve consists of the three line segments connecting x, $x + p_c$, $x + p_N$, $x + np_N$, where $x + p_c$ is the Cauchy point, the minimum of the quadratic model in the Cauchy (steepest decent) direction $-D_S^{-2}g$, and $n \le 1$ (see Ch. 6).

Implementation:

Newtlen := $\| D_S p_N \|_2$;

if Newtlen $\le \delta$

    then begin (* x_+ is Newton point *)

        x_+ := $x + p_N$;

        δ := Newtlen

    end

else begin (* Newton step too long -- x_+ on dogleg curve *)

$$\rho := \frac{\| D_S^{-1} g \|_2^4}{\| L^T D_S^{-2} g \|_2^2 (g^T p_N)}; \qquad (* \| L^T D_S^{-2} g \|_2^2 = g^T D_S^{-2} M D_S^{-2} g *)$$

Our PASCALish code is algorithmically *complete*, and syntactically correct with three exceptions:

    a) mathematical variables (Greek, sub or superscripted) are allowed.

    b) input and output statements are informally specified.

    c) basic mathematical and linear algebra operations are allowed.

The big advantage of this description over actual code is that it is very <u>readable</u> and <u>understandable</u>, whereas removing a or c (and to a lesser extent b) renders it *much* harder to understand. In addition, it is almost trivially translatable into any programming language of interest. Our students have programmed much of the system in FORTRAN and its dialects WATFOR and FLEX, as well as PASCAL, with great ease; PL/1 and ALGOL would be equally easy. The reason PASCAL was chosen for the descriptive language is that it is an algorithmic language designed for readability and easy understanding. (PL/1 or ALGOL would also have been appropriate.) Our descriptions do not make use of any features of PASCAL which are not readily convertible into all languages of interest, including FORTRAN.

The PASCALish code completes the algorithmic description of our modular system. However, there turns out to be an interesting consideration in programming language design which arises in converting this description to code. Note that all that has to be done is to translate the PASCAL statements to the corresponding statements in another language if necessary, and eliminate exceptions a, b and c above. Exceptions a and b are trivial to rectify (e.g., δ probably becomes DELTA). Exception c is also quite easy, but what we would really like is for c to be allowed by the programming language. This turns out to be an issue that is well recognized by researchers in programming language design.

Notice that what we have done up to the algorithmic description level is form top-down a modular system, underlying which are certain data types (integer, real, vector, matrix) and a small number of basic data operations on these types. This is shown in Figure 3.2.

Fig. 3.2 -- <u>Overall structure of the software system</u>

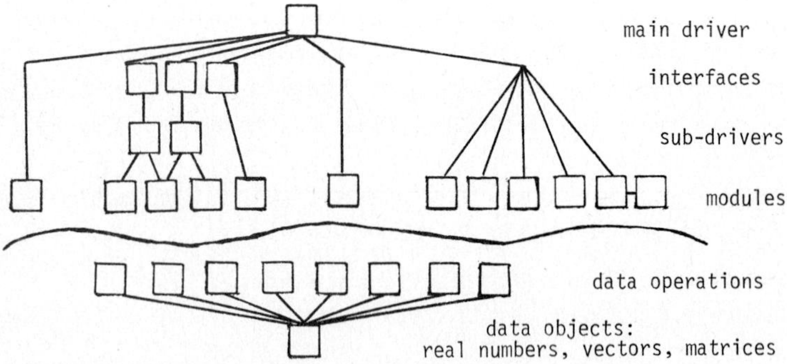

  main driver

  interfaces

  sub-drivers

  modules

  data operations

  data objects:
  real numbers, vectors, matrices

In our whole system, the data operations turn out to be

$$\sum_{i=1}^{n} \alpha_i, \; \max\{\alpha,\beta\}, \; \min\{\alpha,\beta\}, \; \gamma \in [\alpha,\beta], \; v^T w, \; \|v\|_2, \; A \cdot B, \; A \cdot v$$

where α, β, $\gamma \in R$; v, $w \in R^n$; A, $B \in R^{n \times n}$. The two-part hierarchy of Fig. 3.2 (algorithms and data) has become recognized as the form which arises from developing almost any medium or large system (in non-numeric systems the underlying data types and operations are often far more complex than in numeric systems), and there are

starting to be ways to actually write the system so that the underlying data operations can be primitives in the code which are also efficiently executed.

Of course the data operations can simply be made procedures, which is what has usually been done in coding our system. While this is a convenient solution, it is
less efficient than we would like due to the cost of procedure linkage. A more efficient solution is the use of a macro-generator, which allows one to specify simple procedures which are then compiled and executed as in-line code with the proper
parameter substitution. This permits the basic data operations to be executed efficiently. (For an example, see Myers [4].) Perhaps better still, there are several research programming languages which not only enable the expression and efficient use of basic data operations, but also are oriented to supporting a system
with the form of Fig. 3.2. These include SIMULA, MODULA and EUCLID; for a good reference to the work in this field, including references to all of these languages,
see Goos and Kastens [3]. Since it is clear that most numerical algorithms or systems of algorithms, if they are developed modularly, will take a form similar to
Fig. 3.2, we believe that the numerical software community should be interested in
programming languages which allow the specification and use of basic data operations, and perhaps which are also oriented to supporting such modular systems.

Acknowledgement

My thanks to P. Zeiger (Dept. of Computer Science, University of Colorado at
Boulder) for first pointing out the basic data operations issue in my algorithmic
descriptions, and to L. Osterweil (same department) and W. Waite (Dept. of Electri-
cal Engineering, University of Colorado at Boulder) for informing me about some of
the work in this area.

Boulder, Colorado January, 1979

References

[1] J. E. Dennis and J. J. Moré, "Quasi-Newton methods, motivation and theory",
     SIAM Review 19 (1977) 46-89.

[2] J. E. Dennis and R. B. Schnabel, "Quasi-Newton methods for nonlinear problems",
     manuscript in preparation.

[3] G. Goos and U. Kastens, "Programming languages and the design of modular pro-
     grams", in: P. G. Hubbard, S. A. Schuman eds., Constructing Quality Software,
     Proceedings of the IFIP TC2 working conference on constructing quality soft-
     ware (North-Holland, Amsterdam, 1978).

[4] E. Myers, "The BIGMAC user's manual", report no. CU-CS-145-78, Department of
     Computer Science, University of Colorado, Boulder, Colorado (1978).

Performance Evaluation of Numerical Software, Fosdick (ed.)
© *IFIP, North-Holland Publishing Company, 1979*

IMPLEMENTATION AND TESTING OF OPTIMIZATION SOFTWARE*

Jorge J. More'
Applied Mathematics Division
Argonne National Laboratory
Argonne, Illinois

1.   INTRODUCTION

We are concerned with optimization software which must perform satisfactorily on a wide range of machines and compilers. This requirement demands that the development of the software must adhere to strict standards; the purpose of performance testing is to verify that the implementation of an algorithm satisfies these standards.

During the development of optimization software at Argonne National Laboratory we have found that it is necessary to test software at two levels. The first level of testing is concerned with the robustness and reliability of the implementation, while the second level tests the overall performance of the implementation.

A robust implementation extends the domain of the algorithm so that it copes with as many problems as possible without a serious loss of efficiency. A reliable implementation performs well-defined calculations accurately and efficiently.

The implications of these definitions for optimization software are explored in Section 2 and 3. Since we are concerned with performance on a wide range of machines and compilers, we also outline the approach used at Argonne to develop transportable optimization software.

There are many approaches to performance testing, but most of these approaches restrict the test problems, and as a consequence do not really test the overall performance of the implementation. Moreover, performance in optimization is often equated with efficiency, and thus accuracy and reliability have not received sufficient attention.

To address these needs, we suggest guidelines for the implementation of optimization software and show the (often dramatic) effect that these guidelines have on the performance of the implementation. In addition we discuss the weaknesses in the usual approaches to performance testing of optimization software and suggest some remedies.

The background necessary to read this paper is minimal, but at various points in the paper we have used test functions from the optimization literature without reference to the original source. Further information about these test functions is given in the paper of More', Garbow and Hillstrom [8].

2.   IMPLEMENTATION OF OPTIMIZATION SOFTWARE

If optimization software is to perform satisfactorily on a wide range of machines and compilers, then the implementation must adhere to strict standards; we have

---
*Work performed at Cambridge University under the auspices of the U.S. Department of Energy, with partial support from the U.S. Army Research Office (contract DAAG29-78-M-0152) and the Science Research Council.

followed the NATS approach as described by Smith, Boyle and Cody [11]. In this approach, the implementation of each algorithm must be robust, reliable, trans-portable, usable, and valid. We only consider robustness, reliability, and transportability since these attributes are frequently given too little atten-tion in optimization software.

The ideas presented in this section are well known and have been used with great success in other areas of mathematical software. Our contribution is to point out that they can be used profitably in the implementation of optimization software.

Robustness

An implementation of an algorithm is robust if it detects and gracefully recovers from abnormal situations without involuntarily terminating the computer run.

In general it is not possible or reasonable to recover from all abnormal situa-tions, so the aim of robustness is to extend the domain of the algorithm so that it copes with as many problems as possible without a serious loss of efficiency. Optimization software that is not robust exhibits certain symptoms; the main ones are destructive overflows and underflows, and failures due to inadequate termina-tion criteria or invalid arguments.

Destructive overflows and underflows

An overflow or underflow in a well-defined calculation is destructive if it damages the accuracy of the calculation.

One of the simplest examples of the dangers of destructive overflows and under-flows occurs in the computation of the ℓ_2-norm by the formula

$$(2.1) \qquad \| x \| = \left(\sum_{j=1}^{n} x_j^2 \right)^{1/2} .$$

This is a well-defined calculation if $\| x \| \leq$ GIANT, where GIANT is the largest magnitude that can be represented in the machine. In this calculation x_j^2 gen-erates a destructive overflow if

$$|x_j| > (\text{GIANT})^{1/2} .$$

This overflow forces the assignment of the square root of GIANT to $\| x \|$ and since

$$\| x \| > (\text{GIANT})^{1/2} ,$$

this overflow may cause large errors.

In general overflows are destructive, but this is not the case for underflows. The computation of x_j^2 generates an underflow if

$$|x_j| < (\text{DWARF})^{1/2} ,$$

where DWARF is the smallest non-zero magnitude that can be represented in the machine. If some of the other components of x are relatively large, say of order unity, then this underflow is not destructive since it does not affect the accuracy of the computation. On the other hand, if n = 2 and

$$2|x_1| = |x_2| = (\text{DWARF})^{1/2} ,$$

then the underflow generated by x_1^2 is destructive since it leads to an error of about eleven percent.

Destructive overflows and underflows in the calculation of the ℓ_2-norm can be avoided by setting

$$\sigma_1 = \max\{|x_j| : 1 \le j \le n\}$$

$$\sigma_2 = \left\{ \sum_{j=1}^{n} \left(\frac{x_j}{\sigma_1}\right)^2 \right\}^{1/2}$$

$$\|x\| = \sigma_1\sigma_2 \quad .$$

Note that underflows may still occur in the computation of σ_2, but they do not affect the accuracy of the computation. Similarly, an overflow may occur in the product $\sigma_1\sigma_2$, but this overflow is not destructive since in this case the computation of $\|x\|$ is not well-defined.

Avoiding destructive overflows and underflows in optimization software does not impose a significant penalty in terms of efficiency since the critical computations do not occur in the inner loops. It is therefore unfortunate that most optimization software does not attempt to avoid destructive overflows and underflows.

Finally, we emphasize that avoiding destructive overflows and underflows is essential if the software is to perform satisfactorily on a wide range of machines; it is especially important on those machines with small values for GIANT and large values for DWARF (for example, 10^{38} and 10^{-38}, respectively).

Termination criteria

One of the most important components of an optimization algorithm is the termination criteria. The purpose of the termination criteria is to cause an exit from the algorithm if and only if the algorithm finds an approximate solution of the desired accuracy. In addition, the termination criteria should recognize certain abnormal situations such as unreasonably small tolerances, optimization problems without solutions, a continuum of solutions, and infinite solutions.

It is important that the algorithm accept a wide range of tolerance values. For example, consider the unconstrained minimization problem

$$(2.2) \qquad \min\{f(x): x \in R^n\} \quad .$$

If the implementation terminates when

$$\|\nabla f(x)\| \le \text{GTOL} ,$$

then this test is scale dependent and it is therefore improper to require that, for instance, GTOL exceed the machine precision. If this test is being used then all nonnegative values of GTOL should be allowed. The setting GTOL = 0 should be allowed because it is a very convenient way for the user to indicate that he wants maximum accuracy, and in any case, it may be possible to satisfy the user's requirement.

A robust implementation of an optimization algorithm should detect when the tolerances cannot be achieved, inform the user, and exit with the best estimate for the solution to the problem.

The simplest example of an optimization problem without a solution is an unconstrained minimization problem unbounded below. More interesting examples occur in the solution of a system of n nonlinear equations in n variables

$$(2.3) \qquad f_i(x_1,\ldots,x_n) = 0 , \qquad 1 \le i \le n ,$$

or in vector form, $F(x) = 0$. Since nonlinear equations may not have approximate solutions, for example the Chebyquad test functions of Fletcher for $n = 8$ and $n \geq 10$, a robust implementation of a nonlinear equation solver must attempt to recognize when the nonlinear equations do not have a solution and then terminate with an appropriate message to the user.

Now consider the nonlinear least squares problem

$$\min\left\{\sum_{i=1}^{m} f_i^2(x): x \in R^n\right\} ,$$

or the equivalent minimum ℓ_2-norm problem

(2.4) $\min\{\|F(x)\|: x \in R^n\}$

where F is a vector function whose i-th component is the residual f_i. Unlike non-linear equations, nonlinear least squares always have approximate solutions since it is always possible to find a vector x^* such that $\|F(x^*)\|$ is arbitrarily close to (2.4). In certain situations, however, the components of x^* may be forced to assume arbitrarily large values. To make this precise, first consider the unconstrained minimization problem (2.2) and define f to have a critical point at infinity if there is an unbounded sequence $\{x_k\}$ such that $\{\nabla f(x_k)\}$ converges to zero. A consequence of this definition is that if f is continuously differentiable and bounded below then f has a critical point; in particular, smooth nonlinear least squares problems always have critical points. To illustrate this definition in the context of nonlinear least squares, consider Bard's test function which is defined by

$$f_i(x_1,x_2,x_3) = y_i - \left[x_1 + \frac{u_i}{x_2 v_i + x_3 w_i}\right]$$

where $u_i = i$, $v_i = 16-i$, $w_i = \max\{u_i, v_i\}$ and y_i is the i-th data point. This function has a global minimizer near (0.082, 1.13, 2.34), but algorithms started at for instance (10,10,10) are attracted towards a local minimizer near $(x_1^*, +\infty, +\infty)$. To explain the value of x_1^* note that

$$\lim_{\alpha, \beta \to \infty} f_i(x_1, \alpha, \beta) = y_i - x_1 ,$$

and that $y_i - x_1$ is the i-th residual of a (one-dimensional) linear least squares problem with solution

$$x_1^* = \frac{1}{m} \sum_{i=1}^{m} y_i \doteq 0.84 .$$

In most cases infinite solutions do not lead to overflow problems. In the case of Bard's function, $f_i(x) = y_i - x_1$ on a machine with relative precision e_M whenever

$$\frac{u_i}{|x_2 v_i + x_3 w_i|} < e_M |x_1| ,$$

and thus a robust algorithm terminates when x_1 is near 0.84 and, for example, x_2 and x_3 are near e_M^{-1}. Another point concerning infinite solutions is that although they are often a consequence of large initial steps and can usually be avoided by restricting the steps, infinite solutions are valid from the point of view of the optimization algorithm. A robust algorithm tends to avoid infinite solutions but can cope with them.

Invalid arguments

A robust implementation detects when the routine is called with invalid arguments and gives an appropriate error indication. Optimization software tends to have a large number of arguments and in these cases it may not be convenient to indicate which parameter(s) are in error. It is helpful, however, to give the user a list of the possible improper arguments since usually only a small subset of the parameters can be specified improperly (for example, dimensions and tolerances).

It is also desirable for an implementation to accept a wide range of argument values. We have already mentioned this in connection with tolerances. For dimensions, a robust implementation should work for any $n \geq 1$ where n is the number of variables. The only limitation on n should be imposed by the storage available to the user. In the past, many optimization routines failed, often without indication, in the special case $n = 1$. It has been argued that general optimization routines should not be used in the one-dimensional case since more efficient special purpose routines treat this case. But if a user is solving a sequence of problems and some of his problems are one-dimensional, then it is more convenient for the user to use the implementation at hand. A message to the user indicating that $n = 1$ is an invalid argument is not a graceful recovery.

Reliability

An implementation of an algorithm is reliable if it performs well defined calculations accurately and efficiently.

Most optimization calculations are not well-defined since they cannot be guaranteed to terminate with an approximate solution of a prescribed accuracy. For this reason, in this section we only consider the reliability of the linear algebra calculations that occur in optimization algorithms. For these calculations accuracy can be defined in terms of backward error, and efficiency can be measured in terms of arithmetic operations and array references. The question of reliability for a general optimization algorithm is considered in Section 4.

The accuracy of optimization software depends largely on the accuracy of the underlying linear algebra software. However, certain computations are peculiar to optimization software and require special attention. For example, in the Levenberg-Marquardt algorithm it is necessary to compute the quantity

$$(2.5) \qquad \|f\|^2 - \|f+Jp\|^2$$

where J is an m by n matrix, f is an m-vector, and the n-vector p satisfies

$$(2.6) \qquad (J^T J + \lambda D^T D)p = -J^T f$$

for a given nonsingular diagonal matrix D and $\lambda \geq 0$. It can be shown that in exact arithmetic (2.5) must be nonnegative but that in finite precision arithmetic the computation of (2.5) may produce a negative result. To remedy this situation note that

$$\|f\|^2 - \|f+Jp\|^2 = -2f^T Jp - \|Jp\|^2$$

and that (2.6) implies that

$$(2.7) \qquad \|Jp\|^2 + \lambda\|Dp\|^2 = -f^T Jp .$$

Thus

$$(2.8) \qquad \|f\|^2 - \|f+Jp\|^2 = \|Jp\|^2 + 2\lambda\|Dp\|^2 .$$

Clearly, the right hand side expression is nonnegative even in finite precision

arithmetic and is therefore more desirable; it is not entirely satisfactory because it may lead to destructive overflows and underflows. However, a careful
examination of the Levenberg-Marquardt algorithm shows that we only need to work
with the scaled quantity

$$\frac{\|f\|^2 - \|f+Jp\|^2}{\|f\|^2} \; .$$

In view of (2.8), this scaled quantity can be rewritten as

$$\left(\frac{\|Jp\|}{\|f\|}\right)^2 + 2\left(\frac{\lambda^{\frac{1}{2}}\|Dp\|}{\|f\|}\right)^2 \; ,$$

and since (2.7) shows that $\|Jp\| \leq \|f\|$ and that $\lambda^{\frac{1}{2}}\|Dp\| \leq \|f\|$, we
have now eliminated the possibility of overflows and destructive underflows.

In the above example the accuracy of a computation was improved by rewriting an
expression in a more favorable form. This problem was machine independent since
(2.5) may produce a negative result on any machine. Unless we are worried about
the last few bits, accuracy considerations are usually machine independent. In
some computations, however, it may be desirable to obtain maximum accuracy. For
example, in the determination of a Givens rotation it is necessary to compute the
ℓ_2-norm of a 2-dimensional vector. A robust algorithm for this calculation (see
the section on robustness) uses

$$\gamma_1\left\{1 + \left(\frac{\gamma_2}{\gamma_1}\right)^2\right\}^{1/2}$$

where

$$\gamma_1 = \max\{|x_1|, \; |x_2|\} \; ,$$

$$\gamma_2 = \min\{|x_1|, \; |x_2|\} \; .$$

To improve the accuracy of this computation on hexadecimal machines, Cody [2] proposed rewriting this expression as

$$(2.9) \qquad \gamma_1\left(\left\{\alpha_1 + \alpha_1\left(\frac{\gamma_2}{\gamma_1}\right)^2\right\}^{1/2} / \alpha_2\right)$$

where $\alpha_1 = 0.25$ and $\alpha_2 = 0.5$. The numerical results presented by Wisniewski
[12] show that (2.9) is more accurate on non-binary machines and just as accurate
on binary machines.

In optimization software, the efficiency of an implementation is usually dominated
by considerations of robustness and accuracy. For example, (2.6) for $\lambda > 0$ can be
solved by either Cholesky decomposition or by posing (2.6) as the linear least
squares problem

$$(2.10) \qquad \begin{pmatrix} J \\ \lambda^{\frac{1}{2}} D \end{pmatrix} p \overset{\backsim}{=} -\begin{pmatrix} f \\ 0 \end{pmatrix}$$

and solving (2.10) by QR decomposition with column pivoting. The first approach requires about half the number of arithmetic operations of the QR approach but is
less robust and accurate. Unless other considerations come into play, the QR
approach is therefore preferable.

It should not be inferred from the above discussion that efficiency considerations
are not important; the correct inference is that it is important to achieve the
proper balance between efficiency on the one hand, and robustness and accuracy on
the other hand. For instance, although (2.10) could be solved by the singular
value decomposition, this approach is not recommended because it requires a larger
overhead than the QR approach and offers no significant improvement in terms of
robustness and accuracy. Consider now the determination of the Givens rotation
via (2.9). In this case the extra accuracy is obtained at a negligible cost
(since the Givens rotation is invariably used in other computations with a significantly higher overhead) and is therefore justified.

Transportability

An implementation is transportable across a given range of machines and compilers
if it can satisfy specified performance criteria with only a small number of
changes.

There are several ways to achieve transportability, and in general the preferred
approach is the one that requires the least number of changes. The approach used
at Argonne only requires minor changes to one subprogram. The first step in this
approach is to prepare a double precision master implementation. At this stage we
have found it very helpful to follow the recommendations of Smith [10]. The PFORT
verifier of Ryder [9] is then used to check that the master implementation is
written in a subset of ANSI Fortran that is acceptable to most compilers. We note
that PFORT is not designed to check for run-time errors or the appearance of non-
Fortran characters in comments.

The only machine dependencies of the master implementation are the machine parameters. The Argonne optimization programs only require the specification of three
parameters: If the machine has T base B digits and its largest and smallest exponents are EMAX and EMIN, respectively, then these parameters are

    EPSMCH = $B**(1-T)$, the machine precision
    DWARF = $B**(EMIN-1)$, the smallest magnitude
    GIANT = $B**EMAX*(1-B**(-T))$, the largest magnitude.

These parameters are provided by a function DPMPAR in double precision and SPMPAR
in single precision. The values for these parameters were obtained from the
corresponding PORT library function (Fox, Hall, and Schryer [4]) which provides
values for IBM, CDC, Burroughs, Honeywell, DEC (PDP-10 and PDP-11), and Univac
computers; we have added values for the CRAY-1.

The master implementation is tested on IBM equipment for satisfactory performance;
both the H and WATFIV compilers are used. WATFIV is extremely useful to catch
run-time errors such as uninitialized variables and out-of-bounds subscript
references.

Finally, the TAMPR system of Boyle and Dritz [1] is used to produce formatted
single and double precision versions of the master implementation, and the
single precision version is tested on the CDC equipment of Lawrence Berkeley Laboratory via the ARPANET.

The above approach achieves transportability; the only changes necessary to in-
stall the software at a site are to make sure that DPMPAR and SPMPAR provide
appropriate values.

3.   TESTING FOR ROBUSTNESS AND RELIABILITY

We illustrate the performance criteria suggested in the previous section by testing the nonlinear least squares algorithm EO4GAF from the NAG library. Since this

routine is scheduled for withdrawal in the next release of NAG, it will be interesting to test its replacement on the tests described in this paper.

The tests of this section measure the robustness and reliability of the implementation. Since transportability requires that the implementation satisfy specified performance criteria (for example, robustness and reliability) at other installations, the tests below could also be used to measure transportability.

The purpose of E04GAF is to solve nonlinear least squares problems by the Levenberg-Marquardt method as described by Fletcher [3]. This subroutine has 16 parameters; for our purposes only the following 3 parameters are of importance.

XTOL. Vector of tolerances. Algorithm terminates successfully if

$$|\Delta x(i)| \leq XTOL(i), \qquad i = 1,\ldots,n$$

where Δx is the difference between the last two iterates.

METHOD. Integer variable which specifies the scaling technique. If METHOD = 1, then the (implicit) scale factors are set to the diagonal elements of $J^T J$ at the initial point (J is the Jacobian matrix).

IFAIL. Integer variable. On entry IFAIL must be assigned a value. The recommended value is 0. Unless E04GAF detects an error, IFAIL contains 0 on exit.

All of the runs used METHOD = 1 and were performed at the University of Cambridge IBM 370/165 with the Fortran H extended (OPT=2) compiler and in double precision (16 hexadecimal digits). The values of the machine parameters mentioned in the transportability section are

$$EPSMCH \doteq 2.2 \times 10^{-16}$$
$$DWARF \doteq 5.0 \times 10^{-79}$$
$$GIANT \doteq 7.2 \times 10^{75}$$

To test the robustness of E04GAF, we examined the behavior of the implementation when faced with the following situations:

1. Improper arguments.
2. Destructive overflows and underflows.
3. Unusual termination conditions.

E04GAF detects when it is being called with improper arguments and gives a message to the user, but the documentation does not give a list of the possible improper arguments. This turned out to be troublesome because on one of the runs the only improper argument was IFAIL (we had forgotten to set it), and it was necessary to check every one of the arguments before finding out that the only improper argument was the last one. IFAIL gave us further trouble. On the next run IFAIL was set to the recommended value of 0, but this turns out to be the hard fail option and caused our program to terminate after the first unsuccessful exit from E04GAF. We were now forced to read the chapter on error handling to find out that the correct setting was 1 (the soft fail option). However, since we had overlooked the fact that E04GAF resets IFAIL to 0 after a successful exit, the next run again terminated after the first unsuccessful exit. This last point is important because in our experience variables (not arrays!) which are set by the user and then changed by the program often lead to mistakes.

It was fairly easy to show that E04GAF does not avoid the most obvious destructive overflows. As a test function, consider the linear least squares problem defined by

$$f_i(x) = x_i - \frac{2}{m}\left(\sum_{j=1}^{n} x_j\right) - 1 \,, \qquad 1 \le i \le n$$

$$f_i(x) = -\frac{2}{m}\left(\sum_{j=1}^{n} x_j\right) - 1 \,, \qquad n < i \le m \,.$$

It is straightforward to verify that the solution is $(-1,\ldots,-1)$ and that the optimal sum of squares is m-n.

The test consists of solving this least squares problem with starting point

$$x_\alpha = 10^\alpha (1,\ldots,1) \,, \qquad \alpha = 0,10,20,\ldots,70 \,,$$

and with dimensions $n = 5$, $m = 10$ and $n = 5$, $m = 50$. This is a well-defined calculation and a robust (and reliable) nonlinear least squares algorithm should be able to solve these problems.

E04GAF was able to solve this problem successfully for $\alpha = 0,10,20,30$. At $\alpha = 40$ the run was terminated for an excessive number of overflows. If the overflows are masked, then E04GAF can solve the problem for $\alpha = 40$ and 50 but at $\alpha = 60$ the implementation breaks down; the starting point is not changed after 80 function evaluations.

The purpose of this test is to find out the range of numbers that are handled by an implementation, and in the case of E04GAF we have shown that it does not accept function values whose magnitude exceeds the square root of GIANT. Other tests show that the magnitude of the function values and Jacobian elements should be in the interval

$$[(DWARF)^{1/2}, (GIANT)^{1/2}] \,.$$

These restrictions are very bothersome, especially in the early steps. A more important point is that these restrictions are unnecessary; it is possible to implement the Levenberg-Marquardt method in such a way that these difficulties are avoided. See the implementation described by More' [7].

We next tested the ability of the termination criteria in E04GAF to deal with unreasonably small tolerances and infinite solutions.

E04GAF does not deal with unreasonably small tolerances satisfactorily. The documentation specifies that XTOL(i) should not be less than the machine precision, but that is inappropriate if the least squares problem is sensitive to a component x_i with value less than the machine precision. However, E04GAF apparently accepts any value of XTOL(i). It even accepts negative values, but seems to treat them as if they were zero.

The results of running E04GAF on the linear least squares problem defined above with dimensions $n = 5$, $m = 50$, and with starting point $(1,\ldots,1)$ are presented below. We give the number of function evaluations (NFEV) and the number of Jacobian evaluations (NJEV) required by E04GAF as a function of TOL where XTOL(i) $\equiv$ TOL. The documentation states that an exit value of IFAIL = 5 indicates that rounding errors are causing the routine to function abnormally and that repeating the run with larger values for XTOL could avoid this. Although this is indeed correct, the price paid for this information is very high. On such a simple problem a robust nonlinear least squares implementation should detect unreasonably small requested accuracies in 3 or 4 function evaluations.

TOL	NFEV	NJEV	IFAIL
sqrt (EPSMCH)	3	2	0
EPSMCH	5	3	0
sqrt (DWARF)	83	3	0
ZERO	161	3	5

To test the ability of E04GAF to deal with infinite solutions, we used Bard's test function as described in Section 2. The starting points were (10,10,10) and (100,100,100). In both cases E04GAF converged to an infinite solution without any problems; for example, for the starting point (10,10,10) the magnitudes of the final x_2 and x_3 were about 10^{15}.

The reliability (and robustness) of the implementation of E04GAF was tested by the rank-deficient linear least squares problem defined by

$$f_i(x) = i \sum_{j=1}^{n} jx_j - 1 , \qquad 1 \leq i \leq m .$$

The optimal sum of squares is $m(m-1)/(2(2m+1))$ at any point where

$$\sum_{j=1}^{n} jx_j = \frac{3}{2m+1} .$$

We only considered dimensions $n = 5$, $m = 10$ and $n = 5$, $m = 50$. E04GAF obtained the solution to each problem with

$$XTOL(i) \equiv (EPSMCH)^{1/2}$$

in 13 function evaluations and 12 Jacobian evaluations. This lack of efficiency is due to the use of the normal equations in E04GAF. A more reliable implementation can solve this problem in 3 or 4 function evaluations.

4.   TESTING FOR PERFORMANCE

Once an algorithm has been implemented following the guidelines of Section 2, it is necessary to test the performance of the implementation. The purpose of this section is to discuss the weaknesses in the usual approaches to performance testing and to suggest some remedies. We begin by considering the efficiency of optimization software.

Given an optimization test problem and an initial point, the efficiency of the routine may be measured by the number of calls to the user-supplied software or by the execution time. The effectiveness of this approach depends critically on the optimization test problem and on the starting point.

The weakness of the standard starting point provided with most optimization test problems is that the starting point is quite close to a solution and therefore does not test the implementation. To improve this situation, Hillstrom [5] proposed using random starting points chosen from a box surrounding the standard starting point. More', Garbow and Hillstrom [8] proposed an alternative scheme with more flexibility. In this scheme the starting points are of the form

$$\text{FACTOR} * X_s$$

where X_s is the standard starting point and the scalar FACTOR is chosen from the set $\{1,10,100\}$. Other choices of starting points are possible, but the above choice seems to test adequately the implementations. For most problems X_s is an easy starting point, FACTOR = 10 generates a problem of intermediate difficulty, while FACTOR = 100 corresponds to a difficult starting point. Our numerical results for EO4GAF (presented later on) will demonstrate this point.

The weakness of most optimization test problems is that they are well-scaled. To improve this situation More', Garbow and Hillstrom [8] suggested testing for scale invariance. The reason for this test is that the performance of a scale dependent implementation usually deteriorates when it is applied to a badly scaled problem. To emphasize this point, they give an example of an implementation which fails completely (on the standard starting points!) when it is applied to the (badly) scaled problems. In contrast, if a scale invariant algorithm is applied to functions $\hat{F}$ and F related by the change of scale

(4.1)
$$\hat{F}(x) = \alpha F(Dx)$$
$$\hat{x}_0 = D^{-1}x_0$$

where α is a positive scalar and D is a diagonal matrix with positive diagonal entries, then it generates iterates which satisfy

$$\hat{x}_k = D^{-1}x_k , \quad k > 0 .$$

Here F may define the residuals of a nonlinear least squares problem, a system of nonlinear equations, or an unconstrained minimization problem. We note that in addition to the starting point, many algorithms require appropriate changes to other initial conditions. For example, variable metric algorithms are scale invariant if a suitable change is made to the initial variable metric matrix.

The choice of D suggested in [8] is

$$D = \text{diag}(\sigma_1,\ldots,\sigma_n)$$
$$\sigma_i = 10 ** \left[\frac{5(2i-n-1)}{(n-1)}\right]$$

(if n = 1 no scaling is performed). This restricts σ_i to $[10^{-5},10^5]$. In our tests we used this choice of D, and α was chosen from the set $\{10^{-5},1,10^5\}$. These tests show that EO4GAF is numerically scale invariant; in other words, the performance of the subroutine is only marginally affected, although the number of overflows and underflows changes considerably. This is unusual since many optimization routines are highly scale dependent and their performance on badly scaled functions is usually disastrous.

The reliability of optimization software has two main components: the reliability of the (exact arithmetic) algorithm, and the reliability of the termination criteria. The reliability of the algorithm is concerned with the question of whether or not the algorithm converges. The reliability of the termination criteria is based on whether it can produce an approximate solution of the desired accuracy.

These two components of reliability should be differentiated. For example, Lyness and Kaganove [6] have shown that there are no (exact arithmetic) automatic quadrature algorithms which are completely reliable, but since automatic quadrature algorithms can be convergent, the unreliability is due to the termination criteria. In optimization the problem is twofold since many of the interesting optimization algorithms cannot be shown to be convergent under reasonable conditions. For example, unconstrained minimization algorithms which use the BFGS update have only been shown to be convergent if the objective function is convex. Note that even if the algorithm is known to be convergent, the question of adequate termination criteria remains.

To give a definition of accuracy in nonlinear least squares problems, first consider zero residual $(F(x^*)=0)$ problems. Note that the zero residual problem with $m > n$ is ill-posed (arbitrarily small perturbations produce a non-zero residual), and that the zero residual problem with $m = n$ should probably be treated with techniques for systems of nonlinear equations. For non-zero residual problems define

$$(4.2) \qquad \phi_F(x) = \frac{\|F(x)\| - \|F(x^*)\|}{\|F(x^*)\|} \quad .$$

In this definition $\|F(x^*)\|$ refers to a minimum of the nonlinear least squares problem as represented in the computer. This implies that the reasonable values for ϕ_F depend on the precision of the computation of F and on the sensitivity of the problem.

If F is being evaluated with high relative precision, then the computer's $\|F(x^*)\|$ is within a few rounding units of the true $\|F(x^*)\|$, and on a computer with relative machine precision e_M, a reasonably accurate solution satisfies

$$(4.3) \qquad \phi_F(x) \leq e_M^{1/2} \quad .$$

Any x for which $\phi_F(x) \leq e_M$ is a solution to the least squares problem to full machine precision.

If F is being calculated with high relative errors, then the appropriate values of ϕ_F are not easily determined. This is the case for zero residual problems; in these problems the computer's $\|F(x^*)\|$ is usually non-zero and the calculation of F near x^* has high relative errors.

Note that ϕ_F is invariant with respect to the change of scale (4.1). Also note that although x may be accurate according to (4.2), the accuracy of x as determined by $\|x-x^*\|$ may be poor. The reason for this is that for x near x^* there is a constant γ such that

$$\|F(x)\|^2 - \|F(x^*)\|^2 \leq \gamma \|x-x^*\|^2$$

and thus

$$\phi_F(x) \leq \gamma \left(\frac{\|x-x^*\|}{\|F(x^*)\|} \right)^2 \quad .$$

This inequality shows that there is no justification in trying to obtain an x with

$$(4.4) \qquad \|x-x^*\| < \|F(x^*)\| \left(\frac{e_M}{\gamma} \right)^{1/2} \quad .$$

To illustrate the above discussion on performance, let us test E04GAF on the nonlinear least squares problem set presented in [8]. This set contains 18 functions; three of these functions define linear least squares problems, so consider the remaining 15 functions. The number of starting points and the dimensions of the functions are read in as data. The data suggested in [8] for our 15 functions defines 22 problems (a problem is a function together with values for m and n); 13 of these problems use the starting points

$$X_s, \quad 10\,X_s, \quad 100\,X_s \ ,$$

and the remaining 9 problems only use X_s.

The results of running EO4GAF on this problem set are presented below. For these runs we set XTOL(i) to $e_M^{1/2}$ and imposed a limit of $100(n+1)$ function evaluations. For the non-zero residual problerms a run is successful if it terminates with an x which satisfies (4.3). For the zero residual problems in the problem set presented in [8] it is fairly easy to assign a reasonable value to the computer's $\|F(x^*)\|$ (usually of order e_M), and for these problems a run is successful if $\phi_F(x) < 1$. It turned out that our value for the computer's $\|F(x^*)\|$ was such that EO4GAF was always successful on zero residual problems.

		X_s	$10\,X_s$	$100\,X_s$
1.	starting point	X_s	$10\,X_s$	$100\,X_s$
2.	number of starting points.	22	13	13
3.	NFEV median.	17	32	47
4.	infinite solutions	0	1	2
5.	overflows.	YES	YES	YES
6.	failures of algorithm.	0	0	2
7.	failures of termination criteria .	0	1	1
8.	excessive number of function evaluations	0	1	3

We have listed the median of the number of function evaluations because it justifies our claim that the standard starting points do not really test a code.

The number of infinite solutions found by EO4GAF indicates that this code does not restrict the length of the initial step. This is also reflected by the occurrence of overflows for all three types of starting points.

We have listed two kinds of failures. The algorithm fails if it does not terminate with a solution of the desired accuracy within the allotted number of function evaluations. The termination criterion fails if either it terminates prematurely, or if it fails to terminate the algorithm (in the allotted number of function evaluations) after obtaining an x of the desired accuracy. In our tests there were no cases of premature termination; the failures of the termination criteria were due to an attempt to obtain an x which satisfies (4.4).

ACKNOWLEDGMENTS

I would like to thank M. J. D. Powell for his careful reading of the initial version of this paper. His comments prompted a large number of revisions and clarifications. I am also grateful to Burt Garbow and Ken Hillstrom for their contributions to the material presented in Section 4.

REFERENCES

1. Boyle, J. M. and Dritz, K. W. (1974). An automated programming system to
    facilitate the development of quality mathematical software, Proceedings
    IFIP Congress, North-Holland.

2. Cody, W. J. (1974). The construction of numerical subroutine libraries,
    SIAM Review, 16, 36-46.

3. Fletcher, R. (1971). A modified Marquardt subroutine for non-linear least
    squares, UKAEA Research Group Report AERE-R6799, H.M.S.O.

4. Fox, P. A., Hall, A. D., and Schryer, N. L. (1976). The PORT mathematical
    subroutine library, Bell Laboratories, Computing Science Technical Report 47.

5. Hillstrom, K. E. (1977). A simulation test approach to the evaluation of
    nonlinear optimization algorithms, ACM Transactions on Mathematical Software,
    3, 305-315.

6. Lyness, J. N. and Kaganove, J. J. (1976). Comments on the nature of auto-
    matic quadrature routines, ACM Transactions on Mathematical Software, 2,
    65-81.

7. More', J. J. (1977). The Levenberg-Marquardt algorithm: Implementation and
    theory, Numerical Analysis, G. A. Watson, ed., Lecture Notes in Mathematics
    630, Springer-Verlag.

8. More', J. J., Garbow, B. S., and Hillstrom, K. E. (1978). Testing uncon-
    strained optimization software, Argonne National Laboratory, Applied Mathe-
    matics Division Technical Memorandum 324.

9. Ryder, B. G. (1974). The PFORT verifier, Software Practice and Experience,
    4, 359-377.

10. Smith, B. T. (1977). Fortran poisoning and antidotes, Portability of
    Numerical Software, W. R. Cowell, ed., Lecture Notes in Computer Science 57,
    Springer-Verlag.

11. Smith, B. T., Boyle, J. M., and Cody, W. J. (1974). The NATS approach to
    quality software, Software for Numerical Mathematics, D. J. Evans, ed.,
    Academic Press.

12. Wisniewski, J. A. (1978). Some experiments with computing the complex abso-
    lute value, SIGNUM Newsletter, 13, 11-12.

Performance Evaluation of Numerical Software, Fosdick (ed.)
© *IFIP, North-Holland Publishing Company, 1979*

DISCUSSION OF SESSION ON PERFORMANCE EVALUATION
IN OPTIMIZATION AND NON-LINEAR EQUATIONS

Chairman and Discussant
F.A. Lootsma
Department of Mathematics
University of Technology
Delft, Netherlands

J. Rice (to W. Murray): How do you plan to guide the user to make a proper choice
of program from a large set of optimization codes.

W. Murray: Perhaps I could answer the question by describing how the selection is
made in the current NPL Optimization Library which consists of 54 primary routines
together with a number of service routines which check derivatives etc. The Library
documentation contains a general guide which, among other things, identifies the
category of problem. In addition there are selection charts which branch upon
answering "yes" or "no" to simple questions regarding the users problem. At the
end of each path through the chart is the name of the recommended primary routine.
On any selection path the user may be advised to use one or more of the service
routines. If the user ignores the general guide and simply selects the routine he
thinks is most suitable for his purpose, he will be advised about possible alternative
choices within the documentation for the particular routine. If the selected
routine fails, then the user is referred to an alternative routine.

A.H. Sherman: Given the disagreement over how to test optimization software and how
to interpret the results, how can we present material which will be of significant
help to users who have to select among routines for real world problems?

W. Murray: Since I would only recommend a user to select high quality software that
has been specifically developed for multimachine implementation, the choice of
routines is not wide. Indeed the decision concerning which algorithm to develop
into high-level software must be made for the user. Moreover the selection is not
just made on the criterion of performance, as discussed at this meeting. I would
hope that users would be able to choose from a high quality optimization library
in which the routines were not in competition with each other but were complementary.
The process of selection would then follow that described in my reply to John Rice.

Remark by W. Murray: The NAG routine criticized by Moré has been flagged for
withdrawal from the NAG Library. (The NAG Library is released in a series of "Marks".
Users are warned at the issue of a Mark of those routines which are to be withdrawn
at the next.) The criticisms raised in this talk do not apply to the new suite of
routines for nonlinear least squares contained in Mark VII of the Library. Moré
has asked the question "Why do some NAG optimization routines check such matters
as parameters being sensible, etc whilst others do not and where is the demarcation?
I shall answer this question briefly. In its early stages the Library was simply
a collection of material that happened to be available. From the moment that the
National Physical Laboratory took responsibility for the optimization material,
routines have been specifically developed for a multimachine based library. At
Mark VII, the replacement of the early routines was almost completed.

J.K. Reid (to J. Moré): To what extent should we aim for robustness (e.g. avoiding
any possibility of underflow or overflow) at the expense of performance on reasonable
problems? My view is that some loss of performance (say 25%) is acceptable.

<u>J. Moré</u>: There are several aspects of robustness, and avoiding overflows is only one of them. It turns out that this aspect does not seriously affect the efficiency of optimization software since the critical calculations do not occur in the inner loops. Other aspects of robustness may damage the efficiency of the algorithm by increasing the overhead per function evaluation but nevertheless improve the overall performance of the algorithm by decreasing the number of function evaluations required to solve the problem. For example, the use of QR decomposition instead of normal equations in a nonlinear least squares algorithm doubles the overhead per Jacobian evaluation but tends to improve the overall performance of the algorithm.

<u>F.A. Lootsma</u> (to R.B. Schnabel): Why do you recommend a PASCAL type code? Is FORTRAN not usable for modular programming?

<u>R.B. Schnabel</u>: As is stated in the paper, PASCAL is used in the algorithmic <u>description</u> of the code. This description is intended to be readily translatable into code, but also easily readable and understandable. For this reason, an algorithmic language like PASCAL (or ALGOL, PL/1) is used for the description. It can then be translated almost trivially into almost any computer language, and indeed, FORTRAN (or a dialect like WATFOR or FLEX) is the most likely choice. Our students have implemented the code in all of these FORTRAN dialects with no problems. Thus, there is no reason why a modular code cannot be implemented in FORTRAN; however, it is best described in an algorithmic language like PASCAL.

Comment by <u>B. Meyer</u>: I do not think that PASCAL or ALGOL provide significantly better tools for modularization than FORTRAN. The problem is that modular programming involves decomposing a system not only on the basis of processing tasks, but also around data structures. Splitting a program into many routines does not make it modular if data is scattered among them, e.g. by COMMON blocks or global variables. To my knowledge, the only generally available language which provides a solution to that problem is SIMULA 67.

Comment by <u>P. Hemker</u> (on Meyer's comment): One has to distinguish between ALGOL 60 and ALGOL 68. I think that both ALGOL 60 and PASCAL are better equipped than FORTRAN for structuring the activities (tasks) of the code. In addition, ALGOL 68 provides excellent tools for structuring the entities to work on. In particular, features such as the <u>heap</u>, and the possibilities to define ones own data types (<u>modes</u>) and <u>operators</u> working on these modes enable the user to write excellent, clear, and well-structured programs.

<u>J. Rice</u> (to R. Schnabel): You mention that your system will allow the use of certain notations for "sum", "max", "dot product". Why is it, with all the effort in improving programming languages, that facilities for these notations are not being proposed or included in new languages? This would certainly be a great help in the clarity and ease of programming.

<u>R. Schnabel</u>: The answer I get from programming language people is that the use of operations like "sum", "max", and "dot product" is too limited to justify making them built-in functions in programming languages; and that instead, programming languages should have the type of capability discussed at the end of my paper which allows the user definition and efficient execution of primitive data operations. My own opinion is that the primitive data capability would definitely be useful, but also that, while specific mathematical functions may not be justified as built-in primitives in general purpose languages like ALGOL or PASCAL, they should perhaps be built into scientifically oriented languages like FORTRAN.

<u>J. Rice</u>: Will your modular optimization system allow someone to specify a particular algorithm with a short naming of the components to be used?

<u>R. Schnabel</u>: The modular system I have written is a PASCALish <u>description</u>, with the interfaces and guidelines carefully specified. It is up to the user to decide how to implement it. My students, for example, usually code some specific algorithm,

that is, one that chooses one specific module in each place where the system contains alternatives. A library implementation, on the other hand, might contain all of the options, with the choice left to the user at run time, and preferably a default for the "unsophisticated" user. As the system now stands, this could easily be done through the use of two integer-valued input parameters, specifying the global strategy and derivative calculation strategy to be used.

F.A. Lootsma (to K. Schittkowski): Did you employ test problems without gradients, or did you only work with test problems and the gradients of the problem functions in it? How do the codes behave with increasing dimensionality and with increasing difficulty of the test problems?

K. Schittkowski: Until now, all test problems are executed together with their gradients, since most optimization codes do not possess any subprograms to evaluate gradients numerically. In general, there is no direct correlation between the dimension of a test problem and execution time or number of function/gradient evaluations. The number of variables is only one point among many others influencing the efficiency of a code. Finally, it is not possible for me to distinguish between a more or less difficult problem since this question depends on the underlying algorithm, too.

J. Bus: Your randomly chosen test problems are rather special, but with a problem generator it is possible to obtain performance profiles such as the relationship between the performance of codes and the condition of the Lagrangian function (the condition number of its Hessian matrix at a Kuhn-Tucker point). Did you obtain such results?

K. Schittkowski: The construction of test problems is based on signomials. Since we do not know if these problems are too much specialized, we intend to test and compare all submitted optimization programs using the classical or 'real life' test problems as well. The test problems obtained by the generator are chosen for battery tests due to the large amount of programs on one side and of problems necessary to get justified conclusions on the other side. It seems for me more efficient to perform separate tests for special questions like the performance profiles mentioned so far or similar investigations.

At the end of the day, J. Bus explained the position and the objectives of the Committee on Algorithms (COAL) established by the Mathematical Programming Society in 1974: COAL is concerned with computational developments in optimization. There are three major goals: (1) ensuring a suitable basis for the comparison of algorithms, (2) acting as a focal point for computer programs that are available for general calculations and for test problems, and (3) encouraging those who distribute programs to meet certain standards of portability, testing, ease of use, and documentation. Some COAL (related) publications are:

H.J. Greenberg (ed.), Design and Implementation of Optimization Software. Sijthoff and Noordhoff, Alphen aan de Rijn, Netherlands, 1978.

H.P. Crowder, R.S. Dembo, and J.M. Mulvey, Guidelines for Reporting Computational Experiments in Mathematical Programming, in H.J. Greenberg (1978), 519 - 535, and in Math. Programming 15, 316 - 329, 1978.

J.C.P. Bus, A Proposal for the Classification and Documentation of Test Problems in the Field of Nonlinear Programming, in H.J. Greenberg (1978), 507 - 518.

The Committee has a newsletter. Its primary objective is to serve as a forum for the Friends of COAL. Through an informal exchange of opinions, members have an opportunity to share their experiences. Further information is to be obtained from:

<u>Chairman (North American Branch)</u>:

John M. Mulvey
School of Engineering/Applied Science
Princeton University
Princeton, New Jersey 08540, U.S.A.

<u>Chairman (European Branch)</u>:

Jacques P. Bus
Stichting Mathematisch Centrum
2e Boerhaavestraat 49
Amsterdam, The Netherlands

<u>Editor of Newsletter</u>:

Karla L. Hoffman
Center for Applied Mathematics
National Bureau of Standards
Washington, D.C. 20234, U.S.A.

<u>J. Rice</u> (to J. Bus): It is a common experience that software refereeing has been very difficult and far from succesful. Do you feel that the guidelines of COAL will be helpful and that they will be used?

<u>J. Bus</u>: We feel that proper guidelines should be supplied and we hope that they will be used.

SESSION 7 : PANEL DISCUSSIONS

Performance Evaluation of Numerical Software, Fosdick (ed.)
© *IFIP, North-Holland Publishing Company, 1979*

THE USE OF MATHEMATICAL SOFTWARE OUTSIDE
THE MATHEMATICAL SOFTWARE COMMUNITY:
A PANEL DISCUSSION

B. Einarsson
Swedish National Defense Research Institute
Stockholm, Sweden

A. M. Erisman, Chairman
Boeing Computer Services Company
Seattle, Washington, U.S.A.

S. Hitotumatu
Research Inst. for Math. Sciences, Kyoto University
Kyoto, Japan

L. F. Shampine
Sandia Laboratories
Albuquerque, New Mexico, U.S.A.

J. H. Wilkinson
National Physical Laboratory
Teddington, England

N. N. Yanenko
Siberian Division of Academy of Sciences
Novosibirsk, U.S.S.R.

Significant technical progress has been made in recent years
in the mathematical software field. The full benefits of
this progress will be realized only when this software becomes a basic part of applications programs and research projects from many fields. The opening discussion by the chairman suggests that mathematical software has attained only a
fraction of its potential use, and some of the possible
reasons for this are given. The panelists were invited to
support or refute this claim, or to suggest what might be
done to increase the impact of mathematical software. Their
statements follow those of the chairman. Unfortunately, the
contribution of Dr. N. N. Yanenko was not available for publication. Finally, some of the comments from the audience
are recorded.

PROBLEM PRESENTATION BY A. M. ERISMAN

The field of mathematical software has come into being in the last decade to fill
the large area between numerical analysis and software for numerical computation.
As a result of work in this field, mathematical subroutine libraries have improved
significantly, applications software has been modestly influenced, and a new
journal (ACM Transactions on Mathematical Software) has been established. Rice
[1] has surveyed much of the recent progress in this field.

273

A significant problem which we will discuss here is that in spite of the high quality and quantity of work which has been done, there is still a large gap between its actual and potential use. We examine here some of the reasons for this gap, and provide some examples which illustrate it. Understanding the problem may help the mathematical software community have an even greater impact on the user community.

Following are some of the reasons why mathematical software (in the form of subroutine libraries or problem oriented packages) has had limited use in application areas:

Awareness on the part of scientists and engineers.
    Many applications people read their own literature and attend conferences in their own fields (just as mathematicians do) and they are simply not aware of the available software. Within their discipline they are trained in developing solution methods for their models, and this becomes an integral part of the modeling process. In fact, methods are often embedded in the model so that the need for mathematical software is obscured.

Understanding of the problem of developing high quality mathematical software.
    Most mathematical modelers have had an introductory course in numerical analysis where they have programmed Gaussian elimination, Runge-Kutta integration, etc. The vast difference between such a program and a reliable, robust piece of mathematical software is often not understood.

Availability of the software.
    Many researchers, including numerical analysts, are aware of libraries and packages but the software is not readily available to them. Examples include: universities which have IMSL on a tape with only one document in the consultation area, and one DOE lab which had EISPACK on a tape in the desk drawer of the individual responsible for it. It often seems easier to write software than to find it.

Usability of the software.
    There is a difficult tradeoff between the black box software which is easy to use but not flexible for the sophisticated user, and the very complex software which the unsophisticated user cannot handle. Compromises often result in software unsatisfactory to both. An even more significant difficulty in the area of usability is the freedom to use mathematical software in application programs. Proprietary clauses often place restrictions on delivering the software to someone else, or at least make the process very complex.

Appropriateness of the selection of subroutines for a library.
    A general purpose library must be structured so that the unsophisticated user (by far the largest in terms of numbers of users) can readily find what is needed. A library which contains a large number of subroutines, some written simply because the problem was there, makes it very awkward for the user to find what is needed. At Boeing, the most often used subroutine is linear interpolation (a routine not even available on some libraries). The more popular areas from a math software viewpoint (e.g., polynomial rooting, generalized eigenvalue problem) have much less use at Boeing.

Other problem areas (of comparable importance) include
    reliability: software was oversold and then discarded when difficulties were encountered. A bad experience often lasts for a long time;
    efficiency: users perceive that the software cannot be efficient because it is general and cannot take advantage of problem characteristics;
    portability: what happens to the subroutine when the application program is moved to another computer?;

<u>trust</u>:   user has always written his own software and does not trust any other. Unless the user can manipulate the source code, this is the first suspect area when something goes wrong;
<u>responsiveness</u>: two years (or even more) are required to put new capability into a library. Proper help is not always available when something goes wrong.

The gap between users and potential users may be illustrated by reviewing scien tific application programs in electric power dynamic analysis, structural analy- sis, nuclear safety, and circuit analysis, for example. Old methods (fixed step Euler's method for numerical integration, Gauss-Seidel iteration for nonlinear equations, etc.) are not uncommon. Neither are poor implementations of good methods (e.g., recalculation and factorization of the Jacobian at every iteration of a Newton solution to nonlinear systems of equations). At the same time, math software efforts have often missed very important problems (e.g., numerical inte- gration with very <u>low</u> accuracy requirements). Significant progress in sparse matrix computation was made in operation research, power systems analysis, and structural analysis long before mathematical software researchers became inter- ested in the problem.

All of this points to one conclusion: if mathematical software is to meet its potential, much stronger communication ties must be built with the applications communities. These ties should influence the teaching of mathematical modeling, with emphasis on separating models and solution algorithms, using mathematical software at least at the early stages of model building. The communication should include the presentation of mathematical software to the user community at appli- cations conferences. And it should change the mathematical software community's views toward the structure and content of libraries and packages.

Some of the other problems (level of sophistication of software, proprietary conditions, etc.) are more than communications difficulties, and also require careful examination.

DISCUSSION BY B. EINARSSON

I very much agree with the views expressed by Dr. Erisman in his opening remarks. For example, the use of software libraries is quite often impeded by limited local availability of the rather expensive library manuals, or that the local computer center implements a new release with a delay of more than six months. It has also occurred that a version for the wrong kind of equipment has been compiled and loaded by the computer center staff. Also, minor technical details, like differ- ent numbering of input and output units for batch and interactive users, cause problems for the unsophisticated user.

We in the numerical software community can improve the impact of mathematical software by continuously requesting the computer centers to improve their ser- vices in this area, for example, by establishing a competent library advisory service, and simultaneously inform those of our colleagues who write application programs of the potential usefulness of established libraries.

To meet the needs of some users, we would like to be able to offer black box routines in many areas. At present we cannot offer black boxes for finite element solvers of the large general purpose type like NASTRAN, but black boxes for solving ordinary differential equations both efficiently and reliably now exist. After a considerable time of development, it will probably be possible to present also black boxes for the most important partial differential equations.

Another reason for the relatively low use of commercial numerical and statistical libraries at scientific research organizations could be that scientific applica tion oriented computer programs developed at non-academic research institutes can very roughly be divided into two groups:

1) <u>Processing of input data</u>. A lot of input data is processed, often with several transformations in space and corrections for temperature, pressure, humidity, etc. The output consists of a list of values, a graph, or just a few mean values and variances.

These are rather elementary operations from the viewpoint of a numerical analyst. A small number of general purpose subroutines for input manipulation and basic statistics can be used.

2) <u>Producing output data</u>. A few input values are used in a mathematical model to give a lot of output information, like solving a partial differential equation.

These are rather complicated operations from the viewpoint of a numerical analyst. The required subroutines are often <u>not</u> included in the general purpose libraries available so far (lack of routines for partial differential equations and integral equations).

These two groups of programs dominate over the more academic problems usually found at university departments.

There are two other problem areas I would like to mention which impact the use of numerical software.

<u>Problem 1</u>. The different computer centers are not always sufficiently interested in <u>numerical</u> software.
Ex. 1 - One computer center in Sweden requested EISPACK for its computer from another computer center but got a version for another computer, causing the test runs to fail completely because of the wrong numerical value on one machine parameter. (Infinite loop.)
Ex. 2 - It has actually taken nine months to put up the new release of IMSL, although it is a leased product really paid for.
Ex. 3 - Using IMSL on an interactive basis requires the error output to be on the common input/output unit, but using it on the same computer system on a batch basis requires the error output to be on the separate output unit. This fact has not been noticed by the computer center, so if you get an IMSL error output, a system interrupt asking for unit 6 is generated. This is bad human engineering.

<u>Problem 2</u>. A few years ago I wrote a package of spline routines. A colleague at another location obtained slightly changed versions of these codes from a third party. He used the codes in the wrong way and was stopped by error exits. He removed the error <u>exits</u> from the code and obtained almost realistic results before he contacted me and got the documentation of the codes.

Finally, I want to report on the use of local libraries at the Swedish National Defense Research Institute (FOA). FOA is divided into several departments, each of which has one (in one case several) local library. The subroutines are indexed in a common Software Index, now containing 285 subroutines in FORTRAN, SIMULA, and ASSEMBLER, for IBM 360/370 and DEC 10.

A survey of the use of the local libraries was performed for 1971/1972 and showed that most of the uses were from the same department that had originally produced the program. Positive results of the survey were that a majority of the users expressed the view that the routine had been of essential service to them, and that no errors were reported. The negative result was the relative low use of our Software Index.

DISCUSSION BY S. HITOTUMATU

Generally speaking, mathematical or numerical software is not yet familiar to naive users in Japan. There are several reasons. Most of them have already been indicated by our chairman, but I would like to mention a few more.

1. Portability problems. Such problems come from differences in operating systems, such as initialization. Sometimes decimal-binary transformations may cause trouble. In UMS (discussed later) we use rational approximation of fixed constants such as the number pi or the weights of the Gauss integration formula to avoid difficulties from different word lengths.

2. Failure in trivial cases. Sometimes a program for zero finding of a nonlinear equation may fail for the quadratic or linear case.

3. Bad education. It is said that users use only what they have learned. The following are typical examples.
 (i) Until recently, A^{-1} routine was highly used in most computing centers. Probably most users used it to solve the linear equations $Ax = b$.
 (ii) As is shown in table 1 below, Simpson's rule is still most commonly used for quadrature, even when it is not suitable for the case.

4. Selection of algorithms: speed versus reliability. One of my colleagues pointed out that for eigenvalues in the asymmetric case, LR is usually 3 times faster than QR, although often LR causes trouble. Which shall we prefer?

5. There is a general tendency for most computer scientists to underestimate the difficulties in numerical computations, or not to fully understand them.

Tables 1 and 2 may provide interesting data. They are the statistics in the Computing Center of the University of Tokyo during the academic year 1977 April - 1978 March. This organization is open for all scientists in Japan as well as the staffs of the University of Tokyo. The numbers indicate how many times each subroutine is called.

Table 1 - Quadrature

	Single	Double	
Trapezoidal rule	33	85	
Simpson's rule	1017	1444	
Gauss	161	250	
Romberg	5	99	
D.E.F. (by M. Mori)	0	2	
Total	1216	1880	(T.3096)

Table 2 - Linear Equations

	Single	Double	
Gauss (no pivoting)	1179	420	
Gauss (with pivoting)	850	1530	
Gauss-Jordan	6	25	
Repeat accu. improve.	142	199	
LU decom. (complex)	622	607	
Choleski	192	196	
Conjugate gradient	42	61	
Gauss-Seidel	6	51	
Least square	212	180	
Others	19	23	
Total	3279	3302	(T.6581)

We can see several interesting patterns. Single precision (4 byte) is insufficient in most computations. Users have a tendency to start with single precision and then move to double precision. Finally, the newly invented D.E.F. (double exponential formula) is still seldom used.

A new example of numerical software in Japan is UMS (Unified Mathematical Software) designed and produced by JUSE (Japan Union of Scientists and Engineers) for
which I took consulting work. The development started in 1976, and use of UMS
started about June 1978. This system is not a complete exhaustive system, but a
compilation of standard procedures. It is a FORTRAN based program package. It is
capable of not only the standard procedures in numerical computations and statistics but also includes some special outputs (tables or histograms) and unlimited
multiple precision arithmetic. Up to now it works well, and I hope this will
serve as a pioneer for numerical software in Japan.

DISCUSSION BY L. F. SHAMPINE*

The Sandia Laboratories have a long history of large scale engineering and scientific computation. They developed their own library of mathematical software and
have been monitoring its use continuously for about a decade. There is a committee charged with overseeing the development and use of the library.

Sandia has two major laboratories, one in Albuquerque, New Mexico (SLA) and one in
Livermore, California (SLL). Although SLA is roughly seven times the size of SLL,
the usage of the library in the two laboratories is comparable. A complete
explanation of the higher rate of utilization at SLL is not known, but it is
believed that more publicity of the library and better availability of codes and
advice are the most important reasons. In part, the situation is due to the zeal
of the staff supporting the library at SLL, but the most significant factor seems
to be that it is easier to establish personal contact with users in a smaller
laboratory. The lesson to be learned is that usage of a library is not merely a
reflection of its quality.

The library committee estimates that only half the scientific computer users are
making regular use of the library. It is believed that usage is inversely
proportional to the size of the job. The disappointing penetration into very
large codes has been the subject of considerable discussion. Some of the expla-
nations proposed are of general interest:

> Because the very large codes have a long life, many that are of current
> importance were written quite some time ago. It is of the greatest signifi-
> cance that they were written in the style of the time and so are not properly
> modularized. This makes it very difficult to replace obsolete computational
> methods by mathematical software. Furthermore, the sheer size of the codes
> and a natural reluctance to change a successful code discourage any effort to
> rewrite completely the codes. It is to be hoped that better programming
> practice will reach future code writers through their university studies and
> in-house courses. There is some evidence that new employees are more aware
> of improvements in software technology and of the associated advantages of
> libraries.

> Scientists are concerned with their physical problems, and are no more inter-
> ested in numerical questions than they have to be. Thus, their knowledge of
> good numerical practice may be somewhat out of date. There is a tendency to
> believe that crude physical models need only crude numerical methods for

*This work was supported in part by the U.S. Department of Energy (DOE) under
Contract No. AT(29-1)-789 and in part by the U.S. Army Research Office.

their solution. Numerical analysts talk a lot about efficiency, but the applied scientist often finds the task solved by mathematical software to be an inexpensive part of his job. For these reasons, the scientist often underestimates the difficulty of getting <u>reliable</u> numerical results. Consequently, he often does not realize how much work he would be saved by relying on a piece of mathematical software nor the enhanced reliability he would get by turning to the product of intensive study by a specialist.

Perhaps the most important observation from experience with the Sandia library is that mathematical software must be easy to use. Writers of mathematical software are often very conscious of the fact that expert judgment might find one technique a little more suitable for a specific task than another. Those assembling libraries wish to provide a comprehensive, powerful library. The result of these feelings can be a long list of alternative codes. The most common user is not an expert in numerical methods, and does not want to become one. He invariably chooses the code easiest to use which seems to apply to his problem. This observation has many important implications for software design and library management.

DISCUSSION BY J. H. WILKINSON

When Dr. Erisman showed me his manuscript I had the distinct impression that he expected me to react adversely. In fact, I found myself in almost complete agreement with it, though had I written it myself it might have contrived to soften the impact. Dr. Erisman has rightly drawn attention to the gap between the actual use of mathematical software and the potential use, but anyone not acquainted with what has been done might well gain the impression that the effort has so far been a failure. The truth is that the provision of high quality mathematical software is a very difficult task and it was not to be expected that all the problems of communication would be solved in a few years.

I am not qualified to give a survey of the overall mathematical software effort in the UK and I shall content myself with illustrating some of the difficulties in a rather more limited setting. When electronic computers came into operation, the first industrial users in the UK were the aircraft companies, and our customers already had extensive experience of solving the problems of aircraft design on desk machines. At that time there were only a few working machines in existence and a comparatively small number of aircraft companies interested in using them. This was because each of our customers presented his requirements in a different form in spite of the fact that they all had much the same problems.

After a year or so , Dr. Bowden (now Lord Bowden) suggested that a panel should be set up with representatives from the companies on the one hand and the computer centers on the other. The object was to systematize the formulation of the problems and to produce a small number of "standard forms". It worked remarkably well and in quite a short time it completely revolutionized the effectiveness of the computing effort. A great park of the work could be classified as linear algebra, and certainly as far as I was concerned, the problems from the aircraft industry provided a major stimulus for my research in that area. Indeed, although I am probably best known for my work in linear algebra, up to that time I had no special interest in the topic.

Two of the central problems in the aircraft industry were the computation of the normal modes of vibration of an airframe and the determination of its flutter characteristics; both of these involve the computation of the eigensystem of a matrix. As far as the normal modes are concerned, the underlying matrix is essentially real and symmetric so that the eigenvalues are real. In the flutter problem, one is concerned with the solution of the problem $(A\lambda^2 + B\lambda + C)x = 0$, where A, B and C are functions of the velocity and loading factor, and the eigenvalues are complex.

When we first started this work, little was known about the numerical stability of matrix algorithms, but we soon found that many of the methods described in the literature were quite useless for matrices of even quite modest size. In particular, we found that methods for computing the normal modes based on the explicit determination of the characteristic polynomial were fundamentally unstable. In these days I was frequently to be found descending on an unfortunate client with the full weight of my advanced ideas, desperately eager to convince him that there was no future for _any_ algorithm based on such an approach. However, in the intimacy of this conference and knowing that my audience combines good nature with discretion, I shall make a confession. Although the matrices for determining the normal modes were often of quite high order, those associated with the flutter problem were seldom of an order higher than 20. Moreover, the eigenvalues of the latter were invariably complex (conjugate pairs). The characteristic polynomials associated with the flutter problem were quite well conditioned; certainly a condition number equal to the reciprocal of the machine precision on our computer was seldom, if ever, attained, whereas the condition numbers of the roots of the polynomial associated with the normal modes problem almost invariably included some which were well in excess of this value. I found that if one reduced the matrix to Hessenberg form working in single precision, computed the characteristic polynomial explicitly working in double precision and then found its roots working in double precision, these roots, judged as roots of the Hessenberg matrix, were invariably correct to more than single precision! Hence, for the flutter problem this method was not only the fastest I had at my disposal, but it was also the most accurate.

Virtually every flutter problem accepted by me at NPL was solved by an algorithm of which I was perhaps the most determined opponent. I used it knowing the dangers and taking adequate precautions against the possibility that I might, some day, encounter an ill-conditioned flutter polynomial. I never allowed this program to go outside NPL and the idea of giving it, say, to Brian Ford at NAG could not possibly be entertained. I think this illustrates a fundamental problem associated with the provision of robust software.

This example can also be viewed in a different light. Suppose one of my customers had himself developed the algorithm I have just described and had used it successfully for a long period. Imagine, then, his probable reaction to my descending on him and in a heavy-handed manner trying to convince him of the error of his ways. Inevitably it would produce in him a lack of confidence in my judgment! I am sure purveyors of high quality software often find themselves in such a situatioin. I have had examples of this recently in control theory when I have been given a small order problem involving highly defective matrices which has been solved satisfactorily using an unstable method; the method has worked because on the small order example most of the arithmetic has happened to be exact. Using a very stable method, which nevertheless involves some rounding errors, much poorer results have been obtained. Here again one has this fundamental problem of establishing confidence.

In my opinion, the aircraft design panel was successful because it brought together the two sides in a common endeavor. Before the formation of the panel, it had been very much "we" and "they". The gap to which Dr. Erisman referred cannot be closed by the mathematical software community alone; it will need the cooperation of all those working in numerical analysis. Over the next decade I expect to see a substantial growth in cooperative endeavors such as N.A./statistics, N.A./operational research, N.A./control theory, etc. In the last year or two this has certainly been happening in statistics and we have evidence of this in the presence of Dr. Nelder at this meeting. Numerical analysts will certainly find that they have a great deal to learn, but if we are willing to make the effort, I am sure we shall be much more successful in communicating the fruits of our own labors. Only in this way can the gap be closed.

AUDIENCE PARTICIPATION

I. Duff:

Following on from the last comment that Jim made, I would like to comment from recent personal experience on the difficulties of organizing combined meetings with scientists from other disciplines. This clearly ties in with Al's AWARENESS category.

This summer the NRCC (National Resource for Computational Chemistry) organized a 3-1/2 day interface meeting between quantum chemists and numerical linear algebraists at the University of Santa Cruz. Largely because of reciprocal intellectual arrogance and intolerance, the two sides nearly came to blows by the evening of the second day. People did not stop talking but they did stop listening. The chemists did not want to be told how to program or that some of their methods were numerically dangerous, and the numerical analysts resented the arrogance of the chemists and were frustrated by the apparent continual redefinitions of the chemists' problems. The difference in language (for example, diagonal dominance does not mean the same to chemists and numerical analysts) did not make communication any easier.

Suddenly, after splitting into four subgroups (the total attendance was about 40), common sense began to prevail. We stopped thrusting pivoting truisms, etc., down their throats and they opened up to the more palatable suggestions being offered. I think that, in the end, both sides gained something. We began to appreciate that here was a class of problems nearly intractable to most current techniques (matrices of order 10^4 with 10^7 non-zeros) and they are presently examining several techniques suggested which they were not familiar with before the meeting. The moral of this tale is that it is just about possible to make interface conferences work but it requires a lot of hard work, patience, and diplomacy, and there should be some time early in the meeting (which should be at least four days long) for people to meet informally and appreciate the humanity and intellectual integrity of the other side(s).

J. R. Rice:

The typical scientist no longer worries about evaluating sines, logarithms, etc., because these functions are included in the standard languages. Other numerical software could be made easily and widely available if new languages (or preprocessors) included standard mathematical procedures (integration, differentiation, matrix algebra, etc.) as part of the language. Do you foresee any significant developments in this direction and do you think this approach would be successful?

J. A. Nelder:

Users unskilled in programming are hampered in using libraries by having to write driving programs in crude general purpose languages, deficient in vector and matrix arithmetic and with rigid-parameter-passing mechanisms. Besides better languages, we need better education in the structures of experimental and derived data, and standards for the external representation of data structures to allow simple linking of programs and packages.

B. Einarsson:

In response to the complaint of Dr. Nelder regarding unsuitability of present programming languages, I mentioned the intention of WG 2.5 to arrange an IFIP TC 2 working conference on "The Relationship Between Numerical Computation and Programming Languages" in Boulder, Colorado, August 1981.

E. L. Battiste:

On the topic of black boxes: these are exactly what is required by the engineering community, where most numerical computation takes place. This community, the members being embedded in other problem sets, is inclined to utilize numerical procedures as black boxes. The education of such users in the mathematics required for such use is a much harder problem, in which numerical software people are seldom involved. Perhaps they should not be concerned on this point, given the low potential for acceptable solution by them. The optimal method for encroaching our ideas into the non-mathematical community is probably that of adequate treatment of the main numerical software problems, that is, the facilitating of the development of reliable, robust, and easy-to-use kernels.

W. Kahan:

How do you identify or characterize people who could or should benefit from scientific software libraries? How do you find them? How do actual beneficiaries discover their libraries? How do they perceive their benefits? What fraction of potential beneficiaries are actual beneficiaries? How do you know? What procedures will increase that fraction?

W. S. Brown:

Wherever there is scientific computing, and John Rice will tell us that scientific computing is a large fraction (perhaps half) of the total, there are potential users of numerical libraries. However, the need for education cuts both ways. If we get personally involved with applications, we will learn that our tools are not always up to the task. Active collaboration with scientists and engineers will not only teach them the usefulness of our software, but will also teach us how to improve it.

C. W. Gear:

One way to reach the engineering community is through the numerical analysis courses that we teach to engineering students.

W. S. Brown:

To follow up on Bill Gear's suggestion, I would like to remark that Norm Schryer of Bell Labs recently taught an introductory numerical analysis course at Stevens (Stevens Institute of Technology, Hoboken, New Jersey), organized around the Port Mathematical Subroutine Library. For each portion of the library (e.g., spline approximation), he taught both the theory and the use of the relevant Port procedures. He then used that theory and those procedures freely throughout the remainder of the course. With this approach he was able to cover more material than in a conventional course, and also to cover it more thoroughly. Any of the major libraries could be used for such a course, and a suitable text would be the recent book by Forsythe, Malcolm, and Moler (Computer Methods for Mathematical Computations, Prentice-Hall, 1977).

P. Fox:

With reference to the difficulty of getting users in an industrial laboratory setting to learn about available numerical software, we at Bell Laboratories have used the following approach:

A series of seminars on numerical methods is set up and announced in the generally distributed research calendar. Each of the seminars deals with a different topics, e.g., solution of differential equations, partial differential equations, approximation, FFT, linear algebra, optimization, etc. One session includes an overall view of the different software libraries and collections on hand.

Each seminar is given by an expert in that field, and the material covered spans theory to actual programs and their particular applicability.

The sessions have been well attended by both research types and programmers and have served to spread the word about our numerical software capabililties.

A.Erisman

At Boeing we also issue a quarterly newsletter describing new routines, discussing a particular numerical area (e.g., linear algebra or spline capability) and suggesting effective ways of using library software. This has proved to be very popular with users, and a very effective communication vehicle.

H. Stetter

The gap between the numerical analyst and the scientist/engineer is also displayed by journals like "International Journal for Numerical Methods in Engineering", "Journal for Computational Physics", and a few others. A different terminology is used, the authors are largely distinct from those in classical numerical journals, and either group does not seem to read the journals of the other.

I. Duff:

I feel that one way of keeping up a dialogue while at the same time improving the standard of papers appearing in numerical/engineering journals is for numerical analysts to submit papers (or comments on published papers) to these journals. The kudos in our community may not be great but this ennobling gesture should not be unrecognized.

On another subject, I am alarmed at Larry Shampine's apparent comment that routines should be dropped from libraries immediately when they are superseded by better software. I feel that library users expect a certain stability in library composition and would be upset if suddenly their programs didn't work because of a reference to a deleted subroutine. Could you describe your criterion for dropping subroutines in a little more detail?

L. Shampine:

On our library we actually remove subroutines very slowly. When a new routine is made available on the library which is clearly superior to an old one, the new routine is publicized, while publicity for the old one stops. The old routine remains on the library for six months, however, and usage of that routine is monitored. Users get a message in their day file telling about the new routine and warning that the old one will be deleted. Those few users that remain when the routine is about to be deleted are called and advised to use the better routine, but are given an option of having a card deck of the old routine. Card decks on all routines are kept indefinitely. The point is that we do not believe in maintaining two routines for the same task when one is obviously inferior to the other.

Reference

1   J. R. Rice, "Software for Numerical Computation", in Research Directions in Software Technology, compiled from workshop held at Providence, R.I., October 1977. To appear under MIT Press, 1979.

Performance Evaluation of Numerical Software, Fosdick (ed.)
© *IFIP, North-Holland Publishing Company, 1979*

PERFORMANCE EVALUATION OF NON-LINEAR PROGRAMMING

CODES FROM THE VIEWPOINT OF A DECISION MAKER

F.A. Lootsma

Department of Mathematics
University of Technology
Delft, Netherlands

ABSTRACT

The present paper shows how Saaty's priority theory provides a framework for the selection of non-linear programming codes in a particular environment. First, priorities are assigned to the relevant performance criteria, and thereafter some well-known comparative studies are used to compare a number of codes under each performance criterion separately. The procedure outlines what performance evaluation can do and what it cannot do. It demonstrates how priority theory can be used for an integrated assessment of codes taking into account both the performance criteria of a decision maker and the results of performance evaluation.

1. INTRODUCTION

Since the pioneering work of Colville (1968), several authors have compared the performance of non-linear programming codes on a variety of test problems and computers. Hence, a substantial amount of material is available for computer managers and scientists who want to decide which code (or codes) to use for their particular purposes. Nevertheless, the picture is gloomier than we might expect at first sight. There is a widespread disagreement on the performance criteria and the test problems to be used, and the reported results are so condensed or incomplete that a decision maker will frequently run into troubles when he tries to draw conclusions from the comparative studies dealing with the performance of non-linear programming codes.

It is our impression that the authors do not always anticipate the questions that might be asked by decision makers. We therefore find that measurements under certain performance criteria (particularly robustness and efficiency) are intermixed in a confusing manner, whereas other criteria (such as simplicity of use and program organization) are completely neglected. Moreover, attempts are made to estimate the relative performance of codes with an accuracy which is out of proportion with respect to the interests of decision makers: in many cases, the order of magnitude of the relative performance is sufficient to support a decision.

It is the purpose of this paper to outline the role of performance evaluation by sketching the decision of an imaginary optimization specialist who will be responsible for non-linear optimization in an industrial research and development organization. It is his intention to concentrate on a particular methodology and on the related software: he knows that there are various strategies to solve non-linear programming problems, but it is out of the question that he should have the time and energy to become master of all of them.

As a vehicle for discussion we use the priority theory of Saaty (1975, 1977) which has recently been developed to weigh the significant factors in a decision problem via pairwise comparison. Each ratio expressing the relative significance of a pair of factors is displayed in a matrix. Finally, the weights (the so-called priorities) of the factors are obtained by an eigenvalue analysis. The optimization specialist in question has to face a two-level decision problem. First, the significant factors (the performance criteria which are relevant in the given situation) must be identified, and weights (priorities) must be assigned to them, either in a qualitative or in a quantitative manner. Second, some codes are to be compared, and their relative performance must be established under each of the performance criteria. At both levels, priority theory can be used: first, to assign suitable priorities to the significant performance criteria, and thereafter to assign suitable priorities to the codes. At the first level, priority theory is applied only once, at the second level for each performance criterion separately. Finally, adding the priorities per code weighed by the priorities of the performance criteria, one obtains a score for each code. The highest score designates the code to be selected.

This paper will briefly outline the codes to be considered in the given situation. Thereafter, we sketch Saaty's priority theory and its application during the process whereby a code is selected for use in the industrial research and development organization. Finally, there are some concluding remarks to highlight the role of performance evaluation.

2. STRATEGIES FOR NON-LINEAR PROGRAMMING

In its general form, a non-linear programming problem can be written as

$$\text{minimize} \quad f(x)$$
$$\text{subject to } g_i(x) \geq 0 \; ; \; i = 1, \ldots, m,$$
$$h_j(x) = 0 \; ; \; j = 1, \ldots, p.$$

The objective function f and the constraint functions $g_1, \ldots, g_m, h_1, \ldots, h_p$ are real-valued functions of the n-vector x. Real-life problems that can be cast in this mathematical form occur in a variety of circumstances, and although it is dangerous to generalize, we feel that the applications of non-linear programming in industry fall into three distinct categories.

a) <u>Technological applications</u>. There is a variety of design problems in research, development, and engineering departments (Bracken and Mc Cormick (1968)). These problems have a highly non-linear nature and can be formulated with a relatively small number of variables (two to twenty), although larger models (with fifty to hundred variables) are not uncommon. A significant proportion (some 70%) of the problems reduces to the minimization of a sum of squares arising from curve fitting.

b) <u>Business applications</u>. These are mainly found in the area of production planning (long-term production allocation, medium-term capacity planning, or short-term production and transportation scheduling). The size of the models tends to be vary large, both in the number of constraints (hundreds) and in the number of variables (thousands). In this area, non-linear optimization is mostly a refinement of linear programming. The non-linearities, due to economies of scale and difficult to determine precisely, can often be handled via local linearization and repeated application of linear programming, or via separable programming (see Beale (1968)).

c) <u>Discretized problems</u>. Several attempts are made to approximate the solution of optimal-control problems by means of non-linear programming techniques (see Tabak and Kuo (1971)). An optimal-control problem, discretized in time, will lead to a non-linear programming problem with tens or hundreds of variables, each associated with a specification point on the time interval under consideration.

In this paper, we shall be mainly concerned with problems of the first category which arise in industrial research and development laboratories. Today, there is a confusing variety of algorithms to solve these problems. A complete bibliography is beyond the scope of the present paper so that we only mention here some well-known strategies and papers: the <u>reduced-gradient method</u>, particularly developed by

Abadie and Carpentier (1969), Abadie (1975), and Lasden e.a. (1975), see Greenberg (1978); the gradient-projection method of Rosen and Kreuser (1972); the interior and exterior penalty-function methods of Fiacco and Mc Cormick (1968) and Zangwill (1967); the moving truncations of Staha (1973); the augmented-Lagrangian methods described by Fletcher (1975) and Powell (1978); and the geometric-programming method described and tested by Ryckaert (1974, 1978). Many variants of these methods have been programmed, and Table 1 briefly summarizes the number of codes and test problems that have been employed by various authors to test the relative performance of the codes (and of the underlying algorithms).

Table 1

Number of codes and number of test problems used in comparative studies of software for non-linear programming.

Author(s)	Year	Codes	Problems
Colville	1968	34	8
Staha and Himmelblau	1972	4	24
Dembo	1976	11	8
Ryckaert and Martens	1977	16	24
Sandgren	1977	35	30
Schittkowski	1978	17	80

3. PERFORMANCE CRITERIA

The following list of performance criteria for non-linear programming codes has been drawn up in series of discussions with industrial colleagues and with members of the Committee on Algorithms (COAL) of the Mathematical Programming Society (see Lootsma (1976)). The list is not exhaustive, but it seems to contain the criteria which are predominant.

a) Domain of applications. It is important to realize for which type of problems the code has been designed, and under which conditions the underlying algorithm converges to the desired solution. It is also worthwhile to know whether the code is applicable to special problems such as sums of squares.

b) Robustness (reliability and accuracy). A considerable amount of numerical experience on a variety of real-life or randomly generated test problems is necessary to establish whether a code calculates the desired solution with the required accuracy. Moreover, the underlying algorithm should have a sound mathematical basis: established convergence for well-behaved problems, theoretical estimates of the rate of convergence, etc.

c) Efficiency. The relative efficiency of a code with respect to other codes should be measured under the same termination criteria on a variety of real-life or randomly generated test problems. The efficiency, usually measured in terms of (equivalent) function evaluations or CPU time, is sometimes critically dependent on the tolerances of an iterative sub-process (the linear search, for instance). This is a significant obstacle for the comparison of non-linear programming codes. Another complication is, that many algorithms involve the first-order and sometimes the second-order derivatives of the problem functions. The efficiency of a code depends materially on the manner in which the provision of the functions and their derivatives has been organized.

d) Capacity. The maximum size of the problems that can generally be solved by the code, as well as the storage requirements, are important criteria to determine the range of possible applications for a code.

e) Simplicity of use. It is difficult to formulate an objective criterion to decide whether a code is a simple, effective tool in the hands of unsophisticated users. Nevertheless, simplicity of use may be decisive for the successful utilization of a code. Much attention should therefore be given to its user-oriented features such as extensive output provisions (to warn against possible errors and to follow the course of the computations), numerical differentiation (to avoid supplying analytical derivatives or to check the user-supplied derivatives), and documentation of high quality. Finally, it may be important to ask whether the underlying algorithm is conceptually simple, thus enabling both the user and the specialist to discuss the calculated results.

f. Program Organization. Error checking, maintenance, and transfer to other computers are greatly simplifed if the code has a clear (possibly modular) structure. It is desirable, of course, that the code should be written in a higher-level language. Finally, the length of the code (the number of statements, the number of lines of coding) is an important feature of a code, indicating that a considerable amount of work had to be done to obtain a workable tool in a variety of circumstances.

Obviously, it is out of the question that there should be a unique criterion to review non-linear programming codes.

4. PRIORITY THEORY

We consider n factors (performance criteria, decision criteria) which are significant in a decision problem; their respective significances may be unequal in the situation at hand. A well-known procedure to put their significance on a numerical scale is ranking and rating. The factors are ranked in ascending order of significance, and a further refinement is to assign to each factor a numerical value (the weight or priority) between a given, positive lower bound (for the most insignificant factor) and a given upper bound (for the most important factor). Frequently, however, it is not easy to do this for a multiple of factors; a decision maker does not readily accept the reduction to a one-dimensional scale. He sometimes has the awkward feeling that there may be several inconsistencies in his judgement. A single score for each factor on a one-dimensional scale, although eventually useful, is initially felt to be a gross over-simplification.

Priority theory starts with the idea that it is easier to consider each pair of factors separately and to decide whether they are equally significant, or whether one of them is somewhat more significant than the other in the given situation. Quantification of the relative significance for each pair of factors produces a matrix from which suitable priorities can be extracted via an eigenvalue analysis.

We start from the assumption that the significant factors $F_1, \ldots, F_n$ do have positive priorities $w_1, \ldots, w_n$ which are acceptable to the decision maker; for simplicity, we take these priorities to be normalized so that they sum up to unity. Now, the elements a_{ij} of the matrix A of priority ratios can be written as $a_{ij} = w_i/w_j$. The matrix A is clearly a matrix of rank 1 since the rows $2, \ldots, n$ are multiples of row 1. Thus, the matrix A has only one non-zero eigenvalue. Taking w to denote the vector with components $w_1, \ldots, w_n$, we can easily obtain $Aw = nw$. Hence, the non-zero eigenvalue of A is equal to the dimension n, and w is the corresponding eigenvector. The sum of the matrix elements in row i is given by

$$w_i \sum_{i=1}^{n} \frac{1}{w_j} \quad ,$$

so that the row sums provide a multiple of the vector w. The sum of the matrix elements in column j can be written as

$$\frac{1}{w_j} \sum_{i=1}^{n} w_i \quad .$$

Hence, the inverse column sums also provide a multiple of the vector w. These observations will be used in the subsequent analysis of perturbations of A.

Let us now return to the decision maker who estimated the relative significance of each pair of factors. Let r_{ij} denote the numerical value which he assigned to the relative significance of the factors F_i and F_j. If they are felt to be equally significant, then $r_{ij} = 1$. If F_i is more important than F_j, then $r_{ij} > 1$; the values assigned to r_{ij} if F_i is slightly more, rather more, or much more important than F_j will be discussed in the next section. Of course, we have $r_{ij} < 1$ if F_i is less important than F_j, and it must be true that

$$r_{ij} \cdot r_{ji} = 1 . \qquad (1)$$

Finally, we set $r_{ii} = 1$. Obviously, the matrix R is a reciprocal matrix (this is due to (1)) with positive elements. Now, the theorem of Perron and Fröbenius (established in 1908 and 1909; see also Bellman (1960)) guarantees that the absolute largest eigenvalue λ_{max} of R is real and positive, and that there is an eigenvector z with positive components corresponding to it. This is the basis of the priority theory. We normalize the eigenvector z so that its components sum up to unity. Now, the leading idea is to conceive R as a perturbation of A, and to use the normalized Perron and Fröbenius' eigenvector z as an approximation of the normalized weight vector w with components (priorities) $w_1, \ldots, w_n$.

In actual decisions, the pressure of time may be so heavy that facilities for a rapid, handy approximation of the priorities $w_1, \ldots, w_n$ is indispensable (an approximation which is easy to calculate). Taking R again as a perturbation A, we can use the row sums and the column sums of R. Thus, we approximate w_i by the normalized row sums

$$\sum_j r_{ij} \Big/ \sum_i \sum_j r_{ij} ,$$

or by the inverse column sums

$$\left(\sum_{k=1}^{n} r_{ki} \right)^{-1} ,$$

also normalized so that they sum up to unity.

Let us take a simple example to illustrate the theory. Suppose we have three factors F_1, F_2, and F_3. The factors F_1 and F_2 are considered to be equally significant, and both more important than F_3. Thus, we may take $r_{12} = 1$, $r_{13} = r_{23} = 2$, so that the matrix R takes the form

$$\begin{pmatrix} 1 & 1 & 2 \\ 1 & 1 & 2 \\ \frac{1}{2} & \frac{1}{2} & 1 \end{pmatrix} .$$

The reader may verify that both the row sums and the column sums yield the desired eigenvector. Thus, the priorities 0.40, 0.40, and 0.20 are assigned to F_1, F_2, and F_3 respectively whereas the absolute largest eigenvalue λ_{max} of R equals the non-zero eigenvalue 3 of the matrix A.

Priority theory is still under development. It is clear that there are several interesting mathematical problems to be investigated, such as the identification of inconsistencies and the interpretation of the difference between the absolute largest eigenvalue λ_{max} of R and the non-zero eigenvalue n of A. It is possible to show (see Saaty (1977)) that $\lambda_{max} \geq n$. Moreover, $\lambda_{max} = n$ if, and only if, the positive reciprocal matrix R is consistent, in the sense that the matrix elements r_{ij} can be written as $r_{ij} = z_i/z_j$. An appropriate measure for the degree of inconsistency appears to be the expression

$$\frac{\lambda_{max} - n}{n-1} .$$

Further details are beyond the scope of the present paper. We refer the reader
to Saaty (1977).

5. PRIORITIES OF PERFORMANCE CRITERIA

As we announced in the introduction of sec 1, we shall consider the decision
problem of an optimization specialist in a research and development organization.
He will be responsible for non-linear programming in the sense that he will be
asked to solve non-linear programming problems or to help others in solving these
problems. This might be carried out via standard software; alternatively, he may
adopt and modify standard software so that it can be used for special problems;
and it may be necessary to use pieces of standard software for incorporation in a
particular user's program. The type of users will also vary; some of them may be
scientists with a thorough training in numerical mathematics and computer science,
and they may regularly have non-linear programming problems; others will
occasionally run up against these problems, whereas their mathematical skills are
moderate. Hence, although there are many strategies for solving non-linear
programming problems, we assume that the specialist will concentrate on a particular
strategy and on the related software. In order to choose a strategy he has been
advised to use the comparative studies of Colville(1968), Staha and Himmelblau
(1973), Sandgren (1977), and the preliminary results of Schittkowski (1978). The
studies by Dembo (1976), Ryckaert and Martens (1977) pay much attention to geometric
programming, but in the situation at hand a more general strategy seems to be
preferable. In these four comparative studies, reduced gradients and penalty
functions were heavily tested. Moving exterior truncations were present in some
of these studies and exhibited remarkable properties. The gradient-projection
methods and the augmented Lagrangians, however, received less attention; for the
time being, they will therefore be left out of consideration.

Basically, reduced gradients, penalty functions, and moving truncations have the
same domain of applications: the objective functions and the constraints are
supposed to have continuous second derivatives; convexity properties are desirable
for global convergence, although local convergence has been observed for many
non-convex problems. It is therefore acceptable to ignore the domain of applications
as a performance criterion in the given situation. We shall accordingly be concerned
with the following factors (the performance criteria of sec. 3):

F_1: robustness;
F_2: efficiency;
F_3: capacity;
F_4: simplicity of use;
F_5: program organization.

For each pair of factors, the relative significance may be set to one of the
values recommended by Saaty. Thus,

r_{ij} = 1 if F_i and F_j are felt to be equally significant;
r_{ij} = 3 if F_i is felt to be somewhat more important than F_j;
r_{ij} = 5 if F_i is felt to be much more important than F_j.

The intermediate values of 2 and 4 are to be assigned in cases of doubt between
two adjacent values. The relative significance r_{ij} may be set to a value higher
than 5 accordingly as F_i is felt to be predominant with respect to F_j.

The reader will acknowledge that it is not easy to consider the above,
imcomparable factors and to assign a numerical value to the relative significance.
Nevertheless, we will see that it greatly enhances a structured analysis of the
decision. We assume that the decision maker does not hesitate and that he assigns
the values displayed in Table 2. The calculations to approximate the priorities
of the performance criteria are also shown in Table 2. Let us finally assume that
after some deliberations the priorities are adjusted as follows:

Table 2.

Matrix R of Relative Significance, and Approximations to Priorities of Performance Criteria by Normalized Row Sums, Normalized Inverted Column Sums, and Eigenvector corresponding to $\lambda_{max} = 5.18$.

FAKTOR	DESCRIPTION	F_1	F_2	F_3	F_4	F_5	ROW SUMS	NORMALIZED	EIGENVECTOR
F_1	ROBUSTNESS	1	3	3	2	4	13.00	0.40	0.42
F_2	EFFICIENCY	1/3	1	2	2	2	7.33	0.22	0.21
F_3	CAPACITY	1/3	$\frac{1}{2}$	1	$\frac{1}{2}$	$\frac{1}{2}$	2.83	0.09	0.09
F_4	SIMPLICITY OF USE	$\frac{1}{2}$	$\frac{1}{2}$	2	1	1	5.00	0.15	0.15
F_5	PROGRAM ORGANIZATION	$\frac{1}{4}$	$\frac{1}{2}$	2	1	1	4.75	0.14	0.13
	COLUMN SUMS	2.42	5.50	10.00	6.50	8.50			
	INVERTED NORMALIZED	0.43	0.19	0.10	0.16	0.12			

F_1: robustness   0.40
F_2: efficiency   0.20
F_3: capacity   0.10
F_4: simplicity of use   0.15
F_5: program organization   0.15

It is possible to understand that robustness should have the highest priority: unsolved problems do more harm to the reputation of a specialist than inefficient computations (and the subsequent attempts to overcome the failures may be more expensive than inefficient computer utilization). Simplicity of use and program organization, however, cannot be neglected in the area of non-linear programming: experience has shown that there is no black box to solve any problem in this field, and specialists will frequently find that a detailed knowledge of the software is necessary to help the users. Finally, there are many small problems in research and development organizations (with 2 to 20 variables) so that capacity may have a lower priority.

6. PRIORITIES OF CODES

At the second level of the decision process, where priorities must be assigned to codes under various performance criteria, it is sometimes easier to quantify the relative significance: in the comparative studies, we find observations which are based on measurable phenomena (the number of unsolved problems, the elapsed CPU time, etc.). The performance of the codes has been established on a limited number of test problems, however, and this is a major stumbling block: There are real-life and randomly generated test problems, but we do not know whether they constitute a reasonable sample from the problems that occur in real life. Moreover, it is questionable whether the distribution of problems in the given research and development organization coincides with the general distribution of non-linear programming problems in practice. Information about the performance of codes can only be found in these comparative studies, however, and although the results must be considered with some caution we can only draw conclusions from these tests.

Obviously, the comparative studies are concerned with non-linear programming codes, but the underlying algorithm appears to play a significant role. Many reduced-gradient codes and penalty-function codes have been compared, but the four studies mentioned in sec. 5 reveal that there is a considerable amount of consistency in the results. We shall therefore assume that the specialist does not concentrate on a particular code in the categories of reduced gradients and penalty functions. He has in mind a generalized reduced-gradient code (to be designated by GRG) as developed by Abadie (1975) or Lasdon e.a. (1975), and a penalty-function code (to be designated by PENF) like the SUMT code of Fiacco and Mc Cormick (1968) or a FORTRAN version derived from the author's ALGOL 60 procedure (1972) which is published in Greenberg (1978). The COMET code of Staha (1973) for constrained optimization via exterior truncations is the only code in its category. Let us now consider the relative performance of the codes GRG, COMET, and PENF under the performance criteria of robustness, efficiency, capacity, simplicity of use, and program organization as described in sec. 3.

Robustness. The studies of Staha (1973) and Sandgren (1977) show that GRG and COMET are equally robust: they solve practically the same number of test problems under the given termination criteria. A PENF code is definitely lagging behind; the number of unsolved problems is roughly three times higher. It should be noted, however, that the required accuracy is sometimes unreasonably high. In practice, a relative and absolute accuracy of 10^{-2} or 10^{-3} are sufficient. The comparative tests were carried out with accuracies ranging between 10^{-4} and 10^{-8}. It would have been instructive for a decision maker to find comparative results with low and high accuracy. For the time being, the specialist decides to express his opinion in Table 3a, which results into priorities 0.43, 0.43, and 0.14 assigned to GRG, COMET, and PENF respectively.

Table 3a

Robustness (factor F_1)

	GRG	COMET	PENF	
GRG	1	1	3	0.43
COMET	1	1	3	0.43
PENF	1/3	1/3	1	0.14
	0.43	0.43	0.14	1.00

Table 3b

Efficiency (factor F_2)

	GRG	COMET	PENF	
GRG	1	2	3	0.53
COMET	$\frac{1}{2}$	1	2	0.31
PENF	1/3	$\frac{1}{2}$	1	0.16
	0.55	0.29	0.17	1.00

Table 3c

Simplicity of Use (factor F_4)

	GRG	COMET	PENF	
GRG	1	2	2	0.50
COMET	$\frac{1}{2}$	1	1	0.25
PENF	$\frac{1}{2}$	1	1	0.25
	0.50	0.25	0.25	1.00

Table 3d

Program Organization (factor F_5)

	GRG	COMET	PENF	
GRG	1	1/7	1/6	0.07
COMET	7	1	1	0.49
PENF	6	1	1	0.44
	0.07	0.47	0.46	1.00

Efficiency. In the comparative studies, the GRG codes appear to be roughly three times faster than PENF codes. In Staha (1973) the COMET code appears to be competitive with GRG if analytical gradients are supplied, and equivalent to PENF if the gradients are calculated numerically. The last named phenomenon is confirmed by Sandgren (1977) who only operates with difference approximations to derivatives. Schittkowski (1978) considers analytical gradients only, but COMET is not included in his study. Generally speaking, the comparative studies do not properly distinguish between the behaviour of codes if analytical derivatives are present, and if they are not supplied by the user. This is a glaring omission for the decision maker at hand since he will frequently be confronted with attempts to formulate and to test mathematical models. In such a situation the users should not be concerned too much with the provision of derivatives. Properly speaking, there are good reasons to introduce two performance criteria: one for efficiency in the presence of analytical derivatives, and one for efficiency if derivatives are not supplied by the user. The comparative studies, however, do not supply sufficient information for such a refinement. The optimization specialist therefore decides to set the relative significance of the codes as in Table 3b, and finally he sets the priorities to 0.54, 0.30, and 0.16 respectively.

Capacity. The comparative studies do not properly evaluate the codes for increasing dimensionality of the test problems. There is no information about the storage requirements either, so that the specialist decides to assign equal priorities (0.33) to each of the codes.

Simplicity of Use. The comparative studies do not pay much attention to this criterion. Staha (1973) reports some difficulties with GRG codes, but Sandgren (1977) points at the phenomenon that the GRG codes are not very sensitive to variations in parameters which control the computational process. Hence, the specialist decides to express his opinion in Table 3c whereby he obtains the priorities 0.50, 0.25, and 0.25 respectively.

Program Organization. This criterion is also neglected in the comparative studies, but the additional information that the GRG codes take more than 4000 lines of coding, the COMET code some 600 lines, and PENF codes some more than 600 lines, leads to the assignments in Table 3d. After some adjustments the priorities are set to 0.07, 0.48, and 0.45 respectively.

The final results are displayed in Table 4 where we find the priorities of the codes for each factor, separately (columns 1-3) and multiplied by the priority (column 4) of the performance criterion concerned (columns 5-7), as well as the final score per code. We do not expect that the specialist will immediately obey the numerical results and choose a GRG code. Table 4, however, can be used for a structured analysis of the situation and for a structured preparation of the decision. It clearly demonstrates the impact of the performance criteria as well as the strengths and weaknesses of the codes, thus guiding the deliberations towards the decisive argements. This is typically the objective of priority theory.

7. CONCLUDING REMARKS

The above example clearly shows what performance evaluation can do, and what it cannot do. The comparative studies have been used to assign priorities to the codes under various performance criteria (sec. 6), but it is up to a decision maker to identify and to weigh the relevant performance criteria (sec. 5). Priority theory provides the framework for an integrated assessment of codes.

There are several deficiencies in the comparative studies, since the authors do not properly anticipate the performance criteria that might be used. We missed, for instance, an evaluation of how the codes behave under increasing dimensionality of the test problems. Similarly, there is no proper distinction between their behaviour in the absence or presence of analytical derivatives. Simplicity of use

Table 4.

Final Score of Codes GRG, COMET and PENF. The columns 1-3 contain the priorities of the codes under each performance criterion separately, the columns 5-7 the same priorities multiplied by the priority (column 4) of the criterion concerned.

	GRG	COMET	PENF	priority	GRG	COMET	PENF
Robustness	0.43	0.43	0.14	0.40	0.17	0.17	0.06
Efficiency	0.54	0.30	0.16	0.20	0.11	0.06	0.03
Capacity	0.33	0.33	0.33	0.10	0.03	0.03	0.03
Simplicity of Use	0.50	0.25	0.25	0.15	0.08	0.04	0.04
Program Org.	0.07	0.48	0.45	0.15	0.01	0.07	0.07
Final Score					0.40	0.37	0.23

and program organization are mostly ignored in the studies.

As in many decision probems, there is a general vagueness in the significance of
the decision criteria, and the calculated priorities should accordingly be used
as a guideline for the deliberations. Obviously, one is only interested in orders
of magnitude, not in accurate values of the priorities. For performance evaluation
of codes, this implies that rough estimates of the relative performance must be
established under various performance criteria.

There is no ideal code for non-linear programming problems. This was already known,
but Table 4 shows how easily the preference for a GRG code may disappear if the
priorities of the performance criteria are modified. Hence, it is not the task of
performance evaluation to designate the best code in a variety of circumstances.
It only has the humble task to provide the material for the selection of codes in
various cases.

REFERENCES

| 1| J. Abadie and J. Carpentier (1969), Generalization of the Wolfe reduced gradient
     method to the case of non-linear constraints. In R. Fletcher (1969) ed.,
     Optimization. Academic Press, London, 37-47.
| 2| J. Abadie (1975), Méthode du gradient réduit généralisé: le code GRGA.
     Electricité de France, Service I.M.A., Note HI/1756/00, Paris.
| 3| E.M.L. Beale (1968), Mathematical Programming in Practice. Pitman and Sons,
     London.
| 4| R. Bellman (1960), Introduction to Matrix Analysis. Mc Graw Hill, New York.
| 5| J. Bracken and G.P. McCormick (1968), Selected Applications of Non-linear
     Programming. Wiley, New York.
| 6| A.R. Colville (1968), A comparative study of non-linear programming codes.
     IBM New York Scientific Center, Technical Report 320-2949.
| 7| R.S. Dembo (1976), The Current State-of-the-Art of Algorithms and Computer
     Software for Geometric Programming. Working Paper 88, School of Organization
     and Management, Yale University, New Haven, Conn. 06520, USA.
| 8| A.V. Fiacco and G.P. McCormick (1968), Non-linear Programming: Sequential
     Unconstrained Minimization Techniques. Wiley, New York.
| 9| R. Fletcher (1975), An ideal penalty function for constrained optimization.
     J.I.M.A. 15, 319-342.
|10| H.J. Greenberg (1978), Design and Implementation of Optimization Software.
     Sijthoff and Noordhoff, Alphen aan de Rijn, Netherlands.
|11| L.S. Lasdon, A.D. Waren, M.W. Ratner, A. Jain (1975), GRG System Documentation
     (Technical Memorandum CIS-75-01) and GRG User's Guide (Technical Memorandum
     CIS-75-02). Cleveland State University, Cleveland, Ohio, USA.
|12| F.A. Lootsma (1972), The ALGOL 60 procedure minifun for solving non-linear
     optimization problems. Report 4761, Philips Research Laboratories, Eindhoven,
     Netherlands. Published in H.J. Greenberg (1978), pp. 397-445.
|13| F.A. Lootsma (1974), Convergence Rates of Quadratic Exterior Penalty-Function
     Methods for Solving Constrained-Minimization Problems. Philips Res. Repts. 29,
     1-12.
|14| F.A. Lootsma (1976), Non-linear Optimization in Industry and the Development
     of Optimization Programmes, in L.C.W. Dixon (ed.), Optimization in Action,
     Acadamic Press, London, pp. 252-266.
|15| M.J.D. Powell (1978), Algorithms for Nonlinear Constraints that use Lagrangian
     Functions. Math. Programming 14, 224-248.
|16| M.J. Rijckaert (1975), A comparison of Generalized Geometric Programming
     Algorithms. Dept. of Chemical Engineering, Catholic University of Louvain,

Belgium.

|17| M.J. Rijckaert and X.M. Martens (1977), A Comparison of Generalized Geometric Programming Algorithms. Dept. of Chemical Engineering, Catholic University of Louvain, Belgium.

|18| J.B. Rosen and J.L. Kreuser (1972), A Gradient Projection Algorithm for Non-linear Optimization. In F.A. Lootsma, ed. (1972), Numerical Methods for Non-linear Optimization. Academic Press, London, 296-300.

|19| Th. L. Saaty (1975), Hierarchies and Priorities-Eigenvalue Analysis. Internal Report, University of Pennsyvania, Wharton School, Philadelphia, Penn. 19174, USA.

|20| Th. L. Saaty (1977), A Scaling Method for Priorities in Hierarchical Structures. J. Math. Psych. 15, 234-281.

|21| E. Sandgren (1977), The Utility of Nonlinear Programming Algorithms. Thesis, Purdue University, Dept. of Mechanical Engineering, West Lafayette, Indiana 47907, USA.

|22| K. Schittkowski (1978), Randomly Generated NLP Test Problems with Predetermined Solutions. Preprint 35, Math. Institut der Julius-Maximilians-Universität, Würzburg, Germany.

|23| R.L. Staha (1973), Constrained Optimization via Moving Exterior Truncations. Thesis, The University of Texas at Austin, Texas 78712, USA..

|24| R.L. Staha and D.M. Himmelblau (1973), Evaluation of Constrained Non-linear Programming Techniques. Report, The University of Texas at Austin, Texas 78712, USA.

|25| D. Tabak and B.C. Kuo (1971), Optimal Control by Mathematical Programming. Prentice-Hall, Englewood Cliffs, New Jersey.

|26| W.I. Zangwill (1967), Nonlinear Programming via Penalty Functions. Man. Sci. 13, 344-358.

SESSION 8 : GENERAL ASPECTS OF
PERFORMANCE EVALUATION

Performance Evaluation of Numerical Software, Fosdick (ed.)
© IFIP, North-Holland Publishing Company, 1979

METHODOLOGY FOR THE ALGORITHM SELECTION PROBLEM

John R. Rice*

Mathematical Sciences
Purdue University
West Lafayette, Indiana, USA

1. INTRODUCTION

The problem of selecting an effective or good or best algorithm arises frequently and the context of the situation often obscures the basic framework for
this selection problem. A detailed abstract framework for this problem is given by
Rice (1976). It is essential in the performance evaluation of numerical software
that such a framework be recognized and that the evaluation be made within the
framework. The purpose of this paper is to discuss the methodolgy to be used in
an experimental evaluation of software.

The abstract framework is briefly summarized. Then experimental procedures
are discussed followed by a proposed approach to publishing and summarizing the
usually voluminous results of experiments in a way that is both concise and scien-
tifically complete. A key difficulty in the experimental evaluation of software
performance is the definition of the population of input to the programs being
evaluated. The approaches to specifying these populations depends heavily on the
particular software being evaluated; however we consider two cases in some detail
(quadrature and partial differential equations) which are typical for numerical
software evaluation. The explicit use of features is recommended as a way to obtain satisfactory problem populations.

2. THE ALGORITHM SELECTION PROBLEM

The basic abstract model for the algorithm selection problem is given in Fig.
1. The items in this model are:

$\mathcal{P}$ problem space or collection of problems x to be solved
$\mathcal{A}$ algorithm space or collection of programs A to be used
$\mathcal{R}^n$ space of n-vectors p which measure performance
$\mathcal{S}$ mapping from $\mathcal{P}$ to $\mathcal{A}$ which selects an algorithm
p mapping from $\mathcal{P} \times \mathcal{A}$ to $\mathcal{R}^n$ which determines performance
$||\ ||$ norm which produces one final number to evaluate performance

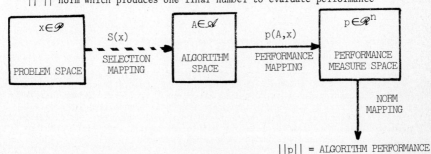

Figure 1. Diagram of the basic abstract model for the algorithm selection
problem. The objective is to have S(x) give high performance.

* This work was supported in part by NSF Grants MCS 76-10225 and MCS 77-01408.

Associated with this model are several questions of best or good selection mappings S.   The most obvious is that of:

Best Selection.   Choose the mapping B(x) which gives maximum performance for each problem, that is

$$||p(B(x),x)|| \geq ||p(A,x)|| \quad \text{for all } A \in \mathscr{A}$$

It is normally unrealistic to expect to obtain B(x) and performance evaluation is usually aimed at obtaining the best algorithm for a subclass of mappings and problems to be solved.   This is explicitly stated as:

Choose the selection mapping $S^*(x)$ from a subclass $\mathscr{S_0}$ which minimizes the performance degradation over the problem subset $\mathscr{P_0}$

See the paper of Rice for other selection problems and other criteria for selection.

Experience shows that people almost always use problem features in their visualization of problem spaces and this should be explicitly recognized in the framework as in Fig. 2.

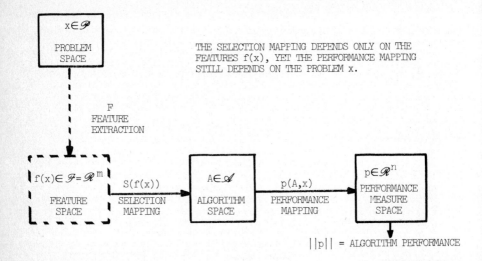

THE SELECTION MAPPING DEPENDS ONLY ON THE FEATURES f(x), YET THE PERFORMANCE MAPPING STILL DEPENDS ON THE PROBLEM x.

Figure 2.   Diagram of the abstract model with selection based on features.

Problem spaces in numerical computation are often very large, amorphous and hard to define.   Simple mathematical definitions are unsatisfactory because they do not exclude impossible or pathological functions.   It is well known that there is no program which can accurately integrate all entire functions (or even all polynomials if numerical integration is used) on a real computer.   Restrictions on the problem space which remove these impossible and nearly impossible problems inevitably lead to problem spaces without a clean definition.

Features serve two purposes, first they allow one to impose some order and coherence in a complicated problem space.   They impose a low dimensional coordinate system in a very high dimensional space and the coordinates have intuitive meaning. Second, they allow one to specify that certain kinds of problems must be in the population, namely some must have certain features.   This is a constructive approach to defining a problem population which is not feasible by merely putting restrictions on a very large population.

Features are primarily used here in the artificial intelligence sense; good features are problem attributes which significantly affect the performance of the algorithms. It is also interesting to view features in the mathematical sense of imposing a lower dimensional coordinate system. Then the question of the best choice of features becomes analogous to the concept of N-widths in functional analysis.

We list some characteristics of this problem in the case of numerical software performance evaluation:

1. The underlying mathematical problem is often unsolvable
2. The problem space is of very high dimension
3. The overall nature of the problem space is not well understood
4. The algorithm space is a small, discrete set
5. The performance criteria are somewhat subjective, they vary with context and any choice of norm mapping is subject to dispute.
6. Large scale performance evaluations are very expensive.

3. THE EXPERIMENTAL APPROACH

Computational experiments on performance seem to be the only feasible approach for the evaluation of much numerical software. It is obvious that the basic principles of scientific experimentation must be followed, yet this has rarely been done. These principles include:

explicitly stated assumptions and hypotheses
objective criteria and measurements
reproducibility of results
access to the actual experimental data
complete descriptions of the experimental apparatus

Jackson and Mulvey (1977) review over 50 comparisons of algorithms and software for mathematical programming and conclude that <u>none</u> of them meet the standards commonly expected in the experimental sciences, most of them do not even come close. A similar experience is indicated by Rice (1976) in a review of 10 substantial comparisons of quadrature algorithms.

A detailed review of the requirements for reporting on experiments with mathematical software is given by Crowder, Dembo and Mulvey (1979). This review implicitly provides many guidelines for conducting experiments and we do not repeat them here. Instead, a few additional observations are made on certain aspects of the experimental procedure.

Experimental techniques generally rely on a random selection from a population. It is essential that the population be well defined, but this does not help much in many numerical problems because the population contains functions and there is no satisfactory way to make random selections from most interesting function classes. Small, discrete populations are well defined (and often used) but it is usually unclear what populations they represent adequately. Thus problem populations, even discrete ones, must be large enough to allow statistically meaningful results or sequential techniques. That is, if one has a hypothesis to test (e.g. program x is better than program y) one can use a portion of population and still have a large enough population left to either independently test the same hypothesis or to test a new hypothesis that is suggested by the data from the original portion.

Adequate problem populations probably should have 50 to 100 members, many of

which are parameterized (e.g. $x^\alpha \cos \beta x$). The parameters allow one to construct
performance profiles. It is important that some of the parameters directly vary
intuitive features such as oscillation, singularity strength, condition, etc. The
parameters also allow some random sampling to be made by choosing parameter values
at random. One in essence has a semi-discrete population to represent the underlying complete population. Later I will argue that such populations should be
standardized.

Standard statistics and statistical tests for experimental data are useful for
two reasons. First, statistics are nice summarizing quantities for data which is
important for making sense out of a mass of data. Second, hypothesis tests with
confidence intervals will clearly reveal inadequancies in the size of the sample
used in an experiment. One can argue quite convincingly that the assumptions for
the standard statistical tests are not satisfied for numerical software experiments. Even so, it is better to use these tests, keeping in mind their inadequancies, than it is to say "I looked at all this data and I conclude that such and
such is true".

I recommend the use of non-parametic tests such as the Friedman, Kendall and
Babington-Smith test (see Hollander and Wolfe (1973)) for many situations. They
can be especially useful for evaluations involving conflicting or incomparable
performance criteria. Suppose one has 100 problems, 5 programs and 3 incomparable
criteria (e.g. ease of preparing input, execution time and accuracy). One then
examines the performance of the 5 programs on each problem and ranks then 1 through
5 based on a <u>subjective</u> weighting of the three criteria. The above tests then
evaluate the significance of the average rankings of the 5 programs. This approach
does not remove the subjectivety from the conclusions reached, but it localizes it
to a lower level where it is easier to judge bias and harder to unconciously influence the conclusions.

4. PUBLICATION OF EXPERIMENTAL RESULTS

The paper of Crowder, Dembo and Mulvey indicates what one should publish in
reporting experimental results, but there is still the conflict between complete-
ness and conciseness. A substantial experiment can require dozens, even hundreds, of
pages to describe completely (e.g. list the programs evaluated) in addition to the
pages needed for the data obtained. The editors of many journals simply cannot
publish such lengthy papers except in the most unusual circumstances. Yet the results of performance evaluations must be published and in a manner that allows complete criticism by others. I describe a publication approach which I believe will
be both concise and complete.

The first part of the approach is the adoption of standardized components of
experiments. Specifically, the problem populations, the algorithms, the statistics
and the summarizing procedures used are indicated by short references to other publications. A population could be indicated by:

"I use the following problems from Smith: $1(\alpha=2, \alpha=4)$, $3(t=6.2)$, 5,6,7,
$9(\alpha=0.5, \alpha=1.5, \alpha=4.0)$, ..., $97(t=4)$"

Algorithms used are those already published or in widespread use. One does not use
"the RQ algorithm that I programmed in MYTRAN". An algorithm that does well in a
good performance evaluation is worthy of publication. One that does poorly should
be forgotten except for showing to those who cannot quite believe that a particular
method does so poorly when implemented.

The second part of the approach is to use hypothesis statements and testings
to express the conclusions of the evaluation. Such a hypothesis might be:

"LINPACK BAND is 50% faster than BAND SOLVE"

The conclusion would be stated that, based on the data obtained, the hypothesis is true with a 98% level of confidence. The actual experimental data need not be published, especially if it is massive; it need only be made available to the relatively small set of readers who will check, repeat or challenge the experiment. In the future we can expect to have alternative means of publication for massive data sets (e.g. microfiche, magnetic cards or tapes).

Thus the publication of an experiment would consist of three parts:

A. A concise description of the experiment using standardized components

B. Raw data

C. The concise, objective conclusions based on the data.

Parts A and C would be published in a normal journal, part B could be published normally if small or be available from the author or, in the future, published in alternative form.

5. FEATURE SPACE FOR NUMERICAL INTEGRATION

A number of function sets for testing numerical integration programs have been given in the literature, but none of them are adequate as a standard set. Here we propose a feature space that such a set should cover and show how simple operations can be used to enlarge a basic set of functions to a quite adequate set. It would be quite easy to generate an actual set of test functions based on this feature space. However, a really high quality set would require considerable study to be sure that some of them were not redundant or overly difficult and that they are all appropriate for inclusion. This is a substantial task.

We propose a feature space with 6 "integrand" features, 3 "auxilary" features and two others not related to the integrand. These are:

		Examples
Integrand f(x):	mathematical smoothness	entire, analytic
	jump discontinuities	
	singularities	x^α, log x, $\sqrt[\beta]{1/\log(1/x)}$
	local variation	peaks, boundary layers
	oscillation	
	noise	
Auxiliary:	scaling x	$x \rightarrow \alpha x + \beta$
	scaling f	$f \rightarrow \alpha f + \beta$
	computational expense	dummy loop in f(x)
Other:	accuracy requirement	
	domain of integration	

About 40 functions, most of them parameterized, should be sufficient to define an adequate population for experimental performance evaluation. This basic set can be substantially enlarged by using the three operations:

addition   multiplication   composition

These operations not only enlarge the set but they also provide a convenient and simple means to generate test integrands with multiple features.

6. FEATURE SPACE FOR PARTIAL DIFFERENTIAL EQUATIONS

E.N. Houstis and I are currently developing a set of test problems for second order, linear, elliptic, two dimensional partial differential equations on rectangular domains. This section is based on the substantial, but still incomplete,

set in Houstis and Rice((1978). The final set will have 55-60 distinct problems
with about half of them parameterized (some have four parameters). About 150
specific problems will be singled out as a set to represent the simple to moderately complicated problems of this class. Our experience shows that the creation of
such a set is by no means straight forward and that it requires a very substantial
effort if one wants to attempt to represent those problems that occur in real applications. Note that this substantial effort is only for a rather small class of
partial differential equations and that an adequate test set representing general
geometry will probably require much more effort.

The features we use are the problem type plus two properties for each of the
operator, solution and boundary conditions. Specifically they are:

> <u>Problem Type</u>: constant coefficients or variable coefficients
>
>   Laplace, Poisson, Helmholtz, self-adjoint, general
>   Dirichlet, Neumann, mixed
>   Real-world or artificial source
> <u>Operator</u>:   smoothness and local variation
> <u>Solution</u>:   smoothness and local variation
> <u>Boundary Conditions</u>:   smoothness and local variation

The values for smoothness, local variation and source are scaled to 0-100 (see
Houstis and Rice (1978) to see how values are assigned). A final feature, problem
complexity is derived from these features by averaging. Our test set contains,
problems of complexity 50 or less and thus does not represent the large scale
problems for which specialized programs are usually written.

A subset of 30 problems was used by Lynch and Rice (1978) to evaluate the
Hodie method and Fig. 3 shows the "profile" of this problem set in the feature
space. This diagram seems to give a nice, if rough, intuitive feeling for the
class of problems used in the evaluation.

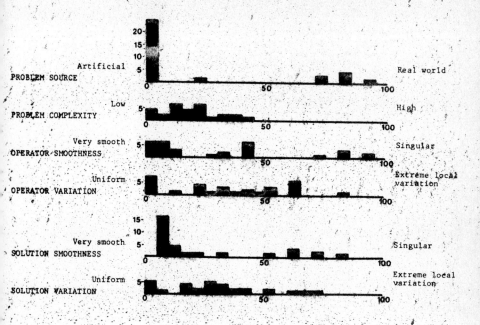

Figure 3. Features of the set of 30 problems used in the performance evaluation of the Hodie method. Heights of bars indicate number of problems from the set.

References

|1| H. Crowder, R.S. Dembo and J.M. Mulvey: On reporting computational experiments with mathematical software, ACM Trans. Math. Software; 5 (1979) to appear.

|2| M. Hollander and D.A. Wolfe: Nonparametric Statistical Methods, John Wiley, New York (1973)

|3| E.N. Houstis and J.R. Rice: A population of partial differential equations for evaluating methods, Computer Science Dept., Purdue University, CSD-TR263, May 15, (1978) 76 pages.

|4| R.H. Jackson and J.M. Mulvey: A critical review of methods for comparing mathematical programming algorithms and software: 1951-1977; presented at TIMS XXIII, Athens, July (1977)

|5| R.E. Lynch and J.R. Rice; The Hodie method and its performance, in Recent Advances in Numerical Analysis (C. deBoor,ed). Academic Press (1978) 143-179

|6| J.R. Rice: The algorithm selection problem, in Advances in Computers, Vol. 15 (Rubicoff and Yovits, eds.), Academic Press, New York (1976) 65-118.

Performance Evaluation of Numerical Software, Fosdick (ed.)
© IFIP, North-Holland Publishing Company, 1979

EXPERIMENTAL DESIGN AND STATISTICAL EVALUATION *

J.A. Nelder
Statistics Department
Rothamsted Experimental Station
Harpenden, Hertfordshire,
United Kingdom

Experiments are characterized as the choice of levels of
stimulus variables (treatments) and the measurement of
response variables. Given a statistical model for the
data, the choice of optimal design can be reduced to a
problem of constrained optimization. Problems connected
with the selection of treatment factors, response
variables and the design matrix are discussed, also the
selection of the systematic part of the model and the
characterisation of error. Comments are made on
factorial experiments (including fractional replication),
mixture experiments, the place of standard batteries of
tests, and the choice between random and systematic
levels for treatments.

It is concluded that existing theory on experimental
design and analysis has application to the testing of
numerical software, but requires the accumulation of
experience as in, e.g., agriculture, for its
effective use.

0.   INTRODUCTION

The appearance at a conference on the performance evaluation of numerical software of a speaker from an agricultural research institute requires some explanation. The only justification I can provide (and you may or may not think it
adequate), is that much of the work on a subject now known as the design and
analysis of experiments originated from Rothamsted, primarily from my two pre-
decessors, R.A. Fisher and F. Yates. The original application, to field
experiments in agriculture, was greatly extended when the basic ideas proved
relevant to experimentation in many other fields. For example, we find
Schatzoff and Tillman (1975) using the idea of fractional replication (developed
by Finney at Rothamsted in 1945) to validate the modelling of the dispatching
algorithm of a time-sharing system.

Though considerable work has been done on modelling aspects of operating systems,
such as paging algorithms, there appears to be very little in the literature
relating to numerical software. I cannot, therefore, give a summary of the state
of the art, but can only use the general framework of experimental design to
raise certain questions, in the hope that these will give rise to useful
discussion.

---

*Supported in part by the U.S. Army Research Office.

1. THE BASIC FRAMEWORK

In an experiment there will be measured variables of two kinds, **stimulus variables** which represent input to the experiment, and **response variables** which represent output. (For simplicity we shall assume a single response variable in what follows.) Stimulus variables are under the control of the experimenter and are usually called treatment factors. A treatment factor may be quantitative (e.g. the size of a matrix to be inverted) or qualitative (e.g., machine range on which program is run). An experiment consists of a series of runs (the word is taken from chemical engineering), in each of which the treatment factors are assigned certain values or levels and the corresponding output is measured. The matrix of levels is called the design matrix.

1.1 Design as a constrained optimization problem

If the k stimulus variables for an experiment with n runs are denoted by the column vectors $x_1, x_2, \ldots, x_k$, and the response variable by the vector y, then we can write the classical linear model for the system in the form

$$E(y) = X\beta$$
$$\text{var}(y) = \sigma^2 I$$

where β is the $(k \times 1)$ vector of parameters to be estimated, $X = [x_1 \; x_2 \; \ldots \; x_k]$ is the $(n \times k)$ design matrix, and σ^2 is the (constant) variance of the ys, which are assumed independently distributed. For any problem there will be constraints on the possible values of the stimulus variables and these constraints will determine a feasible treatment region. A typical formulation of the design problem is to choose X to maximize the generalised information $|X'X|$ subject to the x-values lying in the feasible region. [Other criteria than $|X'X|$ are possible – see Fedorov (1972) for a general account.]

This reduction of the design problem to one of constrained optimization is of undoubted importance and interest. Nevertheless the very process of abstraction involved in its formulation can disguise the fact that it assumes that several difficult questions have already been resolved. In practice it is the resolution of these questions that is likely to be both difficult and time-consuming, and it is the manner of their resolution that may well determine the difference between a good and a poor experiment. We now consider some of these questions.

2. QUESTIONS TO BE RESOLVED

2.1 Selection of stimulus variables

It is useful to distinguish between quantitative and qualitative stimulus variables, hereafter termed variates and factors, in the terminology of Genstat (Nelder et al. (1977)). Variates may be of integer type (e.g. size of matrix in an inversion algorithm) or continuous (e.g. convergence criterion for an iterative process). Factors are limited to a discrete number of levels, which may be unordered (e.g. machine range) or ordered (e.g. a set of values of a variate). The variate-factor distinction is closely related to the terms envisaged in the model to be used in the analysis; thus a factor A and variate X would be associated with a model of the form

$$E(y) = a_i + bX$$

where i indexes the levels of A and b is the parameter associated with X. If the levels of A are quantitative then the a_i may be replaced by some suitable response curve based on those levels.

Stimulus variables may be selected for one of two complementary reasons. Either they are expected to be important, i.e. changes in x are expected to produce large changes in y, or variation in x is expected to have little effect on y but verification of this is needed.

There are essentially two stages of the selection process. In the first the set of stimulus variables is chosen: call these u_1, u_2,..., u_k. In the second the actual variables x_i to be used in the experiment are chosen as functions of the us, i.e.

$$x_i = f_i (u_1, u_2, \ldots u_k)$$

The replacement of the original us by the derived xs will have as its main aim the simplification of the model necessary to describe the resulting response surface of y as a function of the xs. In particular we seek xs such that the effects of them on y are approximately

(i) linear

and (ii) independent

Behind these simplifications is the desire to reduce the number of parameters necessary in the model (parsimony) and to increase the amount of information (in Fisher's sense) about the parameters in the data.

The experimenter's ability to choose the functions f well depends on either the existence of a prior theory or empirical evidence from past experiments or both. Two techniques which have proved useful, given previous experimental results, are

(i) linearization, whereby if an x is thought to be related to a u by $x = g(u, \theta)$, then both x and $\partial x / \partial \theta$ are included in the model to estimate θ. Thus if x^θ is thought to give a linear response rather than x, then both x and $x^\theta \log x$ are included for some trial value of θ, from which an improved value may be derived (see Box and Tidwell (1962));

(ii) canonical variate analysis, whereby linear combinations of the xs most correlated with y are sought, the object being the replacement of many xs by a few combinations of them.

2.2 Selection of response variables

This is governed mainly by a priori considerations of what is required from the experiment. When many measurements are made, some on variables with much greater variability than others, then it may be possible to design an experiment whereby some responses are measured on only a sample of the runs, while retaining the condition that this sample should be a balanced subset of the whole design (see Section 3.1). At the analysis stage, having chosen y, it is still important to consider whether the analysis should be carried out on y itself or some function g(y). There are two distinct reasons for considering an analysis in terms of a function of y, rather than of y. One is that error properties of g(y) may be closer to the desirable form of being (approximately) Normally distributed with constant variance, and the other is that the treatment effects may be more nearly linear or additive (or both) on that scale. Recently models have been developed in which the transformation to produce desirable variance properties may differ from that producing linearity, and suitable algorithms now exist for fitting such models (see, for example, Baker and Nelder (1978)).

2.3 Selection of design matrix

Design points must be chosen from the feasible treatment region. Such a region

may be limited, when testing numerical software, by the hardware (in particular
word size and core space), or by the objectives. If, for example, we are testing the ability of an algorithm for unconstrained optimization to negotiate
curved valleys, then the starting point, considered as a treatment factor, will
be best confined to the neighbourhood of the valley at some distance from the
lowest point.

Designs derived from optimality criteria consist of points on the boundary of the
feasible region, the number of distinct points being in general equal to the
number of parameters to be estimated. Consider, for example, the following
parametrisation of the class of positive-semi-definite matrices of order k.
These can be parametrised (Nelder (1968)) into $k - 1$ 'variance angles',
$k(k - 1)/2$ 'covariance angles' and a scale parameter λ. This last will be
confined to a range (ε, N), where ε is a lower limit and N is an upper limit,
both limits being set by the range of possible real numbers in the machine.
Optimal designs will contain only points for which λ takes the values ε or N.
The main objection to such designs is that they depend on the model being known
with certainty and provide no check on the adequacy of that model. Thus we would
certainly in practice want to include intermediate values of λ to check any model.

3.   FACTORIAL DESIGNS

These designs, introduced by Fisher in the 1920s as an alternative to designs
where one variable (factor) at a time was varied, correspond to the use of a
rectangular grid of points in the treatment space. A two-dimensional example is
given below, where the 9 crosses indicate the 3×3 factorial, and the circled
points the one-factor-at-a-time design.

An advantage of factorial designs is that if joint response to, say, $\underline{x}_1$ and $\underline{x}_2$
is additive then the table can be summed up in terms of its two margins, and all
yields are used (in different linear combinations) to estimate the effects of
each factor. If the response is non-additive then this will be discovered, as
it would not be with the one-at-a-time design. In contrast to optimal designs
which require exact knowledge of the model, factorial designs allow minimal
initial assumptions to be made about the form of response.

3.1   Fractional replication

An apparent disadvantage of factorial designs is the rate at which the total
number of treatment combinations increases with the number of factors. Thus 10
factors at only 2 levels each produces $2^{10} = 1024$ combinations, while the same
number of factors with 3 levels gives $3^{10} = 59049$ combinations, which usually far
exceeds what an experimenter can afford. A solution is to use only a fraction of
the possible combinations, i.e. to do a fractional replicate of the complete
experiment. The simplest example is the half replicate of the 2^3 experiment
whereby the 8 treatments (representable as 3 digit binary numbers) are split into
2 groups

$$(000,011,101,110) \quad \text{v} \quad (100,010,001,111)$$

and either half is chosen for the experiment. The contrast between the 2 groups
is the ABC interaction of the 3 factors A, B, and C, say. It follows that in
interpreting the yields from such an experiment, the average (or main) effect of
A will be measured by the same contrast as the two-factor interaction BC. Such

pairs are said to be <u>aliased</u>, and we have for this design

$$A \equiv BC, \quad B \equiv CA, \quad \text{and} \quad C \equiv AB$$

In a $1/4$ replicate of the 2^6 design defined by an <u>alias sub-group</u> with elements

$$I, \quad ABCD, \quad ABEF, \quad CDEF$$

where I is the identity element, each main effect has 2 3-factor-interaction aliases and one 5-factor-interaction alias. Thus A is aliased with

$$BCD, \quad BEF, \quad \text{and} \quad ACDEF$$

Unambiguous interpretation depends on being able to discard all aliases but one as being of unlikely importance. Fortunately the frequent occurrence of one or two null treatments often allows this to be done.

3.2   Mixture experiments

When, for example, a quadrature algorithm is to be evaluated, severe problems arise in defining the class of functions on which it is to be tested and on which subsequent recommendations are to be made. A possible procedure consists in choosing a basic set of test functions

$$f_1, \; f_2, \ldots, \; f_k$$

and performing runs with derived functions

$$g_1, \; g_2, \ldots, \; g_k$$

where
$$g_j = \sum_i p_{ji} f_i, \text{ with } \sum_i p_{ji} = 1 \text{ and } p_{ji} \geqslant 0.$$

The design for a set of such runs can be represented by points on or within a simplex, and designs analogous to factorial experiments can be defined (see Scheffé (1958) and (1963)). Cox (1971) has commented on the corresponding models for the analysis of such designs.

4.   WHAT IS ERROR?

The classical account of experiments presupposes the existence of variation in the response of a nature expressible in terms of a random variable having some distribution (usually assumed Normal). Now for an algorithm the response variable is strictly a deterministic function of the inputs, in the sense that the exact response on a given machine could be calculated given the machine coding of the algorithm. Nonetheless it may still be entirely adequate to describe part of the variation of such responses in statistical terms, in other words to treat such variation as if it were random. However, statistical description still needs to be accurate, and blind assumptions of Normality must be avoided. Note that accumulated rounding and truncation errors in an algorithm of any appreciable length are almost a perfect source of errors in the classical sense; however, much greater irregularities in the response surface are likely to be superimposed by things such as branches in the algorithm which depend on ranges of input parameters, and these irregularities may have much less smooth cumulative effects. If the experimenter merely assumes that all local variation after removal of trends can be treated as independent random variables with a tractable distribution, he may go seriously astray in his analysis. He needs to beware of at least three things:

(i) the existence of subclasses of runs with biased responses. Such subclasses should be modelled by a systematic component, which if omitted, will bias the error estimate. The bias may be in either direction;

(ii) empirical distribution functions that vary widely with grid size. If an experiment with a grid of factor levels, distance δ, produces a set of apparent errors with a certain distribution function, then another experiment with a grid at distance $\delta/2$, should not produce a greatly different result. If it does then the assumption of independence of errors is suspect:

(iii) distributions with very long tails. These distributions have frequencies of extreme values that tend to zero much more slowly than those of the Normal distribution. Inferences from such distributions based on tendency to Normality of means, etc., may then be misleading.

5.   THE TEST-BATTERY CONTROVERSY

There has been considerable discussion, both at this conference and elsewhere, about the value of standard batteries of tests in evaluating software. Now any test problem corresponds to the choice of a point in some treatment space. It may not be explicitly stated what that treatment space is and how its factors and levels are being defined, and hence it may be difficult to recognise a battery of tests as being a set of points of a design, but some kind of design they certainly comprise.

Those opposing batteries of tests seem to be saying that the tests (i.e. points in the treatment space) are arbitrary, and that we should be more systematic. Those supporting claim that the tests correspond to important subsets of problems, frequently met, and ask how can one be systematic? For example, how does one sample the class of integrable functions for a quadrature algorithm?

I suggest that the controversy reflects uncertainty about the definition of the treatment space, and about suitable models for analysis. Thus there would be no argument about doing experiments covering a large treatment region if it could be shown that the predictions of response from such an experiment were sufficiently accurate over the whole region to be useful for particular sub-regions corresponding to commonly met problems.

6.   RANDOM OR SYSTEMATIC TREATMENT LEVELS?

In selecting levels for a factor in an experiment, there is always the possibility either of selecting at random from the range of interest, or of taking values on a systematic grid (not necessarily evenly spaced). When treatment levels are selected at random it is always possible to use sampling theory to assign limits of error to predicted mean values. However, such an interpretation assigns all the variation produced by that factor to error, and only mean predictions are attempted, rather than some attempt to describe how the response changes with input. For such a description systematic designs are always more informative, though it may be necessary to randomize the origin of the grid to avoid the inclusion of special subsets. In general, if factors producing heterogeneity of response are known, the design should be systematised with respect to them, being randomized with respect to everything else. All designs require for their analysis an adequate model to describe the macro features of the response surface.

7.   CONCLUSION

I conclude that standard principles of experimental design have a part to play in software evaluation, but that their development in this field requires the accumulation of experience, of the kind that exists in agriculture, in such matters as

(i) factor definition

(ii) response-surface shapes, on both macro and micro scales

(iii) error distribution forms.

The development of suitable designs and techniques of analysis will depend on the intelligent scrutiny of existing data and on careful assessment of the behaviour of standard procedures in this new field.

References

1. Baker, R.J. and Nelder, J.A. (1978) The GLIM System Manual, Release 3 Oxford: Numerical Algorithms Group.

2. Box, G.E.P. and Tidwell, P.W. (1962) Transformation of the independent variables. Technometrics 4, 531-50.

3. Cox, D.R. (1971) A note on polynomial response functions for mixtures. Biometrika 58, 155-9.

4. Fedorov, V.V. (1972) Theory of optimal experiments. Trans. and Ed. W.J. Studden and E.M. Klimko, New York: Academic Press.

5. Nelder, J.A. (1968) Regression, model-building and invariance. J.R. Statist. Soc. A 131, 303-15.

6. Nelder, J.A. et al. (1977) Genstat Manual. Statistics Department, Rothamsted Experimental Station, 469.

7. Schatzoff, M. and Tillman, C.C. (1975) Design of experiments in simulator validation. IBM Journal of Research and Development, 19, 252-62.

8. Scheffé, H. (1958) Experiments with mixtures. J.R. Statist. Soc. B 21, 344-60.

9. Scheffé, H. (1963) Simplex-centroid design for experiments with mixtures. J.R. Statist. Soc. B 25, 235-63.

Performance Evaluation of Numerical Software, Fosdick (ed.)
© *IFIP, North-Holland Publishing Company, 1979*

EVALUATION OF NUMERICAL SOFTWARE INTENDED FOR MANY MACHINES - IS IT POSSIBLE?

Brian Ford, Graham S. Hodgson and David K. Sayers
The Numerical Algorithms Group Limited
7 Banbury Road
Oxford OX2 6NN
England

To evaluate software intended for many machines one requires
a design and operational specification against which its
performance should be judged. The specification must take
full account, for example, of the arithmetic assumptions upon
which the algorithm is based. Further when evaluating
software we should ensure that essentially the same code is
being tested for each configuration. We show by means of an
example that even in the newer languages it is difficult to
preserve such similarity of source code.

1   INTRODUCTION

Evaluation of numerical software means different things to different people
but most would agree that the construction of numerical software consists of
four phases:

a) the design of the algorithm
b) its realisation as a source language subroutine or program
c) the testing of the compiled code on a given configuration
d) the writing of the supporting documentation.

Any performance evaluation is in some way an attempt to measure the degree of
success, or otherwise, obtained in each phase of the construction. Some
people place all the emphasis on one phase of the construction. Typically the
theoretician will consider evaluation at the algorithmic level only and take
into account such factors as stability and convergence proofs, existence of
error bounds and guaranteed termination in a finite number of steps.
Broadening our view to consider phases b) and c) we see that other factors
become involved, such as program size, efficiency of coding and the
reliability and robustness of the code.

It is a moot point how much weight should be attached to each property of
the software and this is one of the major stumbling blocks to universal
agreement about codes. Certainly some circumstances demand overwhelming
emphasis on particular aspects of the software: otherwise excellent software
which is so large that it cannot fit into a particular machine is useless for
users of that machine. One can imagine situations in which accuracy is of
paramount importance and others in which speed of response, typically in a
real-time situation, is vital.

This problem is alleviated if the software has an operational specification,
in the context of which its performance can be measured. Thus a software
writer might design his software to be extremely accurate and it would be
churlish to criticise the routine too severely for being slow. Were the
software inaccurate however, then harsh condemnation would be in order.

Nearly all attempts at the evaluation of numerical software to date have
concentrated on the performance of software on one particular machine. (For
example, Hull [11] did his work on the evaluation of O.D.E. software on an
IBM machine.) There are at least two reasons for this. Firstly authors only
had access to the one machine and secondly the importance of 'portability' was
not fully appreciated until recently. The growth of multi-machine sites and

networks, the heavy costs of producing good software and the desire to communicate information in the form of programs have led to the desire to produce software which will run and perform well on many different machines. Under these circumstances we argue that the detailed testing of a piece of software on one machine is incomplete. Such evaluation should be over many different configurations.

In section 2 we review the additional problems of multi-machine evaluation, consider the manner in which assumptions about arithmetic can affect the adaptability [6] of algorithms and their effectiveness as the basis of transportable software [10] and attempt an answer to the question 'when are programs comparable?'. In section 3 we examine how the attempted development of transportable source-text is affected by prelude mechanisms in the newer high-level languages, and by the absence of defined mechanisms of library incorporation. Our conclusions are inevitably far from encouraging.

2   EVALUATION OF SOFTWARE

2.1 Additional Problems of Multi-Machine Evaluation. Obviously the software will have to compile correctly on all the machines and hence strict adherence to a standard language in the coding is necessary. 'Tricks' which rely on particular compilers for their success are obviously bad [20]. In practice the mere adherence to a standard, such as ANSI, proves insufficient and subset languages such as PFORT [18] are necessary [1].

The reason for this is not merely the inadequacies of the standards, of which the failure of the Algol 60 report [15] to define input/output is a classic example, but also of the compiler writers. For instance Algol 60 allows dynamic own arrays, however few compilers implement these. Sometimes the standards are vague and compiler writers have taken different courses of action according to their own individual circumstances. Typical compiler variations are:

  a) handling of array parameters to a subroutine by address or
     by copying,
  b) whether a DO statement is always evaluated once or not,
  c) different levels of optimisation causing expressions to
     be evaluated in a different order.

It is clear then that the adherence of the code to a suitably defined subset language is of importance in testing over many machines. But this is not all we require, for some machines have facilities in hardware which others have only in software. For instance the ICL 1906A can implement a double length accumulation of an inner-product by hardware but the ICL 1904A needs to resort to software. As a result accumulation of inner-products is relatively expensive on the ICL 1904A compared to the ICL 1906A. A routine, designed to be fast but not prepared to meet this situation, would thus fare badly on the ICL 1904A although the author may have thoroughly tested it on the ICL 1906A.

It is self-evident that an absolute yardstick by which we measure a routine's performance on several machines is unrealistic. The yardstick must vary from machine to machine to reflect the different computing resources, be they central processor power or available storage. To emphasise the obvious, a routine which is 'fast' on two machines A and B may take very much longer on machine B. It is 'fast' on machine B relative to rival candidates on that machine.

Nevertheless we would like to follow the overall concept of evaluation on

one machine, as described in the introduction, and see how this extends to
several machines. We argued that the testing was a means of evaluating how
closely the software met its specification. For this to be meaningful over
several machines the extent of the specification needs to be enlarged.

The specification of the software should address two requirements: the
design goals of its developer in terms of algorithmic properties and software
performance; and an operational specification with a description of the
precise purpose of the software, its intended environments of use and the
users to whom it is directed.

For the evaluation of software on one machine the design goals might be
described by reference to its actual performance in solving a particular set
of problems ([9], [14]). Factors considered would include accuracy, speed,
reliability and robustness [3]. The operational specification would
describe the numerical area addressed by the software, the particular
configuration on which it has been certified [4] and whether it is for
general use or directed to a particular community (through the nomenclature
and vocabulary employed in the user interface, source text and documentation
[8]).

For use on many machines the design goals and the operational specification
of software are significantly more complicated. Definition of the problems
upon which evaluation might be based poses difficulties because of the use of
different arithmetics producing different 'solutions'. The accuracy the
routine can attain, the speed of the software, its reliability in solution of
the problem set and its degree of robustness must all be described in terms
which are specific for each configuration, and yet may be meaningfully
combined to give a general measure of performance.

For the operational specification the precise purpose of the software will
need to be all embracing (e.g. sequentially operating mainframe and mini
computers) but specific with regard to exceptions (e.g. not for general use
in paged environments with storage of arrays by row) and limitations (e.g.
not for machines with less than 2K actual store). It must delineate the
configurations upon which the software will run correctly and describe the
documentation available for users of each environment.

Specifications covering these requirements are of course very rare [19],
and aspects of the operational specifications are as warmly contested as are
features of the design goals.

In 2.2 we shall consider various assumptions made by algorithm designers
about arithmetic and related features and discuss how their choice (and its
effects on algorithm design) may determine the ability of software to meet a
specification for generally used transportable software [6].

2.2 Arithmetic Assumptions Made by Algorithm Designers. To formulate a
specification for a piece of software we need to know all of the assumptions
that have been made in the creation of the software. Often these are part of
the specification. Dekker [5] for instance is careful to state the
properties he requires of the floating point arithmetic. An evaluation of his
software on a configuration which did not satisfy his criteria would not only
be unfair, but give the misleading impression that the software was in some
sense wrong. Other authors make assumptions, either consciously or
subconsciously, which are not so clearly stated. It is convenient to discuss
these under five headings, ordered in terms of increasing sophistication.

2.2.1 All Machines Do the Same Arithmetic. Instinctively one feels that

this assumption is a beginner's error but a moment's thought reveals that this
is not necessarily so. The specification of the software might be to return
a low accuracy approximation to a function within a suitable range of
argument. For example

$$1-\tfrac{1}{2}(1+c_1 X+c_2 X^2+c_3 X^3+c_4 X^4)^{-4}$$

with $c_1=0.196\ 854$, $c_3=0.000\ 344$

   $c_2=0.115\ 194$, $c_4=0.019\ 527$

approximates the normal or Gaussian probability function

$$P(x) = \frac{1}{\sqrt{2\pi}} \int_{-\infty}^{x} \exp(-t^2/2)\,dt$$

with error $< 2.5x^{-4}$. Within this bound the formula might serve as the basis of
an algorithm meeting the required specification; the machine arithmetic is
unlikely to affect the achieved accuracy significantly.

Commonly however heuristics creep into software as it is being developed.
Rarely are all such heuristics flagged, yet a failure to adapt them when the
software is moved to another machine can prove disastrous. For example a
termination criterion of 10^{-8} may be reasonable on a CDC 7600 but is unlikely
to be so on a single precision DEC System 10 or IBM 370 FORTRAN system.

2.2.2 Computers Have Finite Arithmetic which Preserves Mathematical
Properties. It might be appreciated that the computer has finite arithmetic
but natural, and outwardly reasonable, assumptions might be made concerning
computer arithmetic which prove to be untrue in some environments.

As an illustration real numbers A,B,C, and D exist such that A/B < C/D in a
mathematical sense yet on the CDC 7600 using the FTN compiler with round
options the computed test (A/B)>(C/D) yields the value TRUE. [One set of
such numbers is $A=1$ $B=1+2^{-46}$ $C=1-3x2^{-4}$ $D=1-3x2^{-4}+3x2^{-48}$].

As we depend on machine arithmetic, so we often depend upon the compiler
functions library. Unless such functions are carefully developed, programs
employing them will experience spurious discontinuities in the
function values. It is also natural to expect the standard library functions
to return the expected accuracy, but often these hopes are not fulfilled.

Such factors cause consternation at the explicit assumptions made by
algorithm developers, such as Rutishauser [17], that mathematical properties
such as monotonicity are preserved on computers and hence can be employed in
the construction of termination conditions for iterations. It is true that
Rutishauser often noted the effect if a particular property was not
preserved. Unfortunately other workers are rarely so scrupulous. And
whatever its pedigree, software based upon such assumptions has neither the
reliability nor the robustness we would expect of general purpose
transportable [6] codes.

2.2.3 Simplistic Model of a Machine. A very simple model of a computer
accepts the finite arithmetic and its properties but defines specific actions
in the event of overflow and underflow. A number of software developers
define the action on overflow to be that control is returned to the operating
system and the program aborted. On underflow, zero is accepted and the
computation continued. It is assumed that these actions are common to most
computers and this may well be the case. There is however a significant
number of machines not covered by this model. Amongst these we list

Honeywell systems under GCOS for underflow and some IBM systems for overflow.

If software is intended for general use it should be robust. Further if the algorithm has been unable to solve the problem, the code should advise the user of the point at which the computation terminated and the suspected reason for the failure.

Unfortunately the simplistic model of a machine does not provide sufficient information to enable such an operational specification to be met. If the user's program computes an overflow (or an underflow) he may receive so little information from the program (since he cannot be told what caused the event; why it has occurred; what should be done) that he does not know how to proceed.

2.2.4 Parameters of a Conceptual Machine. In this approach [7] the assumption is made that a simple model of a computer, covering the essential features for numerical computation, can be developed and described in terms of related parameters, which may be given particular values for a particular configuration. Algorithms are designed employing these parameters, enabling them to adapt [6] to the various configurations within which they compute. The parameters are used in the transportable source-language routines and given specific values during compilation or execution.

In FORTRAN these numbers may be made available by function call, as values in a COMMON block or substituted by a pre-processor. More modern languages such as Algol 68 have a limited facility for environmental enquiry [21] built into them, which can be augmented by the use of a library prelude.

Experience has shown that transportable software can successfully be developed using this approach.

2.2.5 Parameters of a Conceptual Machine - with Arithmetic Axiomatically Based. A recently developed model of floating-point computation involves a set of arithmetic parameters (similar in concept and substance to those defined above) and a set of axioms which describe floating-point computation in terms of these parameters. In abstract this approach appears to be a significant improvement in numerical software technology since it permits an error analysis of the computation. Early work by Brown [2] has demonstrated the elegance of the strategy but has also made clear the problems associated with its implementation. Machine anomalies cause difficulties in defining parameter values. A relaxation of parameter values is necessary on many machines in order to ensure that the arithmetic obeys the axioms of the model. The essential assumptions made are that this model relates closely to the real world on all machines and that this relaxation of parameter values does not adversely affect software performance; this assumption may well fail on machines with 'quirky' arithmetic.

2.3 When are Programs Comparable? Suppose a piece of software runs on several different machines and has a specification clearly stating the design aims and assumptions made by the original author. The question naturally arises, "How different, at the source text level, can two versions of the program be, before we regard them as essentially different pieces of software?" For example a single precision CDC program might transfer without change to an IBM 370. However for reasonably accurate results double precision might be necessary. Do we regard the double precision IBM version as essentially the same piece of software as the CDC version? Yes - provided that we have a systematic and reliable means of transforming the source-text from one precision to another.

For we note that representation differences are frequently necessary when transforming programs between configurations, if not normally in FORTRAN, then at least in other languages. In Algol 60 for instance 'not equals' is frequently represented by # or 'NE'. To regard working programs on different machines as essentially different because one used # and the other 'NE' does not seem reasonable. So we certainly must allow some representation differences in the source-text.

On the other hand there are compilers which provide an option to generate double precision code directly from single precision source text. Whether the double precision compiled code is obtained in this way or indirectly via double precision source text, is not a significant distinction from the point of view of performance evaluation.

Indeed we should consider the scope of our performance evaluation as consisting not simply of many machines, but of many machine environments (i.e. configurations); single precision computation on an IBM 370 and double precision computation on the same machine are two different environments in which we may very reasonably wish to evaluate the same piece of software.

However, the fundamental question is: how do we evaluate program performance in each configuration, given their different attributes? A limited illustration of what can be done is given by the procedure for implementing the NAG Library in different configurations. At this stage routines have already been selected for inclusion in the Library and the limited operational requirement is that the routines should always perform with the expected reliability, robustness and accuracy in each configuration (different degrees of accuracy will of course be expected in different configurations). A comprehensive suite of implementation test programs and data sets is provided, which can be systematically transformed for use in each configuration. These programs are run and the results checked to ensure that the requirements have been met. To check for accuracy, we need a measure of the error in the computed results, expressed in terms of the relative machine precision, the base of the floating-point arithmetic and other configuration-dependent factors: this measure can then be compared with a theoretically derived bound or expected value, which depends on the type of problem, its dimensions and degree of ill-conditioning. To check for robustness we need to include test problems which will present the routine with all the exceptional conditions that are anticipated (noting that some exceptional conditions will only occur on certain machines, e.g. on those where the reciprocal of the largest floating-point number would underflow). And to check reliability we need to ensure that all the compiled code is exercised in each machine (ideally in such a way that any compilation errors will be detected). To achieve these aims rigorously presents varying degrees of difficulty in different areas of numerical analysis covered by the library and so far we have been more successful in some areas than others. Nevertheless we are confident that this limited degree of performance evaluation is achievable, in practical terms, across a wide range of different configurations.

Reverting to the question of the representation of programs, and our pursuit of transportable software for general use, we would hope to be able to compare simply at the source-text level in the more modern languages such as Algol 68 which have much greater sophistication. In the following section we shall explore this at greater depth and illustrate that, because of the absence of a standard library pick-up mechanism, a diversity of different pick-up mechanisms has forced NAG to make important changes at the source text level in some of its Algol 68 routines. This situation only arises where some of the more sophisticated features of the language have been used, but nevertheless where this does occur the resulting codes are semantically very different.

3   THE ENVIRONMENT PROVIDED BY NEWER LANGUAGES

In an attempt to cocoon the user program and present it with a standard environment, the newer programming languages have defined the concept of a prelude. By achieving the goal of a standard environment, a source code which is invariant across differing configurations becomes possible (only representational changes as in FORTRAN and Algol 60 are then necessary). This concept has grown from the idea of providing a very limited environment for a FORTRAN program by a set of intrinsic and basic external functions. Languages such as Algol 68 [21] and PASCAL [12] define standard preludes to attempt to provide a machine-independent environment. In Algol 68 this may be further extended by library preludes to declare and define (in Algol 68) additional facilities to be available to the user program.

The properties of either the standard or a library prelude have nothing logically to do with the problems of providing a mechanism for separate compilation, as is demonstrated by the complete separation of these concepts in the recently published standard for modules and separate compilation mechanisms in Algol 68 [13]. However this standard was published too late to influence currently available compilers, and in this section we discuss the difficulties experienced in providing an Algol 68 library on two machine ranges (CDC (see 3.2.1) and ICL 1900 (see 3.2.2)) each with different (and nonstandard) separate compilation mechanisms.

It will of course be possible to provide a library in source text form for the user to incorporate directly in his program; a separate compilation mechanism then becomes unnecessary. However we have discounted this approach. Not only does it pose a significant security and software maintenance problem but the Library routines cannot be fully tested once and then confidently expected to perform correctly on all future occasions. Each minor modification of the compiler and its libraries puts the integrity of the source-text subroutines at risk. The compiler interface to precompiled code is not threatened to the same extent.

We will discuss some of the problems associated with preludes by referring to the Algol 68 standard and library preludes since this is currently the more advanced and precisely defined prelude mechanism. PASCAL has a rather weaker standard prelude, but as yet no universally defined library mechanism. Despite its lack of generality PASCAL too suffers from the defects which we are about to demonstrate.

We are not advocating the abolition of the prelude, far from it, but we wish to show that it alone is not enough - a facility for separate compilation is necessary if we are to achieve a truly machine-independent environment (and hence an invariant source code). We hope to encourage all compiler writers to regard a facility for separate compilation as an integral part of the language; and for Algol 68 to refer them specifically to [13].

The advantages for performance evaluation of having a machine-independent environment and hence, except for representational changes, a source code which is invariant across all configurations is self evident.

3.1 Standard and Library Preludes.

In Algol 68 the standard prelude provides the environment common to all programs. It contains a facility for environmental enquiries. These may either define the extent of the standard environment (for example, *real lengths* gives the number of extra lengths of real numbers) or define implementation dependent features of the environment (for example, *max real* gives the precision of a real number). The prelude also contains standard declarations and definitions (of, for example complex numbers and arithmetic).

The library prelude defines extensions BUT NOT REDEFINITIONS of the standard prelude.

Further, preludes are intended to provide facilities on a global basis; the concept of facilities being provided more locally is excluded.

At first sight the prelude mechanism may seem ideal for the construction of portable code but unfortunately a number of prelude dialects have arisen; some facilities may not be implemented or may be implemented differently. We can categorise the differences as

   a) implementational errors for which we give two examples:
      ÷ (integer divide) on ICL 1900(68R)
      rounds to −∞ and not to 0.
      SHORTEN (LONG REAL rounding) on CDC
      truncates rather than rounds.
   b) variations explicitly encouraged by the report -
      the number of different lengths which an
      implementation distinguishes is at the discretion
      of the implementor. (In consequence, LONG REAL
      operations may be treated as REAL operations
      which may confuse users who fail to check the
      environmental query *real lengths*.)
   c) variations (allowed and) not defined by the report.
      (For example, the type of rounding for the
      arithmetic operations +,−,∗ and / is at the
      discretion of the implementor.)

Any variation in the standard prelude cuts at the heart of our aim of an invariant source text. As language dialects have caused great difficulties for the preparation of transportable programs in FORTRAN and rendered meaningful performance evaluation across different configurations problematic, so prelude dialects in Algol 68 can render subroutines inoperable or inefficient in particular environments. For example if LONG REAL operations are treated as REAL operations by a compiler, iterative refinement for the accurate solution of simultaneous linear equations simply will not work.

Unfortunately there is no easy way to correct either of the variations a) or c) above of the standard prelude. In particular a library prelude may not be used for this purpose since the syntax forbids redefinitions and what we do must apply not only to the internal objects of a library prelude but also to the particular program. The report does not permit the nesting of preludes in a block structured way. Hence a program accessing preludes does not have the required structure available to it for redefining the standard prelude through the use of library preludes.

It is possible to include LONGer (and SHORTer) precision versions of operators in a library prelude (see b) above , but this facility should be used with caution. If such operators are provided piecemeal and not as a self-coherent whole (that is all of the usual arithmetic operations or some reasonable subset of them), the standard prelude is required to do the best it can to fill the gaps (and provide identifiable operators implemented in lesser precision). In consequence, some operations can yield less accuracy than the unwary user might expect.

3.2 Mechanisms for Separate Compilation. What then do we require of a
library structure? A library may be composed of different types of objects:
subroutines, operators, data structures and communication variables. For
large libraries, such as NAG [16] we need subdivision into separately
compiled parts (modules) to obtain

     a) manageable modules for compilation,
     b) manageable modules for loading and running,
and c) protection of one module from other modules
       (or even the program).

Such subdivision was not defined by the Algol 68 report [21], although a
standard for separately compiled modules [13] has been promulgated recently by
WG2.1 (August 1978). Meanwhile due to the absence of a standard library
mechanism in the report at least two drastically different library mechanisms
have evolved (neither of which conforms to the new standard). The effect
for the portability of Algol 68 source text and for performance evaluation of
the code for different configurations is devastating.

3.2.1 A Method based on Preludes. The first mechanism is that used by the
CDC Algol 68 compiler; it is a method based on preludes. Only routine texts
may be separately compiled. Other objects must be global and put into a
library prelude.

   This therefore requires that mode declarations, priority declarations,
variables and operator and routine definitions must be either global to the
whole library (and to the particular program) or must be local to a separately
compiled module (and not available to the particular program). It also means
that storage is reserved by either

     a) local generation in a routine,
     b) local generation in the prelude
       (which is global to every particular program),
or c) generation on the heap in order to extend
       its scope beyond a routine text.

   For example if we separately compile this storage generating procedure:

```
PROC gen1000 = REF [ ] REAL:
    BEGIN ¢ local generators are local to the
        procedure body, heap generators
        are therefore needed ¢
        HEAP [1 : 1000] REAL x;
        FOR i TO 1000 DO x[i] : = 0.0 ; x
    END;
```

and with a library prelude

```
    REF REF [  ] REAL vector = HEAP REF [  ] REAL;
    PROC gen1000 = REF [ ] REAL:
        PR a system name for the separately
            compiled routine text PR SKIP;
    PR prog PR SKIP;
```

then the assignment   vector : = gen 1000
would generate the appropriate initialised vector, but on the heap. To
selectively generate values for some (but not all) particular programs, we
must therefore use the heap rather than the stack of the compiler. Although
the use of the heap (which must have an associated garbage collector) is not
inefficient on this particular implementation, this does not apply universally.

Hence an attempt to maintain invariant source code by using the heap, can mean large overheads on some implementations and therefore significantly affects performance evaluation.

3.2.2 A Method based on Ranges. We can avoid the need for the heap generators used above with a method based on ranges (the analogue of a block in Algol 60). This is essentially the ICL 1900 68R system, but with the notation of the new standard [13]. Any range may be separately compiled and we will then call it a module. Such modules then surround the particular program.

Using this library mechanism we must recast the previous example into modules. Note how significantly different the source code has become:

MODULE A = DEF
¢ a public declaration ¢ PUB [1:1000] REAL gen1000;
                        FOR i TO 1000 DO gen1000[i] : = 0.0
          FED;

then ACCESS A (REF [ ] REAL vector : = gen1000)
would generate the appropriate initialised vector, but NOT on the heap.
Not only does the source code look very different (note the absence of procedures) but its semantic interpretation is also very different (due to the use of the stack rather than the heap).

3.2.3 The Proposed Standard Method [13]. The method based on ranges (3.2.2) is the more general of the two methods. It avoids serious restrictions being imposed on source text which is to be separately compiled; we have shown above that such restrictions are imposed with the method based on preludes (3.2.1). The method based on ranges has been generalised and its effects more rigorously defined in the new standard.

One must be aware that interrelations between ranges can produce unexpected results since these ranges form a concentric structure. By separately compiling:

MODULE G = DEF PUB REAL g; ..... FED;
MODULE H = DEF PUB REAL g,h; ... FED;

the order of accessing G and H becomes important. The resulting programs have the following structures:

            accessing G then H   accessing H then G

BEGIN REAL g; ....   BEGIN REAL g,h; ....
   BEGIN REAL g,h;   BEGIN REAL g;
   ¢ which redeclares g of module G ¢   ¢ which redeclares g of module H ¢
   ....
      BEGIN   BEGIN
         {particular program}   {particular program}
      END   END
   END   END
END   END

Ideally a user should not need to worry about this. Hence great care must be taken in the design of such modules to ensure that their order of accessing is unimportant. This can be achieved without insisting that all modules are completely independent since the library mechanism allows only some of the declarations of a module to be made public. (The variable g should not be

made public in both of the modules G and H above.) Only public declarations
are made available to other modules for use or for republishing. Hence the
declarations available to the particular program can be carefully controlled.

However this does not solve the problem where completely independent
libraries are involved - the choice of which declarations can be made public
is then impossible since the library writers are unaware of one another.
This is solved by making clashes between objects from different libraries
illegal unless a module hierarchy is explicitly defined by the user. By
writing

ACCESS G ACCESS H ({program text})

a user defines such a heirarchy; by writing

ACCESS G,H ({program text})

he does not.

This standard incorporates the features which we consider necessary in any
library mechanism.

   a) A truly machine-independent environment can be constructed
      by tailoring a given standard prelude.
   b) The user should be able to choose whether he invokes the
      tailored or standard environment on his configuration.
   c) The library mechanism should not determine the modular
      division of a library. Hence, the actual division of a
      library into modules should not need to correspond with
      the apparent division indicated by user accessible
      modules.
   d) The result of interdependencies must be well-defined.

While diverse library mechanisms remain, an algorithm has no unique 'best'
source code - it must be modified for the different library mechanisms. The
existence and adherence to a defined standard satisfying the above
requirements for the library mechanism is just as important to library writers
and users as is adherence to the definition of the language itself.

4   CONCLUSION

In evaluating numerical software over many machines one must judge it upon
the claims made in its design and in its operational specification.

These claims must take full account of the arithmetic assumptions upon
which the software is based. Further, to evaluate software fairly, one must
be sure that on each configuration essentially the same software is being
tested.

The newer languages exclude this possibility given the present state of
development, by failing to define and implement a satisfactory library pick-up
mechanism. However the recent proposals in Algol 68 should rectify this
situation in the future and may give a lead to other languages.

This paper has described some of the problems associated with evaluation
over many machines. It is important that these problems are solved so that a
meaningful evaluation can be performed.

5 REFERENCES

1 J. Bentley and B. Ford
   On the Enhancement of Portability within the NAG Project -
   A Statistical Survey (1977)
   Portability of Numerical Software, Ed. W. Cowell,
   Springer-Verlag.
2 W.S. Brown
   A Realistic Model of Floating-Point Computation
   Private Communication.
3 W.J. Cody (1974)
   The Construction of Numerical Subroutine Libraries
   SIAM Review, 16, 1 pp. 36-46.
4 W.R. Cowell and L.D. Fosdick (1977)
   Mathematical Software Production
   Mathematical Software III, Ed. J.R. Rice
   Academic Press, pp. 195-224.
5 T.J. Dekker (1971)
   A Floating-Point Technique for Extending the Available Precision
   Numer. Math. 18, pp. 224-242.
6 B. Ford (1977)
   The Evolving NAG Approach to Software Portability
   Software Portability, Ed. P.J. Brown
   C.U.P.
7 B. Ford (Editor) (1978)
   Parameterisation of the Environment for Transportable Numerical Software
   A.C.M. TOMS, Vol.4 2, pp. 100-103.
8 B. Ford and J. Bentley (1978)
   A Library Design for all Parties
   Numerical Software - Needs and Availability (ed. D.A.H. Jacobs)
9 P.E. Gill and W. Murray (1979)
   Performance Evaluation for Optimization Software
   This volume.
10 S.J. Hague and B. Ford (1976)
    Portability - Prediction and Correction
    Software Practice and Experience, Vol.6, pp. 61-69.
11 T.E. Hull, W.H. Enright, B.M. Fellen, A.E. Sedgwick (1972)
    Comparing Numerical Methods for Ordinary Differential Equations
    SIAM J. Numer. Anal., 9, pp..603-637.
12 K. Jenson, N. Wirth (1978)
    Pascal User Manual and Report (2nd Ed.)
    Springer-Verlag.
13 C.H. Lindsey and H.J. Boom (1978)
    A Modules and Separate Compilation Facility for Algol 68
    Algol Bulletin AB43.3.2.
14 J. Lyness (1979)
    Performance Profiles and Software Evaluation
    This volume.
15 P. Naur (Editor) (1963)
    Revised Report on the Algorithmic Language Algol 60
    Computer Journal.
16 Numerical Algorithms Group Limited (1978)
    NAG Algol 68 Library Manual, Mark 2
    Numerical Algorithms Group Limited.
17 H. Rutishauser (1966)
    The Jacobi Method for Real Symmetric Matrices
    Numer. Math. 9, pp. 1-10.
18 B.G. Ryder and A.D. Hall (1973)
    The PFORT Verifier.
    Bell Labs. Computing Science Technical Report 12.

19 B.T. Smith et al, (1974)
    Matrix Eigensystem Routines - EISPACK Guide
    Springer-Verlag.
20 B.T. Smith (1977)
    Fortran Poisoning and Antidotes
    Portability of Numerical Software (Ed. W.R. Cowell)
    Springer-Verlag, pp. 178-256.
21 A. van Wijngaarden et al, (1977)
    Revised Report on the Algorithmic Language Algol 68
    SIGPLAN Notices, Vol.12, 5, pp. 1-70.

Performance Evaluation of Numerical Software, Fosdick (ed.)
© *IFIP, North-Holland Publishing Company, 1979*

DISCUSSION OF THIRD SESSION ON GENERAL ASPECTS
OF PERFORMANCE EVALUATION

L.M. Delves
Department of Computational and Statistical Science
University of Liverpool
Liverpool

1. METHODOLOGY OF TESTING

The three papers in this session, and also the earlier paper of James Lyness and
Hans Stetter, have a common theme. This is that much (most) of the comparative
testing of numerical software being carried out is far too unstructured; and the
judgements made as a result of the tests far too subjective.

Any comparative testing of two or more routines designed to do the same job
involves two stages :

1) Solving a number of test problems with the routines
2) Analysing the results.

1.1 CHOOSING THE TEST PROBLEMS

To date, both phases have often been carried out haphazardly. Collections of test
problems have grown by accretion in those fields (numerical quadrature, o.d.e.'s)
where an individual run is relatively cheap; or have remained small (two or three
sparse matrices; four or five time - hallowed non-linear objective functions) in
fields where an individual test is relatively expensive. The number of tests
required to adequately distinguish between two routines, is a priori unlikely to
be small just because a single run is expensive; nor is cheapness an elegant
substitute for effective test design. Rather, there seems to be general agreement
that families of test problems should be viewed statistically, as a sample from an
underlying population. The major difficulty faced in testing software is that
this underlying population is ill defined; and contains very steep hills. The
paper by Rice, and that by Stetter, expand on the first of these difficulties :
which classes of function do we wish to test? A typical test package for, say,
quadrature will contain integrals each displaying some difficulty : a rapid
oscillation, or a sharp peak, for example. Rice suggests that these features be
explicitly recognised, and presumably graded; formal recognition, as opposed to
subjective recognition by inclusion of examples in the test battery, then leads
naturally to an experiment properly designed to test the efficiency of algorithms
in dealing with one or more features, while the features themselves give a
(hopefully) low-dimensional but adequate description of the population. Such a
classification also allows easy recognition of the fact (often, alas, overlooked
in tests) that some routines are designed specifically for (or, not for), problems
with particular features; to test them on a completely broad battery of problems
is then almost certainly misleading. As an example, the NAG library contains
(currently) two routines for the solution of Fredholm integral equations of the
second kind; one for smooth kernels, and the other for kernels of Green's function
type. To complain that the latter is inefficient for smooth kernels, or the
former inaccurate for Green's function kernels, is to miss the point of having
both; yet both would behave poorly overall if tested on a battery containing both
kinds of problem.

The most extreme use of features in testing is that of Lyness, whose methodology
for testing quadrature routine tests (currently) only one feature at a time, and
ranks routines separately for each such feature. Such an approach is almost
certainly more expensive than necessary; the paper by Nelder serves as an

excellent reminder to us all of how economic tests should be designed. Lyness'
procedure has however the very great advantage of taking explicit account of the
second difficult aspect of the testing problem : the steep 'difficulty gradients'
which exist even amongst test problems from an identical feature family. These
gradients have been displayed explicitly by Lyness in the field of numerical
quadrature. To summarise one such example : a given routine, used to integrate
two functions having peaks of identical height and width, but peaking at slightly
different points in the region of integration, may integrate one very accurately
and the other very poorly. If a battery of tests contains one example of such an
integral, the routine will appear good, or bad, for problems with this feature
depending on which one was chosen for inclusion. The only remedy for this is to
check the average performance of the routine over a family of problems; in this
case, a family with a parameter governing the position of the peak; and it seems
now to be generally accepted that problems in a quadrature test battery should
where possible represent parameterised families of this type. It is not, as far
as I can see, so generally accepted in other fields of testing; such as
optimisation or sparse matrix solvers. Parameterised classes in these fields will
certainly lead to expensive tests; they will not be necessary if problems in
these fields do not show the same steep difficulty gradients as can be displayed
in quadrature. How many readers will care to bet that this is the case?

1.2 ANALYSING THE RESULTS

The tenor of the discussion above reflects a general opinion that the test
problems should be chosen as far as possible objectively, to represent
statistically the classes of problems, whether restrictive or wide, for which a
particular routine was designed. Subjective judgements remain, in the choice of
'features' to be tested; but they should be displayed for all to see. Similar
opinions seem to be general for the analysis of the test results : it is quite
undesirable to study a set of test results by eye; form a subjective opinion as to
the 'quality' of the routine; and publish the opinion, but not the test results.
Maybe beauty is in the eye of the beholder, but one should publish both the
definition of beauty being used, and the results on which the verdict is based.
Specifically, there seems now to be general agreement that to achieve significant
comparisons, and despite the possible costs, it is necessary to go through the
statistical procedures common in other fields : that is,

> 1) To decide, in advance of the analysis (if not of the tests themselves)
>   on one or more hypotheses which are to be tested. For example :
>
>   Routine A is quicker than B for problems having feature C
>
> 2) To test this hypothesis on the data collected.

It is most unlikely that suitable hypotheses are known in advance of the testing;
indeed, the argument for scanning the test results, subjectively and in detail,
is that such a scan suggests the relative merits and demerits of routines, and
hence suggests suitable hypotheses. But it is important that these be tested
statistically on further runs [so split your tests into two : one part to suggest
ideas, the other to verify them].

2. MULTI-MACHINE TESTING

Most tests are still carried out on a single machine; increasingly, libraries
such as NAG and IMSL must seek routines which perform well with (ideally) no
changes, in many different machine environments. Despite some wistful discussions
on the possibility of developing a specific library for each machine range, there
is general recognition that this is not financially possible. There also seems
to be reasonably general agreement that, given a suitable parameterisation of
these environments (and the difficulty of achieving sufficient detail in this
parameterisation varies from area to area), contributors are willing and able to

write parameterised code in standard FORTRAN (or whatever) for which in most cases the gain available from a rewrite on a given machine would be only modest (20%?). Certainly NAG now achieves an almost complete transportability for its FORTRAN library, to most machine ranges; however, inevitably, difficulties remain with some machines, the chief of which are :

Compilers implementing non-standard dialects

Machines with quirky   arithmetic

The latter comprise those which fail to fit the parameterised arithmetic model being used, or which fit it only with obviously silly values of the parameters. There was very general support for the efforts of Kahan and his coworkers in pursuing a standardised arithmetic specification.

Many of the portability aids which have been gradually developed within the various library groups to achieve portable FORTRAN code, are present in the language design of new languages such as Algol 68; further, the spread of incompatable dialects of these languages should be (and on the whole is) prevented by the (almost) watertight specification of the language by its designers. However, the original design failed to specify a library mechanism, with the result that while individual routines may be completely portable, the library as a whole may not be; the difficulties being experienced by NAG are displayed in the paper by FORD et al. Algol 68 now has a standard library mechanism, which hopefully will contain this problem in the future.

AUTHOR INDEX